HEALTHCARE TECHNOLOGIES SERIES 59

Deep Learning in Medical Image Processing and Analysis

IET Book Series on e-Health Technologies

Book Series Editor: Professor Joel J.P.C. Rodrigues, College of Computer Science and Technology, China University of Petroleum (East China), Qingdao, China; Senac Faculty of Ceará, Fortaleza-CE, Brazil and Instituto de Telecomunicações, Portugal

Book Series Advisor: Professor Pranjal Chandra, School of Biochemical Engineering, Indian Institute of Technology (BHU), Varanasi, India

While the demographic shifts in populations display significant socio-economic challenges, they trigger opportunities for innovations in e-Health, m-Health, precision and personalized medicine, robotics, sensing, the Internet of things, cloud computing, big data, software defined networks, and network function virtualization. Their integration is however associated with many technological, ethical, legal, social, and security issues. This book series aims to disseminate recent advances for e-health technologies to improve healthcare and people's wellbeing.

Could you be our next author?

Topics considered include intelligent e-Health systems, electronic health records, ICT-enabled personal health systems, mobile and cloud computing for e-Health, health monitoring, precision and personalized health, robotics for e-Health, security and privacy in e-Health, ambient assisted living, telemedicine, big data and IoT for e-Health, and more.

Proposals for coherently integrated international multi-authored edited or co-authored handbooks and research monographs will be considered for this book series. Each proposal will be reviewed by the book Series Editor with additional external reviews from independent reviewers.

To download our proposal form or find out more information about publishing with us, please visit https://www.theiet.org/publishing/publishing-with-iet-books/.

Please email your completed book proposal for the IET Book Series on e-Health Technologies to: Amber Thomas at athomas@theiet.org or author_support@theiet.org.

The Institution of Engineering and Technology

Deep Learning in Medical Image Processing and Analysis

Edited by
Khaled Rabie, Chandran Karthik, Subrata Chowdhury and Pushan Kumar Dutta

The Institution of Engineering and Technology

Published by The Institution of Engineering and Technology, London, United Kingdom

The Institution of Engineering and Technology is registered as a Charity in England & Wales (no. 211014) and Scotland (no. SC038698).

First published 2023

The Institution of Engineering and Technology
Futures Place
Kings Way, Stevenage
Hertfordshire SG1 2UA, United Kingdom

www.theiet.org

British Library Cataloguing in Publication Data
A catalogue record for this product is available from the British Library

ISBN 978-1-83953-793-6 (hardback)
ISBN 978-1-83953-794-3 (PDF)

Typeset in India by MPS Limited

Cover Image: Andrew Brookes/Image Source via Getty Images

Contents

About the editors

Khaled Rabie is a reader at the Department of Engineering at Manchester Metropolitan University, UK. He is a senior member of the Institute of Electrical and Electronics Engineers (IEEE), a fellow of the UK Higher Education Academy and a fellow of the European Alliance for Innovation (EAI). He is an area editor of IEEE Wireless Communications Letters and an editor of IEEE Internet of Things Magazine.

Chandran Karthik is an associate professor of mechatronics engineering at Jyothi Engineering College, India. He is a member of the Association for Computing Machinery (ACM), the ACM Special Interest Group on Computer Human Interaction (SIGCHI), and a senior member in IEEE, member in the IEEE Robotics and Automation Society. His research interests include medical robots, sensors, automation, machine learning and artificial intelligence-based optimization for robotics design.

Subrata Chowdhury is with the Sreenivasa Institute of Technology and Management Studies, Chittoor, Andhra Pradesh, India. He has edited five books in association with the CRC Press and others. He has published more than 50 articles in international and reputed journals. His research interests include data mining, big data, machine learning, quantum computing, fuzzy logic, AI, edge computing, swarm intelligence, and healthcare. He is also an IEEE member.

Pushan Kumar Dutta is an assistant professor at Amity University Kolkata with experience in book editing, proofreading, and research publication. He has published in IEEE conference and Scopus journals. He received the Best Young Faculty in Engineering award from Venus International Foundation Awards (2018) and the Young Researcher Award from IOSRD, India (2018). He is a senior member of IEEE and IET. His research interests include AI, machine ethics, and intelligent systems for biomedical applications.

Chapter 1

Diagnosing and imaging in oral pathology by use of artificial intelligence and deep learning

Nishath Sayed Abdul[1], Mahesh Shenoy[1], Shubhangi Mhaske[2], Sasidhar Singaraju[3] and G.C. Shivakumar[4]

Over the past few decades, dental care has made tremendous strides. Recent scientific discoveries and diagnostic tools have allowed for a sea change in the practice of conventional dentistry. Medical imaging techniques, including X-rays, MRIs, ultrasounds, mammograms, and CT scans, have come a long way in helping doctors diagnose and treat a wide range of illnesses in recent decades. Machines now may imitate human intellect via a process called artificial intelligence (AI), in which they can learn from data and then act on those learnings to produce outcomes (AI). AI has several potential applications in the healthcare industry. The use of AI in dentistry could improve efficiency and lower expenses while decreasing the need for specialists and the likelihood of mistakes being made by healthcare providers. Diagnosis, differential diagnosis, imaging, management of head and neck diseases, dental emergencies, etc. are just some of the many uses of AI in the dental sciences. While it is clear that AI will not ever be able to fully replace dentists, understanding how this technology might be used in the future is crucial. Orofacial disorders may be diagnosed and treated more effectively as a result of this. A doctor's diagnostic ability and outcome may be jeopardized by factors like increased workload, the complexity of work, and possible fatigue. Including AI features and deep learning in imaging equipment would facilitate greater productivity while simultaneously decreasing workload. Furthermore, they can detect various diseases with greater accuracy than humans and have access to a plethora of data that humans lack. Recent AI advancements and deep learning by use of image analysis in pathology and possible applications in the future were discussed in this chapter.

[1]Faculty of Oral Pathology, Department of OMFS and Diagnostic Sciences, Riyadh Elm University, Kingdom of Saudi Arabia
[2]Department of Oral Pathology and Microbiology, People's College of Dental Sciences and Research Center, People's University – Bhopal, India
[3]Department of Oral Pathology and Microbiology, Rishiraj College of Dental Sciences – Bhopal, India
[4]Oral Medicine and Radiology, People's College of Dental Sciences and Research Center, People's University – Bhopal, India

1.1 Introduction

There has been talk of a 48% annual increase in the volume of medical records. In light of the data deluge and the difficulties in making good use of it to enhance patient care, several different AI- and ML-based solutions are now in development (machine learning (ML)). The field of artificial intelligence (AI) known as ML has the potential to give computers human-level intelligence by allowing them to learn independently from experience without any human intervention or programming [1]. Thus, AI is defined as the subfield of computer science whose central research interest is the development of intelligent computers able to execute activities that traditionally have required human intellect [2].

Researchers all across the globe are fascinated by the prospect of creating artificially intelligent computers that can learn and reason like humans. Even though it is an application in dentistry is still relatively new, it is already producing impressive outcomes. We have to go back as far as 400 BC when Plato envisioned a vital model of brain function [3]. AI has had a significant influence in recent years across many areas of dentistry, but notably oral pathology. When used in dentistry, AI has the potential to alleviate some of the difficulties currently associated with illness detection and prognosis forecasting. An AI system is a framework that can learn from experience, make discoveries, and produce outcomes via the application of knowledge it has gleaned [4]. The first stage of AI is called "training," and the second is called "testing." In order to calibrate the model, it is first fed its training data. The model backtracks and utilizes historical instances, such as patient data or data with a variety of other examples. These settings are then used on the test data [3]. Oral cancer (OC) prognostic variables, documented in a number of research, may be identified using AI and a panel of diverse biomarkers. Successful treatment and increased chances of survival both benefit from detecting malignant lesions as soon as possible [5,6]. Image analysis of smartphone-based OC detectors based on AI algorithms has been the subject of several investigations. AI aids in OC patients' diagnosis, treatment, and management. By simplifying complicated data and alleviating doctors' weariness, AI facilitates quicker and more accurate diagnoses [7,8].

The word "AI" may have a certain meaning, but it really encompasses a vast variety of methods. For instance, deep learning (DL) aims to model high-level abstractions in medical imaging and infer diagnostic interpretations. Keep in mind that "AI" covers a wide range of technologies, including both classic "classical" machine learning and more recent "deep" forms of the same. Through the use of pre-programmed algorithms and data recognition procedures, conventional machine learning provides a quantitative judgment on the lesion's type and behavior as a diagnostic result [8]. Supervised and unsupervised approaches are subcategories of classic machine learning techniques. The diagnostic input is checked against the model's ground truth, which is validated by the training data and outputs in the supervised method. Unsupervised methods, on the other hand, are machine learning models that are not based on a set of predetermined values, and therefore they use techniques like data extraction and mining to discover hidden patterns in the data or specimen being studied. Using nonlinear processing units with numerous hidden layers, DL (also known as neural networks) is a collection of computational algorithms used in ML for learning and comprehending input and correlating it with output. To analyze massive datasets, DL is preferable since it can cope with data abstraction and complexity, unlike traditional ML [9,10].

There has been a recent uptick in the study of AI-based medical imaging and diagnostic systems. The potential for AI to enhance the precision and efficiency of disease screenings is motivating its adoption in the dental industry. Oral, pulmonary, and breast cancers, as well as other oral disorders, may all be detected with the use of AI technology [11–13]. These methods are now being tested in order to determine whether or not they should be included in diagnostic frameworks, especially for use in disease screening in areas with limited access to healthcare professionals. Utilizing AI may ease the burden of screening and analyzing massive data sets for cancer spot identification. There has to be more study done on the use of AI in the process of illness diagnosis. First and foremost, early detection must be measured against the accuracy and efficiency of AI in detecting a specific illness in contrast to a skilled doctor [14].

1.1.1 Application of Artificial Intelligence in the Field of Oral Pathology

It is widely accepted that microscopic morphology is the gold standard for making diagnoses in the area of pathology [8]. In order to analyze a pathology specimen, it must go through many processes, including formalin fixing, grossing, paraffin embedding, tissue sectioning, and staining. The majority of pathology diagnoses are made by human pathologists who examine stained specimens on glass slides under microscopes. One major drawback of morphologic diagnosis, however, is that results might vary depending on the individual pathologist. Thus, it is crucial to incorporate AI in the pathology area for consistent and more accurate diagnosis. Numerous recent efforts have been attempted, including scanning and saving a digital picture of a whole histopathology slide (whole slide image) [15].

According to the most up-to-date data from the World Health Organization (WHO), OC accounts for over 85% of all cancer-related deaths worldwide, impacting 4.5 million people. The mortality rate may be reduced by 70% if early diagnosis is implemented [15]. Oral epithelial dysplasia may be diagnosed and graded based on the presence of a number of distinct histological characteristics, as well as the presence and severity of corresponding structural alterations. Immature cell proliferation causes a loss of polarity, which manifests as an increase in the size and shape of nuclei, an increase in the ratio of nuclei to the cytoplasm, an uneven distribution of chromatin within nuclei, and an increased number of mitotic figures. This takes time and effort on the part of pathologists, and the results might differ from one observer to the next owing to differences in perspective. "Consequently, a computer-aided image classification technique that includes quantitative analysis of histological features is required for accurate, quick, and complete cancer detection." There have been studies into automatic cancer detection utilizing classifiers and enhanced features for quite some time now as a way to get around problems like a lack of clinicopathological competence or a lack of specialized training among oral oncopathologists. Labeling the distinct layers in histological slices of polystratified tissues was revolutionized by Landini and Othman [16] in 2003. Despite its lack of dimensions beyond two (or 2D), this approach may be useful for explicitly describing the relevant spatial arrangements. The same scientists have previously formalized the geometrical organization of benign, precancerous, and malignant tissues in 2D sections using statistical metrics of graph networks. When comparing normal, premalignant, and malignant cells, discrimination rates of 67%, 100%, and 80% were reported, respectively, demonstrating reliable and objective measurement [17]. "The goal of Krishnan *et al.*'s study was to increase the classification accuracy based on textural features by classifying histological tissue sections as normal, oral submucous fibrosis (OSF) without dysplasia, or OSF with dysplasia." The accuracy, sensitivity, and specificity increased to 95.7%, 94.50%, and 98.10%, respectively, when texture was mixed with higher-order spectra. Clinicians may now use their newly developed oral malignancy index to swiftly and simply evaluate whether mouth tumors are benign or malignant [18]. "To enhance the recognition of keratinization and keratin pearl from in situ oral histology pictures, Das *et al.* (2015) created a computer-aided quantitative microscopic approach, i.e. an automated segmentation methodology." When compared to domain-specific facts specified by domain experts, our technique successfully segmented 95.08% of the data [19].

In microscopic pictures, crucial visual markers in OC detection include architectural changes of epithelial layers and the presence of keratin pearls. If a computer program could do the same identification work as a physician, it would be a tremendous help in the interpretation of histology pictures for diagnosis. Das *et al.* proposed a two-step method to visualize oral histology images: first, a deep convolutional neural network (CNN) with 12 layers (7 7 3 channel patches) is used to partition the constituent layers; then, in the second step, keratin pearls are recognized from the partitioned keratin regions using surface-based highlight (Gabor channel) prepared irregular woods. The detection accuracy of a texture-based random forest classifier for identifying keratin pearls was found to be 96.88% [20]. Using a chemically-induced animal model, Lu *et al.* (2016) created a computer-aided technique for diagnosing tongue cancer. After having tongue tissue processed histologically, images of stained tissue sections were taken to use as benchmarks for later classification of tumorous and benign tissue. Most distinguishing was a texture trait characterizing epithelial structures. They found that the average sensitivity for detecting tongue cancer was 96.5%, with a specificity of 99% [21]. By analyzing hyperspectral photographs of patients, Jeyaraj and Samuel Nadar 2019 created a DL algorithm for an automated, computer-aided OC detection method. They were able to achieve a 91.4% accuracy in classification over 100 different picture datasets, with a sensitivity of 0.94 and a specificity of 0.91 [22]. To achieve a desired conclusion, a classification system must be self-learning and able to adapt its rules accordingly. Such a system is future-proof, continues to learn new things, and mimics the human brain in terms of how it gathers information [23].

1.1.2 AI as oral cancer prognostic model

Early diagnosis is critical for OC patients due to the dismal outlook of the advanced illness stage. Incorporating data from cytology images, fluorescence images, CT imaging, and depth of invasion, AI learning technologies may help speed up and improve the accuracy of OC diagnoses. Several studies have shown that OC may start anywhere in the mouth, including the tongue, buccal mucosa, or anywhere else in the mouth, while others have looked for the disease at an advanced stage to see if it can be detected early. The complexity of OC progression stems from its wide range of possible outcomes [24]. Sunny *et al.* did research employing artificial neural networks (ANN) for the early diagnosis of OC in light of the development of tele cytology (TC), the digitalization of cytology slides. Twenty-six different forms of AI were put to the test against traditional cytology and histology, and 11,981 prepossessed photos were loaded for AI analysis using the risk categorization model. TC was shown to be equally as accurate as traditional cytology; however, it was less sensitive in identifying possibly malignant oral lesions. Malignancy detection accuracy was raised to 93% using the ANN-based model, and 73% of lesions were properly classified as possibly malignant. For this study, researchers employed a noninvasive technique called "brush biopsy" to gather samples; this should be taken into account while looking for malignancy. In their work, Jeyaraj *et al.* used a deep-learning method for the definition of oral malignant development that relied on regression to identify OC [22]. One hundred hyperspectral images (HIS) were evaluated as part of the development of a computer-aided OC identifying system using a deep-learning method of CNN. When comparing the findings of the regression-based approach to the standard technique using the same pictures, they found that the former had a sensitivity of 91.4% for identifying malignant tumors. When compared to the standard method, the suggested model of the algorithm yielded higher-quality diagnoses. Uthoff *et al.* investigated the feasibility of employing smartphone photos and AI technologies for the detection of OC [25]. The point-of-care idea served as inspiration for the creation of pictures optimized for use on mobile devices. The images were enhanced using autofluorescence and white light imaging and then fed into AI systems trained to detect OC. The 170 autofluorescent photographs were taken in total. This method was not only more accurate but also easier to implement. However, in order to provide sufficient proof, the research need to be extended to a large population. Nayak *et al.* conducted a similar investigation utilizing autofluorescent spectrum pictures and analyzed the data using principal component analysis (PCA) and ANN [26]. Findings from ANN performance were somewhat better than those from PCA, which is a method of computing based on the principle components of data. The usage of a fluorescence spectroscopic picture is advantageous since it is a non-invasive diagnostic method that eliminates the need for a biopsy. Musulin *et al.* conducted a research utilizing Histology photos to conclude that AI performed better than humans at identifying OC [27]. Similarly, Kirubabai *et al.* found that, when presented with clinical photographs of patients with malignant lesions, CNN performed better than human experts in classifying the severity of the lesions [28].

"In order to detect nodal metastasis and tumor extra-nodal extension involvement, Kann *et al.* used deep-learning computers to a dataset of 106 OC patients [8]. The data set included 2875 lymph node samples that were segmented using computed tomography (CT). Here, we investigated how useful a deep-learning model may be in improving the treatment of head and neck cancer. Deep neural networks (DNN) were rated more accurate with an AUC of 0.91." Measurements of areas under the receiver operating characteristic (ROC) curve (AUC) may be made in two dimensions. Similar results were found by Chang *et al.*, who used AI trained on genomic markers to predict the presence of OC with an AUC of 0.90 [29]. The research compared AI using a logistic regression analysis. There was a significant lack of statistical power since the study only included 31 participants. Future research should include a bigger sample size [3].

Cancer research has long made use of ML. Evidence linking ML to cancer outcomes has grown steadily over the previous two decades. Typically, these investigations use gene expression patterns, clinical factors, and histological data as inputs to the prognostic process [30]. There are three main areas of focus in cancer prognosis and prediction: (i) the prediction of cancer susceptibility (risk assessment), (ii) the prediction of cancer recurrence, and (iii) the possibility of redeveloping a kind of cancer after full or partial remission Predictions might be made using large-scale data including ancestry, age, nutrition, body mass index, high-risk behaviors, and environmental carcinogen exposure. However, there is not enough data on these characteristics to make sound judgments. It has become clear that new types of molecular information based on molecular biomarkers and cellular characteristics are very useful indicators for cancer prognosis, thanks to the development of genomic, proteomic, and imaging technologies. Research shows that combining clinicopathologic and genetic data improves cancer prediction findings. The OC prognosis study conducted by Chang *et al.* employed a hybrid approach, including feature selection and ML methods. Both clinicopathologic and genetic indications were shown to be related to a better prognosis, as shown by their research. [29] Exarchos *et al.* set out to identify the factors that influence the progression of oral squamous cell carcinoma so that they might predict future recurrences. They pushed for a multiparametric decision support system that takes into account data from a variety of fields, such as clinical data, imaging results, and genetic analysis. This study clearly demonstrated how data from several sources may be integrated using ML classifiers to provide accurate results in the prediction of cancer recurrence [31]. Integrating multidimensional heterogeneous data and using various methodologies might provide useful inference tools in the cancer area, as has become obvious [23].

1.1.3 AI for oral cancer screening, identification, and classification

Oral squamous cell carcinoma (OSCC) presents unique logistical challenges, particularly in low- and middle-income countries (LMICs), where there are fewer head and neck cancer clinics and fewer doctors experienced with OSCC. When risk factors and visual examinations are used together, community health professionals may almost half the death rate from OC in high-risk populations. Cost-effectiveness analysis has shown that this kind of screening is beneficial in those at high risk for OC. There was no effect on mortality, morbidity, or cost in previous large-scale OC screening investigations. Despite the fact that conventional OC screening should be beneficial in LMICs, substantial populations in sectors with a high OC risk in LMICs often lack access to healthcare, necessitating alternative techniques adapted to the specific restrictions and features of each region. OC screening accuracy may be improved with the use of many AI-based algorithms and methodologies that have emerged in the recent decade. They may be as effective and accurate as traditional methods of screening, if not more so, while eliminating the requirement for highly trained and regularly retrained human screeners. In 1995, researchers began using AI to predict who would develop OC. Researchers found that a trained ANN could identify oral lesions with a sensitivity of 0.80 and a specificity of 0.77 [32]. Subsequent research confirmed that by screening only 25% of the population with this method, high-risk people could be identified and 80% of lesions could be detected. In 2010, researchers conducted a case-control study that compared the accuracy of prediction models based on fuzzy regression and fuzzy neural networks to that of professional doctors.

AI's ability to facilitate remote healthcare interactions has the potential to increase the speed with which screenings may be implemented, which is particularly important in LMICs, where their influence is most felt. The potential of AI as a tool for remote oral screening has been underlined in recent years, and there has been a surge of interest in AI-based telehealth applications. For high-risk populations in areas with few resources, researchers at many institutions worked together to create a very affordable smartphone-based OC probe using deep learning [13,25,33]. Images of autofluorescence and polarization captured by the test, as well as OSCC risk variables, were analyzed using an innovative DL-based algorithm to provide an evaluative output that provides emergency data for the screener. In the most important clinical preliminary, 86% of cases were found to match the screening calculation and the highest quality level result. After being primed further, the calculation's overall awareness, explicitness, positive predictive value, and negative predictive incentive for spotting intraoral sores all increased from 81% to 95%. The accuracy of automated screening was estimated to be over 85% in a variety of studies, which is much higher than the accuracy of traditional screening by community health workers. These findings are quite promising, especially in light of the fact that there will be 3.5 billion mobile phone users worldwide by 2020. Particularly important in underserved and rural areas, studies like these show that non-expert medical care providers like attendants, general specialists, dental hygienists, and local area well-being laborers can effectively screen patients using man-made intelligence-supported applications integrated into cell phones. When it comes to intraoral photos of mucosal sores taken with a smartphone, the degree of concordance between the image and the clinical evaluation is considered to be moderate to high, whereas it is lower for low-goal shots. Nonetheless, supported AI at low cost and cell phone-based innovations for early screening of oral wounds could serve as a viable and reasonable approach to reducing delays in the master and clinical consideration framework and allowing patients to be triaged to seek okay and ideal treatment. Using a separate approach taking soft inducing into account, intraoral photographs of OSCC, leukoplakia, and lichen planus wounds were correctly identified as OSCC and lichen planus bruises 87% of the time, and lichen planus wounds 70% of the time. Similarly, deep convolutional neural network (DCNN) models achieved performance levels comparable to human professionals in recognizing OC's first phases when trained on a small collection of pictures of tongue sores. A newly designed robotized DL technique, trained on 44,409 pictures of biopsy-proven OSCC lesions and solid mucosa, produced an AUC of 0.983 (95% CI 0.973–0.991), with a responsiveness of 94.9% and an explicitness of 88.7% on the internal approval dataset.

In an early work by van Staveren *et al.* [34], autofluorescence spectra were taken from 22 oral leukoplakia lesions and 6 healthy mucosal areas to see how well an ANN-based ordering algorithm performed. According to the published data, ANN exhibits 86% responsiveness and 100% explicitness when analyzing phantom pictures of healthy and sick tissues [33,34]. Wang *et al.* [35] autofluorescence spectra of premalignant (epithelial dysplasia) and harmful (SCC) sores were separated from those of benign tissues using a half-way least squares and artificial neural network (PLS-ANN) order calculation, with 81% responsiveness, 96% explicitness, and 88% positive prescient value achieved. As explained by others, using an ANN classifier as an exploration method may lead to high levels of responsiveness (96.5% or more) and explicitness (100% particularity). De Veld *et al.*'s investigation on autofluorescence spectra for sore order using ANN indicated that although the approach was effective in differentiating healthy mucosa from disease, it was less effective in differentiating benign tissue from premalignant sores. By performing 8 solid, 16 leukoplakia, and 23 OSCC tests utilizing Fourier-transform infrared spectroscopy (FTIR) spectroscopy on paraffin-inserted tissue slices, we were able to develop an SVM-based strategy for diagnosing oral leukoplakia and OSCC based on the biomarker choice. It was stated that the authors had success in locating discriminating spectral markers that indicated significant bio-molecular alterations on both the qualitative and quantitative levels and that these markers were useful in illness classification. The malignant areas' boundaries were also accurately delineated in both the positive- and negative-ion modes by an ML-based diagnostic algorithm for head and neck SCC utilizing mass spectra, with accuracies of 90.48% and 95.35%, respectively. A DCCN-based algorithm was recently examined for its ability to detect OC in hyperspectral pictures taken from individuals diagnosed with the disease. When comparing photographs of cancerous and benign oral tissues, researchers found a classification accuracy of 94.5%. Recent animal research and another investigation using imaging of human tissue specimens also reported similar findings. Indicative accuracy was increased to an average of 88.3% (awareness 86.6%, explicitness 90%) by using DL methods to assess cell structure with confocal laser endomicroscopy for the location of OSCC. The most basic optical coherence tomography (OCT) models were used, together with a robotized finding computation and an image management system with an intuitive user interface. When comparing the robot-assisted disease screening stage to the histopathological gold standard, it showed a responsiveness of 87% and an explicitness of 83% in distinguishing between healthy and abnormal tissues [8].

1.1.4 Oral cancer and deep ML

The phrases AI and ML are sometimes used synonymously in academic writing, despite their distinct meanings. The phrase "AI" was first used by John McCarthy, sometimes referred to as the "father of AI," to describe robots that might potentially carry out behaviors traditionally associated with intelligence without any human interaction [35]. The data fed into these machines allows them to solve issues. In the realm of AI, machine learning (ML) may be found. The phrase was first used by Simon Cowell in 1959. When given a dataset, ML [36] makes predictions using algorithms like ANN. These networks are modeled after the human brain and use artificial neurons joined together to process incoming data signals [37].

Implementing machine learning effectively requires access to a large amount of data sets. Data may relate to a wide range of information sources, including visuals such as clinical photos or radiography, text such as patient data or information on the patient's symptoms, and audio such as the patient's voice, murmurs, bruits, auscultation, or percussive sounds. AI's capacity to learn from new data is a game-changer for the future of healthcare. Though most applications are still in their formative stages, the research has shown encouraging outcomes. Dentists, in order to thrive in the new healthcare environment, must be familiar with the basic ideas and practical uses of AI in dentistry. AI has recently been proposed as a tool in healthcare, particularly for illness detection, prognosis forecasting, and the creation of individualized treatment plans. AI can help dentists with a variety of tasks, but it excels at those that need quick choices. It may alleviate pressure on dentists by eliminating their need to make split-second decisions and improving patient care overall [37].

The worldwide death toll from cancer as well as the number of new cases has been rising at an alarming rate. In 2015, the WHO stated that cancer was either the leading cause of death or a close second in 91 of the world's 172 countries. Cancers of the mouth and throat are the sixth most common form of the disease worldwide. "Diseases affecting the oral cavity, pharynx, and lips account for around 3.8% of all cancer cases and 3.6% of all cancer deaths. In high-risk countries like India, Pakistan, Sri Lanka, and Bangladesh, OC is the most common disease among men, accounting for as much as 25% of all new cases each year." There is a critical need for individualized strategies for OC prevention, diagnosis, and therapy in light of the disease's increasing prevalence. There is a consensus among experts that patients' chances of survival and the likelihood of their cancer returning after treatment are both influenced by the therapy they receive. In order to enhance the quality of care for patients with OC, they advocated for a system that more accurately classifies individuals into groups according to their condition before treatment begins. Diagnostic, therapeutic, and administrative decisions may all benefit from data mining technologies and analysis when used by medical experts. For example, decision trees, a kind of supervised learning in the realm of data mining, are useful for tasks like categorizing and forecasting. To differentiate between the symptoms shown by patients who died from and survived OC in the past, Tseng *et al.* [23] created a unified technique that incorporates clustering and classifying aspects of data mining technology [38].

The prognosis and survival rate for those with OC is improved with early diagnosis. The mortality and morbidity rates from OC may be reduced with the use of AI by helping with early detection. Nayak *et al.* (2005) employed ANN to classify laser-induced autofluorescence spectra recordings of normal, premalignant, and malignant tissues. This was contrasted with a PCA of the identical problems. The findings demonstrated a 98.3% accuracy, a 100% specificity, and a 96.5% sensitivity, all of which are promising for the method's potential use in real-time settings. CNN was utilized by Uthoff *et al.* (2017) to identify precancerous and cancerous lesions in autofluorescence and white light pictures. When comparing CNN to medical professionals, it was shown that CNN was superior in identifying precancerous and cancerous growths. With more data, the CNN model can function more effectively [25]. Using confocal laser endomicroscopy (CLE) images, Aubreville *et al.* (2017) trained a DL model to detect OC. Results showed that this approach was 88.3% accurate and 90% specific. Comparative research was undertaken by Shams *et al.* (2017) utilizing deep neural networks to forecast the progression of OC from precancerous lesions in the mouth (DNN). DNNs were compared to SVMs, RLS, and multilayer perceptron (MLP). DNN's 96% accuracy was the highest of all of the systems tested. Additionally, Jeyraj *et al.* validated these results (2019). "Using hyperspectral pictures, malignant and noncancerous tissues were identified using convolutional neural networks. CNN seems to be useful for image-based categorization and the detection of OC without the need for human intervention. Research on OC has exploded in recent years." Several studies have achieved their goals by creating AI models that can accurately forecast the onset and progression of OC. Research comparing DL algorithms to human radiologists has produced mixed findings. The accuracy of DL for detecting cervical node metastases from CT scans was evaluated by Ariji *et al.*, 2014. From 45 patients with oral squamous cell carcinoma, CT scans of 137 positive and 314 negative lymph nodes in the neck were utilized. Two experienced radiologists were used to evaluate the DL method's output. In terms of accuracy, the DL network was on par with human radiologists. The researchers also used DL to identify tumors that have spread beyond the cervical lymph nodes. Among the 703 CT scans we obtained from 51 individuals, 80% were utilized as training data, and 20% were used as test data to determine whether or not the disease had spread beyond the nodes. The DL system outperformed the radiologist, indicating it might be utilized as a diagnostic tool for spotting distant metastases. When it comes to diagnosing dental conditions including cavities, sinusitis, periodontal disease, and temporomandibular joint dysfunction, neural networks, and ML seem to be just as good, if not more so, than professional radiologists and clinicians. Using AI models for cancer detection enables the consolidation of disparate data streams for the purpose of making decisions, evaluating risks, and referring patients to specialized care. Indications are promising for the diagnostic and prognostic utility of AI in studies of premalignant lesions, lymph nodes, salivary gland tumors, and squamous cell carcinoma. By facilitating early diagnosis and treatment measures, these approaches have the potential to lower death rates. In order to provide an accurate and inexpensive diagnosis, these platforms will need access to massive amounts of data and the means to evaluate it. These models need to be fine-tuned until they are both highly accurate and very sensitive before they can be successfully adopted into conventional clinical practice. More so, regulatory frameworks are required to put these models into clinical practice [37].

1.1.5 AI in predicting the occurrence of oral cancer

However, even though there are effective methods for treating OC now, the disease often returns. When dealing with oral malignant development, treatment options are dependent on the progression of the illness. An insufficient or unnecessary treatment plan may result from a lack of an evidence-based staging system [38]. There have been several proposals for prognostic biomarkers and therapeutic targets throughout the years, however, they are not reflected in the current cancer staging system. Predictions of OC have previously been made using conventional statistical approaches such as the Cox proportional hazard (CPH) model, which is unsuitable for forecasting circumstances like OC. Taking into account the intricate "dataset" of OC, an AI-based predictive predictor will provide positive results. Using AI for predicting OC has shown promising outcomes in previous research [39,40]. The probability of oral tongue squamous cell carcinoma recurrence was evaluated among four ML algorithms in a research by Alabi *et al.*, which included 311 patients in Brazil. Some AI-inspired machine learning frameworks were employed. Support vector machines (SVMs), naïve Bayes (NBs), boosted decision trees (BDTs), and decision forests (DFs) were among the methods used (DF). The accuracy of diagnosis from all of these algorithms improved, but the BDT approach improved the most. As a result of the limited size of the sample, more information from external algorithms is required. AI and the gene expression profile were utilized by Shams *et al.* to predict the onset and progression of OC from precancerous lesions. About half (51) of those who participated in the study had OC. There were no malignant cells in any of the other 31 samples. We looked at SVMs, CNNs, and MLA to see which would be best for certain tasks (MLP). Machines taught using deep neural networks performed better than those trained with MLPs (94.5% vs. 94.5% accuracy). From a set of four models, Chui *et al.* discovered that BDT provided the most reliable predictions for cancer incidence (linear regression (LR), binary decision trees (BDT), support vector machines [SVM], and k-nearest neighbors (KNN)). The symptoms of patients who died from OC and those who recovered were compared and contrasted by Tseng *et al.* [38]. Results from a comparison of traditional logistic regression, a decision tree, and an ANN were analyzed and presented for 674 OC patients. Survival time, number of deaths, new cancer diagnoses, and spread of disease were employed as prognostic indicators in this study. Decision trees were shown to be simple to understand and accurate, whereas ANN was found to be more comparable to traditional logistic regression. For their study, Rosma *et al.* analyzed the predictive power of AI for cancer in a

Malaysian cohort by factoring in each person's unique demographic and behavioral risk factors. Predictions of OC were evaluated using expert opinion, a fuzzy regression model, and a fuzzy neural network prediction model. Fuzzy regression may be used to build a link between the explanatory and response variables in situations when there is insufficient data. Human experts were unable to match the accuracy of the neural network and fuzzy regression model used in AI-based OC prediction [3].

1.1.6 AI for oral tissue diagnostics

The elimination of subjectivity, the automation of the procedure, and the application of objective criteria to the findings are all ways in which advances in AI technology might make tissue diagnostics more accurate. Researchers used image analysis to determine the mean nuclear and cytoplasmic regions of oral mucosa cytological smears; this calculation demonstrated a responsiveness of 0.76, an explicitness of 0.82, and the ability to distinguish between normal/nondysplastic and abnormal/unsafe mucosa. In order to examine a slide model and capture large standard pictures of stained brush biopsy tests for histological assessment of gathered cell tests, researchers have developed a tablet-based reduced enhancing point of convergence that combines an iPad mini with various optics, Drove light, and Bluetooth-controlled engines. The results showed that the proposed technology, by integrating high-quality histology with regular cytology and remote pathologist interpretation of pictures, would enhance screening and reference adequacy, particularly in rural locations and medical care offices without competent experts [8].

1.1.7 AI for OMICS in oral cancer

New omics technologies (including genomes and proteomics) have made it possible to gather enormous datasets on cancer. Ovarian cancer (OC) and pancreatic cancer (PC) omics studies have used AI to improve prognostic prediction models, identify nodal contribution, discover HPV-related biomarkers [41], and differentiate transcriptome and metabolite markings. The authors, Chang *et al.* [29], using clinicopathologic and genomic data (p53 and p63) from 31 patients with oral disease, found that ANFIS was the most reliable device for predicting oral harmful development surmise, with the highest accuracy achieved by the 3-input components of alcohol use, depth of invasion, and lymph node metastasis (93.81%; AUC = 0.90). Clinicopathologic and genetic data when combined allowed for a more refined guess prediction than was possible with simply clinicopathologic data alone.

In addition, in 2020, researchers analyzed 334 high-level OC patients' entire clinicopathologic and genetic data to review an ML-based framework for endurance risk order. Super-deep sequencing data from 44 quality variety profiles in growth tissue assays, all of which are linked to illness, were used to formulate the method. Patient age, orientation, smoking status, cancer site, cancer stage (both T and N), histology findings, and cautious outcomes were also recorded. By combining clinicopathologic and hereditary data, a more accurate prediction model was developed, outperforming prior models that relied only on clinicopathologic data.

Clinical evaluations, imaging, and articulation quality data were all used by Exarchos *et al.* [42] to identify characteristics that foreshadow the onset of OC and predict relapse. Classifiers were built independently for each dataset, and then they were combined into one cohesive model. Moreover, a dynamic Bayesian network (DBN) was used for genetic data in order to develop disease evolution tracking software. The authors were able to provide more customized therapy by separating patients into those at high risk of recurrence (an accuracy of 86%) and those at low risk (a sensitivity of 100%) based on the DBN data from the first visit. "The association between HPV infection and the presence of apoptotic and proliferative markers in persons with oral leukoplakia has been studied by others using a particular sort of ML called a fuzzy neural network (FNN)." Clinical and immunohistochemical test data, demographics, and lifestyle habits of 21 patients with oral leukoplakia were input into an FNN system, with HPV presence/absence acting as a "output" variable. Researchers used this method to relate a positive proliferating cell nuclear antigen result to a history of smoking and a history of human papillomavirus infection to survival in people with oral leukoplakia. Transcriptome biomarkers in OSCC were found by support vector machine classifier-based bioinformatics analysis of a case-control dataset. Sputum samples from 124 healthy people, 124 people with premalignant diseases, and 125 people with OC sores were studied using conductive polymer shower ionization mass spectrometry in novel research to detect and confirm dysregulated chemicals and reveal changed metabolic pathways (CPSI-MS). Evidence suggests that ML to CPSI-MS of spit samples might provide a simple, fast, affordable, and painless alternative for OC localization since the Rope approach when applied in combination with CPSI-MS, has been shown to yield a sub-atomic result with an accuracy of 86.7%. AI research has shown a connection between alterations in the spit microbiome and the development of dental disease. The salivary microbiome of individuals with OSF was compared to that of individuals with OSF and oral squamous cell carcinoma using high-throughput sequencing of bacterial 16S rRNA by Chen *et al.* (OSCC). When comparing OSF and OSF + OSCC instances, the AUC was 0.88 and the mean 5-fold cross-approval exactness was 85.1% thanks to the ML analysis's effective coordination of elements of the bacterial species with those of the host's clinical findings and lifestyle. Man-made intelligence applications in omics are aimed at completing tasks that are beyond the scope of human capability or conventional fact-based methods of investigation. Through the use of AI and ML methods, which enable the coordinated translation of omics information with clinicopathologic and imaging features, we may be able to enhance clinical treatment and broaden our understanding of OC [8].

1.1.8 AI accuracy for histopathologic images

Histopathological examination is the most reliable method for diagnosing OC. However, as it is based on subjective evaluations, the screening accuracy by the physician is also subjective. Certain traits and characteristics help the pathologist establish if a patient presents with malignancy and the stage when OC histopathologic samples are analyzed. Due to the quantificational nature of manual sample assessment for diagnostic characteristics, there exists the possibility of inaccuracy, which in turn leads to erroneous findings [43]. As a result, cytologic and histologic hallmarks of OC may now be detected with greater speed and precision because of advancements in AI. And because of advancements in AI, huge data sets can be processed and analyzed to find OC. These investigations employed two different kinds of samples: histologic and biopsy samples, and photographic pictures. Biopsy and histologic samples were utilized in six separate investigations. Various studies have looked at the possibility of using cellular changes as a marker for identifying cancer samples as distinct from normal and aberrant cell nuclei [7,44,45]. For this study, Das *et al.* used the suggested segmentation approach to examine epithelial alterations in the oral mucosa of OC patients by identifying keratin pearls. Successfully using their suggested CNN machine, they measured the keratinization layer, a crucial characteristic in identifying the OC stage [14].

1.1.9 Mobile mouth screening

Developed by researchers at Kingston University (United Kingdom) and the University of Malay (Malaysia), the Mobile Mouth Screening Anywhere (MeMoSA) software takes pictures of the mouth and sends them to a server where professionals may look at them remotely. They want to use thousands of images for training a deep learning system that can detect OC symptoms and abnormalities and then include that system into the app. Professor Dr. Sok Ching Cheong of Malaysia, an expert in cancer research, felt that incorporating AI into MeMoSA had a great deal of potential in ensuring that their efforts in early detection continued to overcome boundaries throughout locations where the illness is most common [23].

1.1.10 Deep learning in oral pathology image analysis

Researchers may now investigate the potential of AI in medical image processing since diagnostic imaging has become so commonplace. In particular, DL, an AI approach, has shown substantial accomplishments in solving a variety of medical image analysis challenges, most notably in cancer identification in pathological pictures. Many forms of cancer, including breast cancer, lung cancer, prostate cancer, and others, have been targeted by proposals for large-scale implementations of DL-based computer-aided diagnostic (CAD) systems. In spite of this, research suggests that DL is often missed while examining OSCC pathology pictures. Using a CNN and a Random Forest, Dev *et al.* were able to identify keratin pearls in pictures of oral histology. The CNN model had a success rate of 98.05% in keratin area segmentation, while the Random Forest model achieved a success rate of 96.88% in recognizing keratin pearls [20]. Oral biopsies were analyzed by Das *et al.*, who used DL to assign grades to pictures based on Broder's histological classification system. In addition, CNN was suggested because of the great precision with which it can categorize data (97.5%) [46]. "Using a CNN trained with Active Learning and Random Learning, Jonathan and his coworkers were able to classify OC tissue into seven distinct subtypes (stroma, lymphocytes, tumor, mucosa, keratin pearls, blood, and adipose). We determined that AL's accuracy is 3.26% points greater than RL's [47]. Also, Francesco *et al.* used many distinct DL approaches to classify whole slide images (WSI) of oral lesions as either carcinoma, non-carcinoma, or non-tissue. The accuracy of a deeper network, such U-Net trained using ResNet50 as an encoder, was shown to be superior to that of the original U-Net [48]. In recent research, Rutwik *et al.* (91.13% accuracy) employ ResNet to perform binary classification on images of oral disease." Rutwik *et al.*, in a recent research, used several pre-trained DL models to classify OSCC pictures as normal or malignant. ResNet was able to get the greatest accuracy of 91.13% [49].

The primary focus at the moment is on determining how well AI and its subsets perform in screening for oral illnesses and disorders, particularly mouth cancer, utilizing photographic and histologic images. So far, the vast majority of research has shown that ML algorithms are very effective in detecting OC. Recent developments in ML algorithms have made it possible to identify OC using a method that is both effective and noninvasive, and which can compete with human professionals. However, many tumors are not diagnosed until they have advanced because of how readily accessible the mouth and throat are during a normal inspection. The clinical appearance of the lesion may be a signal for experts to detect OCs. The use of AI to facilitate faster and more precise detection of OC in earlier stages is a potential technique for lowering death rates. The use of AI in oncology is gaining momentum as researchers seek to enhance the accuracy and throughput of cancer lesion detection [14].

1.1.11 Future prospects and challenges

Dental AI is still in its infancy. It is still not widely used in routine dental procedures. There are still many obstacles to overcome before it can be widely used in medical diagnosis and treatment. Institutions and private dentistry practices alike have access to massive data sets that can be used for machine learning. Federal rules and legislation are necessary for addressing data sharing and privacy concerns. Most research has this one fundamental flaw that our solution can fix: an absence of data sets. Both the European Union's General Data Protection Regulation and the United States' California Consumer Privacy Act were enacted by their respective legislatures to safeguard consumers' personal information and mitigate any risks related to data sharing. VANTAGE6, Personal Health Train (PHT), and DataSHIELD are all examples of federated data platforms that make it feasible to exchange data while still meeting privacy and security requirements. Also, with the help of AI, data that is now quite disorganized may be transformed into a unified whole that is simple to use and understand. The majority of the research discussed in this article used supervised image analysis to detect structures or relationships. The data is incomplete and cannot be used to make decisions or offer care. It is necessary to develop AI for unsupervised diagnosis and prognosis of illnesses in order to lessen the reliance on human subjectivity and increase the prevalence of objectively correct conclusions. Rural areas are characterized by a lack of resources and human capital. Healthcare efforts powered by AI have the potential to provide high-quality medical attention to underserved areas. AI's influence on therapy, along with its efficacy and cost-effectiveness, must be assessed via prospective randomized control trials and cohort studies [50,51].

1.2 Conclusion

AI is developing fast to meet a growing need in the healthcare and dental industries. Much of the study of AI remains in its infancy [52]. There are now just a small number of dental practices that have implemented internal real-time AI technologies. Data-driven AI has been proven to be accurate, open, and even superior to human doctors in several diagnostic situations [53]. AI is capable of performing cognitive tasks including planning, problem-solving, and thinking. Its implementation may cut down on archival space and labor costs, as well as on human error in diagnosis. A new era of affordable, high-quality dental treatment that is more accessible to more people is on the horizon, thanks to the proliferation of AI in the dental field.

References

[1] Rashidi HH, Tran NK, Betts EV, Howell LP, and Green R. Artificial intelligence and machine learning in pathology: the present landscape of supervised methods. *Acad. Pathol.* 2019;6:2374289519873088. doi:10.1177/2374289519873088. PMID: 31523704; PMCID: PMC6727099

[2] Alabi RO, Elmusrati M, Sawazaki-Calone I, *et al.* Comparison of supervised machine learning classification techniques in prediction of locoregional recurrences in early oral tongue cancer. *Int. J. Med. Inform.* 2020;136:104068.

[3] Khanagar SB, Naik S, Al Kheraif AA, *et al.* Application and performance of artificial intelligence technology in oral cancer diagnosis and prediction of prognosis: a systematic review. *Diagnostics* 2021;11:1004.

[4] Kaladhar D, Chandana B, and Kumar P. Predicting cancer survivability using classification algorithms. Books 1 view project protein interaction networks in metallo proteins and docking approaches of metallic compounds with TIMP and MMP in control of MAPK pathway view project predicting cancer. *Int. J. Res. Rev. Comput. Sci.* 2011;2:340–343.

[5] Bànkfalvi A and Piffkò J. Prognostic and predictive factors in oral cancer: the role of the invasive tumour front. *J. Oral Pathol. Med.* 2000;29:291–298.

[6] Schliephake H. Prognostic relevance of molecular markers of oral cancer—a review. *Int. J. Oral Maxillofac. Surg.* 2003;32:233–245.

[7] Ilhan B, Lin, K, Guneri P, and Wilder-Smith P. Improving oral cancer outcomes with imaging and artificial intelligence. *J. Dent. Res.* 2020;99:241–248.

[8] Kann BH, Aneja S, Loganadane GV, *et al.* Pretreatment identification of head and neck cancer nodal metastasis and extranodal extension using deep learning neural networks. *Sci. Rep.* 2018;8:1–11.

[9] Ilhan B, Guneri P, and Wilder-Smith P. The contribution of artificial intelligence to reducing the diagnostic delay in oral cancer. *Oral Oncol.* 2021;116:105254.

[10] Chan CH, Huang TT, Chen CY, *et al.* Texture-map-based branch-collaborative network for oral cancer detection. *EEE Trans. Biomed. Circuits Syst.* 2019;13:766–780.

[11] Lu J, Sladoje N, Stark CR, *et al.* A deep learning based pipeline for efficient oral cancer screening on whole slide images. arXiv 2020;1910:1054.

[12] Ehteshami Bejnordi B, Veta M, Johannes van Diest P, *et al.* Diagnostic assessment of deep learning algorithms for detection of lymph node metastases in women with breast cancer. *JAMA* 2017;318(22):2199–2210.

[13] Coudray N, Ocampo PS, Sakellaropoulos T, *et al.* Classification and mutation prediction from non–small cell lung cancer histopathology images using deep learning. *Nat. Med.* 2018;24:1559–1567.

[14] Song B, Sunny S, Uthoff RD, *et al.* Automatic classification of dual-modalilty, smartphone-based oral dysplasia and malignancy images using deep learning. *Biomed. Opt. Express* 2018;10:5318–5329.

[15] Al-Rawi N, Sultan A, Rajai B, *et al.* The effectiveness of artificial intelligence in detection of oral cancer. *Int. Dent. J.* 2022;72(4):436–447. doi:10.1016/j.identj.2022.03.001. Epub 2022 May 14. PMID: 35581039; PMCID: PMC9381387.

[16] Komura D and Ishikawa S. Machine learning methods for histopathological image analysis. *Comput. Struct. Biotechnol. J.* 2018;16:34–42.

[17] Erickson BJ, Korfiatis P, Kline TL, Akkus Z, Philbrick K, and Weston AD. Deep learning in radiology: does one size fit all? *J. Am. Coll. Radiol.* 2018;15:521–526.

[18] Landini G and Othman IE. Estimation of tissue layer level by sequential morphological reconstruction. *J. Microsc.* 2003;209:118–125.

[19] Landini G and Othman IE. Architectural analysis of oral cancer, dysplastic, and normal epithelia. *Cytometry A* 2004;61:45–55.

[20] Krishnan MM, Venkatraghavan V, Acharya UR, *et al.* Automated oral cancer identification using histopathological images: a hybrid feature extraction paradigm. *Micron* 2012;43:352–364.

[21] Das DK, Chakraborty C, Sawaimoon S, Maiti AK, and Chatterjee S. Automated identification of keratinization and keratin pearl area from in situ oral histological images. *Tissue Cell* 2015;47:349–358.

[22] Das DK, Bose S, Maiti AK, Mitra B, Mukherjee G, and Dutta PK. Automatic identification of clinically relevant regions from oral tissue histological images for oral squamous cell carcinoma diagnosis. *Tissue Cell* 2018;53:111–119.

[23] Lu G, Qin X, Wang D, *et al.* Quantitative diagnosis of tongue cancer from histological images in an animal model. *Proc. SPIE Int. Soc. Opt. Eng.* 2016;9791. pii: 97910L.

[24] Jeyaraj PR and Samuel Nadar ER. Computer assisted medical image classification for early diagnosis of oral cancer employing deep learning algorithm. *J. Cancer Res. Clin. Oncol.* 2019;145:829–837.

[25] Krishna AB, Tanveer A, Bhagirath PV, and Gannepalli A. Role of artificial intelligence in diagnostic oral pathology – a modern approach. *J. Oral Maxillofac. Pathol.* 2020;24:152–156.

[26] Sunny S, Baby A, James BL, *et al.* A smart tele-cytology point-of-care platform for oral cancer screening. *PLoS One* 2019;14:1–16.

[27] Uthoff RD, Song B, Sunny S, *et al.* Point-of-care, smartphone-based, dual-modality, dual-view, oral cancer screening device with neural network classification for low-resource communities. *PLoS One* 2018;13:1–21.

[28] Nayak GS, Kamath S, Pai KM, *et al.* Principal component analysis and artificial neural network analysis of oral tissue fluorescence spectra: classification of normal premalignant and malignant pathological conditions. *Biopolymers* 2006;82:152–166.

[29] Musulin J, Štifanić D, Zulijani A, Cabov T, Dekanić A, and Car Z. An enhanced histopathology analysis: an AI-based system for multiclass grading of oral squamous cell carcinoma and segmenting of epithelial and stromal tissue. *Cancers* 2021;13:1784.

[30] Kirubabai MP and Arumugam G. View of deep learning classification method to detect and diagnose the cancer regions in oral MRI images. *Med. Legal Update* 2021;21:462–468.

[31] Chang SW, Abdul-Kareem S, Merican AF, and Zain RB. Oral cancer prognosis based on clinicopathologic and genomic markers using a hybrid of feature selection and machine learning methods. *BMC Bioinform.* 2013;14:170–185.

[32] Kourou K, Exarchos TP, Exarchos KP, Karamouzis MV, and Fotiadis DI. Machine learning applications in cancer prognosis and prediction. *Comput. Struct. Biotechnol. J.* 2015;13:8–17.

[33] Exarchos KP, Goletsis Y, and Fotiadis DI. Multiparametric decision support system for the prediction of oral cancer reoccurrence. *IEEE Trans. Inf. Technol. Biomed.* 2012;16:1127–1134.

[34] Speight PM, Elliott A, Jullien JA, Downer MC, and Zakzrewska JM. The use of artificial intelligence to identify people at risk of oral cancer and precancer. *Br. Dent. J.* 1995;179:382–387.

[35] Uthoff RD, Song B, Birur P, *et al.* Development of a dual-modality, dual-view smartphone-based imaging system for oral cancer detection. In *Proceedings of SPIE 10486, Design and Quality for Biomedical Technologies XI*, 2018. 10486. https://doi.org/10.1117/12.2296435.

[36] van Staveren HJ, van Veen RL, Speelman OC, Witjes MJ, Star WM, and Roodenburg JL. Classification of clinical autofluorescence spectra of oral leukoplakia using an artificial neural network: a pilot study. *Oral Oncol.* 2000;36:286–293.

[37] Wang CY, Tsai T, Chen HM, Chen CT, and Chiang CP. PLS-ANN based classification model for oral submucous fibrosis and oral carcinogenesis. *Lasers Surg. Med.* 2003;32:318–326.

[38] Bowling M, Fürnkranz J, Graepel T, and Musick R. Machine learning and games. *Mach. Learn.* 2006;63:211–215.

[39] Patil S, Albogami S, Hosmani J, *et al.* Artificial intelligence in the diagnosis of oral diseases: applications and pitfalls. *Diagnostics* 2022;12:1029.

[40] Tseng WT, Chiang WF, Liu SY, Roan J, and Lin CN. The application of data mining techniques to oral cancer prognosis. *J. Med. Syst.* 2015;39:59.

[41] Kim DW, Lee S, Kwon S, Nam W, Cha IH, and Kim HJ. Deep learning-based survival prediction of oral cancer patients. *Sci. Rep.* 2019;9:1–10.

[42] Lucheng Z, Wenhua L, Meng S, *et al.* Comparison between artificial neural network and Cox regression model in predicting the survival rate of gastric cancer patients. *Biomed. Rep.* 2013;1:757–760.

[43] Campisi G, Di Fede O, Giovannelli L, *et al.* Use of fuzzy neural networks in modeling relationships of HPV infection with apoptotic and proliferation markers in potentially malignant oral lesions. *Oral Oncol.* 2005;41:994–1004.

[44] Exarchos K, Goletsis Y, and Fotiadis D. A multiscale and multiparametric approach for modeling the progression of oral cancer. *BMC Med. Inform. Decis. Mak.* 2012;12:136–150.

[45] Shahul Hameed KA, Shaheer Abubacker KA, Banumathi A, *et al.* Immunohistochemical analysis of oral cancer tissue images using support vector machine. *Measurement* 2020;173:108476.

[46] Rahman TY, Mahanta LB, Das AK, *et al.* Automated oral squamous cell carcinoma identification using shape, texture and color features of whole image strips. *Tissue Cell* 2020;63:101322.

[47] Rahman TY, Mahanta LB, Choudhury H, *et al.* Study of morphological and textural features for classification of oral squamous cell carcinoma by traditional machine learning techniques. *Cancer Rep.* 2020;3:e1293.

[48] Das N, Hussain E, and Mahanta LB. Automated classification of cells into multiple classes in epithelial tissue of oral squamous cell carcinoma using transfer learning and convolutional neural network. *Neural Netw.* 2020;128:47–60.

[49] Folmsbee J, Liu X, Brandwein-Weber M, and Doyle S. Active deep learning: Improved training efficiency of convolutional neural networks for tissue classification in oral cavity cancer. In *2018 IEEE 15th International Symposium on Biomedical Imaging (ISBI 2018)*, IEEE, 2018, pp. 770–773.

[50] Martino F, Bloisi DD, Pennisi A, *et al.* Deep learning-based pixel-wise lesion segmentation on oral squamous cell carcinoma images. *Appl. Sci.* 2020;10(22):8285.

[51] Palaskar R, Vyas R, Khedekar V, Palaskar S, and Sahu P. Transfer learning for oral cancer detection using microscopic images, 2020, arXiv preprint arXiv:2011.11610.

[52] Rodrigues JA, Krois J, and Schwendicke F. Demystifying artificial intelligence and deep learning in dentistry. *Braz. Oral Res.* 2021;35:1–7.

[53] MacHoy ME, Szyszka-Sommerfeld L, Vegh A, Gedrange T, and Wozniak K. The ways of using machine learning in dentistry. *Adv. Clin. Exp. Med.* 2020;29:375–384.

Chapter 2

Oral implantology with artificial intelligence and applications of image analysis by deep learning

Hiroj Bagde[1], Nikhat Fatima[1], Rahul Shrivastav[2], Lynn Johnson[1] and Supriya Mishra[1,3]

To improve the effectiveness of human decision-making and to lessen the burden of the massive amount of work required of the human race, artificial intelligence (AI) has been created. The first AI-based dental solutions have been developed in tandem with the explosion of digital technology in many facets of modern life (AI). AI has been speculated to have game-changing implications for the healthcare sector, allowing for increased efficiency among healthcare workers and brand-new methods of delivering medical treatment. Numerous research has shown that AI and deep learning (DL) by image analysis are making significant contributions to the field of medicine by identifying previously unknown diseases and pinpointing the most effective therapies for individual individuals. Dentistry calls for the development of novel, inventive procedures that are beneficial to both the patient and the practitioner in terms of achieving the most effective and suitable treatment alternatives. The dentistry industry may benefit from the use of AI if it could help with the early diagnosis and correct prognosis of implant cases. Many medical professionals, including experts and generalists, lack the expertise necessary to properly use the DL system through image analysis, which includes the careful planning and interpretation of anatomical data gleaned from radiographic examinations. That is a major issue. It is a problem for dentists, and there is no current treatment for it. Radiographic interpretation using AI systems provides many advantages to the physician and may help alleviate this problem. As an added bonus, it may help dentists avoid wasting time or energy on incorrect diagnoses and treatment plans brought on by the difficulty of the job, laziness, or lack of expertise. However, intelligent robots will never be able to completely replace people in the medical field, despite the fact that AI has the potential to serve as a supplemental tool to enhance diagnostic and therapeutic treatment. Although the area of AI is still in its infancy, it has made significant progress in the medical and dental sectors in recent years. As a consequence of this, it

[1]Department of Periodontology, Rama Dental College, India
[2]Department of Oral Medicine and Radiology, Rama Dental College, India
[3]Department of Periodontology, Government Dental College and Hospital – Raipur, India

is necessary for dentists to keep in mind the potential consequences that it may have for a prosperous clinical practice in the years to come.

2.1 Introduction

Since the 1950s, researchers have been actively studying artificial intelligence (AI), one of the newest subfields of computer science [1]. The use of deep learning (DL) and AI is spreading across the medical and dental communities. AI, as described by John McCarthy, was one of the field's first pioneers "the science and engineering of making intelligent machines" [2]. It is not hard to find places where AI has been put to use. The use of AI in healthcare has been on the rise in recent years, and its results have been encouraging. AI has already found uses in several areas of healthcare, including human biology and dental implants [1].

AI is a subfield of man-made intelligence in which a framework learns how to use factual examples found in a dataset to make forecasts about the way of behaving of new information tests, while man-made reasoning (man-made intelligence) refers to the review, improvement, and examination of any PC framework showing "wise way of behaving" [3]. Simon Cowell coined this term in 1959 [4]. Machine learning's foundational purpose is to discover regularities in new data (test data) for the purpose of performing tasks like classification, regression, and clustering. Training for machine learning algorithms may be done in two ways: supervised and unsupervised. "Classification (deciding what category a given data point belongs to) and regression are two examples of tasks that are often accomplished through supervised training, in which the learning model is fed a collection of input–output pairs of training data (finding a numerical relationship between a set of independent and dependent variables)." However, unsupervised training is often used for tasks like clustering and dimensionality reduction, in which the goal is to merely collect the essential characteristics in a given data set. Third, ML uses algorithms like artificial neural networks (ANNs) to make predictions based on the data it has been fed. These networks are modeled after the human brain and use artificial neurons joined together to process incoming data signals. The idea was proposed in 1943 by Warren McCulloch and Walter Pitts. A stochastic neural analog reinforcement calculator was then developed by Minsky and Dean Edmunds in 1951 [4].

DL, a specialized subfield of machine learning that employs sophisticated techniques based on ANEs, has seen a surge in popularity in recent years (ANN). Because of its superior generalization capabilities, DL has found usage in other fields outside data analytics, including engineering and healthcare. It was not until 2006 that Hinton *et al.* presented the concept of a convolutional neural network (CNN), now often referred to as DL. When processing information, it employs neural networks with several layers. Using data to examine patterns, DL systems may be able to provide better results. In 1969, the backpropagation algorithm was created, and it was this innovation that cleared the door for DL systems. Important turning points in the development of AI are shown in Figure 2.1.

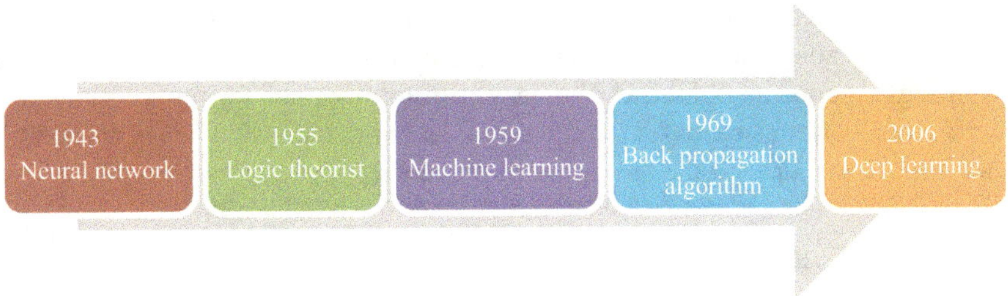

Figure 2.1 Important milestones in the advancement of AI

Recently, AI has been proposed as a tool for healthcare providers to employ in a variety of contexts, including illness diagnosis, prognosis forecasting, and the creation of individualized treatment plans. In particular, AI can help dentists with judgments that must be made quickly yet with great significance. It may alleviate pressure on dentists by eliminating their need to make split-second judgment calls, all while offering patients better, more consistent healthcare [4].

2.2 Clinical application of AI's machine learning algorithms in dental practice

The capacity of computers to store and transmit vast amounts of data has led to a surge in data collection in recent years. The term "big data" is often used to describe this deluge of information. In order to establish reliable standards and precise forecasts, it is now crucial to use novel methods that integrate statistical (mathematical) and computational patterns into data analysis and interpretation. In this light, ML becomes apparent as a subfield of AI focused on data mining. Algorithms using ML may draw meaningful conclusions from historical data, improving their ability to aid in decision-making. ML is used to recognize important data patterns within datasets and to suggest models that best explain the data.

Thanks to recent developments in this field, it is now possible to differentiate between several classes of learning methodologies and algorithms for recognizing patterns in data. It is widely accepted that supervised, unsupervised, and reinforcement learning approaches may help achieve this objective. By constructing mapping functions between the input and output variables, ML facilitates supervised learning. Using labeled variables completes the analysis and produces findings that are more indicative of the actual or intended criteria. Incorporating it into medical practice as an alternative to or in addition to expert opinions is facilitated by this. Supervised learning is one of the most well-known and promising approaches to this problem. Dentists have used these techniques for years to diagnose and classify a wide range of oral and maxillofacial conditions, as well as to forecast the likelihood of disease and other events. When it comes to running algorithms, unsupervised learning merely needs access to data. To achieve this goal, we train the algorithm to identify patterns in the data, to handle non-linear and interactive combinations of multiple predictors, to draw correct conclusions from our analyses, and so on. Unlabeled dental patient datasets may nevertheless be able to detect labels, such as those linked with certain patterns of bone loss owing to periodontal disease. This may help form groups for further study. The accuracy of algorithms based on reinforcement learning has recently been apparent in dental clinical practice via the use of image processing apps [5].

2.2.1 Applications in orthodontics

Most dental ML applications seem to be associated with improved diagnostic capabilities. In orthodontics, ML algorithms' capacity to optimize and increase the use of existing data has tremendously assisted the diagnosis of dental maxillofacial anomalies for the assessment of treatment needs by using training datasets consisting of radiographic images. Extraction of craniofacial landmarks and analysis of cephalometric factors have been used in the past to identify dental abnormalities accurately. As a consequence of frequent variations in observer position, the accuracy of standard methodologies for cephalometric assessments is highly susceptible to numerous mistakes. Modern AI methods help make these fixes more effective. When compared to the accuracy of other classifiers, support vector machine (SVM) algorithms performed best in this proposed technique for automated dental deformity identification. The ML community has shown a growing fondness for SVM-based models. The best way to categorize information in an n-dimensional space is using these classifiers. This technique may be used to improve efficiency in comparison to the dentist in terms of both the amount of photographs needed and the rate at which they are analyzed. The SVM method has been characterized as useful for learning tasks with a high number of attributes. Another aspect that makes SVMs appealing is that their complexity does not change with the size of the training data. Numerous studies have shown the usefulness of applying different neural network algorithms for segmentation, automatic identification, analysis, and extraction of image data in orthodontics, all with the goal of providing a more precise diagnostic tool. Convolutional neural networks (CNNs) techniques were utilized in a recent research with a large dataset, leading to accurate training and a precise model for cranial shape analysis. Model results were consistent with those selected by knowledgeable human examiners, allowing for speedier achievement of the benchmark [6]. It is already common knowledge that CNN methods can greatly enhance picture quality by minimizing blur and noise, hence their usage has spread extensively in the field of dentistry. As their efficacy has been shown in the categorization of dental pictures, these DL techniques are prioritized for use in challenging tasks involving a huge quantity of unstructured data. The use of neural networks to prediction problems has been strongly recommended by recent studies. A genetic algorithm/ANN combination was developed to estimate the eventual size of canine and premolar teeth that have not yet erupted during the mixed dentition period. The effects of orthognathic treatment on facial attractiveness and age appearance have been studied using CNNs, both for typically developing people and those who have had cleft therapy. Further, large-scale hybrid dental data gathered from various healthcare facilities and public registries may be valuable throughout the training phase. If new discoveries are to be made in this area, researchers with an interest should pool their resources and share the information they have gathered [5].

2.2.2 Applications in periodontics

It seems that data on people with periodontal disease, including their molecular profiles, immunological parameters, bacterial profiles, and clinical and radiographic characteristics, may be properly analyzed using ML algorithms [7]. Periodontal disease cannot be diagnosed without first determining bone levels, then detecting bacteria in subgingival fluid samples, and finally analyzing gene expression patterns from periodontal tissue biopsies. ML approaches may be quite useful in situations when this diagnosis is difficult for early practitioners to make during normal exams. Typically, SVM was utilized as the analytical classifier in these research. Other algorithms, including naive Bayes, ANNs, and decision tree algorithms, were utilized in two of the investigations. A decision tree is a kind of tree structure that offers a hierarchical arrangement starting at the top, with a root node. The decision tree is thought to provide a clear analysis with high levels of transparency and explanatory power. Several authors have offered suggestions on how the models' effectiveness and clarity may be improved. Bootstrap aggregated (or bagged) decision trees, random forest trees, and boosted trees are all examples of ensemble approaches that may be used to construct more complex tree models. For example, naive Bayes is advantageous in the medical field since it is easy to understand, takes use of all available data, and provides a clear rationale for the final choice.

2.2.3 Applications in oral medicine and maxillofacial surgery

Maxillofacial cyst segmentation and identification, as well as the detection of other common mouth diseases, are only two examples of the expanding applications of ML-based diagnostics in the field of oral medicine and maxillofacial surgery during the last decade. Although it is often more time-consuming and costly than other diagnostic procedures, histopathological examination is the gold standard for diagnosing these lesions. Therefore, it is very important to pay particular attention to the development of novel approaches that speed and enhance diagnostic procedures. Many studies have shown that the CNNs algorithm is effective for this task. A database of 3D cone beam computed tomography (CBCT) images of teeth was analyzed, and SVM was shown to have the highest accuracy (96%) in categorizing dental periapical cysts and keratocysts. In addition to ANNs, k-NN, naive Bayes, decision trees, and random forests were also explored. The k-NN classifier uses the idea that adjacent data points have similar characteristics. k-NN is said to be very sensitive to superfluous details, which might hamper learning and cloud the model's interpretability. This classifier's transparency, reflecting the intuition of human users, is, nonetheless, what makes it so valuable. The naive Bayes classifiers have this quality as well. Overall, it is not easy to choose the right algorithm for a given goal. Further, the characteristics and pre-processing of the datasets may affect the evaluation of their performances. Selection case-based reasoning (CBR) has been used in analysis, according to other research. CBR gives input by accumulating and learning from past situations. Thus, even if new cases may be introduced, new rules may be established. Similar clinical manifestations are seen in a wide variety of oral cavity disorders, which may make accurate diagnosis challenging. The diagnostic accuracy and patient compliance with the prescribed treatment plan are compromised as a result. The CBR technology has been helpful in creating a thorough and methodical strategy for the one-of-a-kind identification of these illnesses, with the goal being a more precise definition of similarities and differences between them. Although preliminary, the findings demonstrate that the algorithms have promise for enhancing the standard of care and facilitating more effective treatment. These methods have also been put to use in other contexts, such as the identification and segmentation of structures, the classification of dental artifacts for use in image verification, and the classification of maxillary sinus diseases. There has also been a lot of talk about how to apply ML algorithms to anticipate perioperative blood loss in orthognathic surgery. It is feasible that a random forest classifier might be used to estimate the expected perioperative blood loss and so prevent any unanticipated issues during surgery. This forecast has the potential to aid in the management of elective surgical operations and improve decision-making for both medical professionals and their patients. Researchers have employed ANNs to make accurate diagnoses in situations involving orthognathic surgery, with 96% accuracy.

2.2.4 Applications in forensic dentistry

The developments of these contemporary instruments have also influenced forensic dentistry and anthropological tests. With an accuracy of 88.8%, CNNs algorithms were used in one study to categorize tooth kinds using dental CBCT images. Estimating age from teeth is crucial in forensic dentistry, thus naturally there have been a number of research examining the use of automated approaches in this area. Estimating age by the degree of tooth growth is a laborious and intricate manual process. The examination of bone patterns using computerized algorithms was the subject of an intriguing research investigation. Automated methods that help increase the accuracy and speed of ML-based age estimate have a great deal of practical value and need additional research and testing.

2.2.5 Applications in cariology

The results of this line of inquiry have showed promise, with possible applications including the development of image-based diagnostic tools for caries lesions and the prediction of the disease's prognosis. Caries, or cavities, in teeth continue to be a problem for many individuals. Dentists in practice and their patients will welcome the introduction of novel options and technologies that promise to enhance the state-of-the-art methods of diagnosis and prognosis. Out of a massive dataset, researchers were able to find models with remarkable performance for predicting adult root caries. When compared to the other algorithms employed for root caries diagnosis, the SVM-based technique performed the best, with a 97.1% accuracy rate, a 95.1% precision rate, a 99.6% sensitivity rate, and a 94.3% specificity rate. Cross-sectional data were used in the research, which raised several red flags regarding the model's predictive power. The use of longitudinal data in research is recommended for greater generalization and validation of findings. In addition, a study with a solid methodological basis examined the use of graphical regression neural networks (GRNNs) to predict caries in the elderly, and the results were promising: the model's sensitivity was 91.41% on the training set and 85.16% on the test set. A GRNN is a sophisticated kind of nonparametric regression-based neural network. The ability of these algorithms to generate predictions and compare the performance of systems in practice is greatly enhanced by the fact that they need just a few number of training samples to converge on the underlying function of the data. With just a tiny amount of supplementary information, the modification may be obtained effectively and with no more intervention from the user. The cost-effectiveness of these technologies was analyzed for the first time in a ground-breaking research; this is a crucial factor in deciding whether or not to use them in a clinical setting. In conclusion, the research showed promise for the use of automated approaches to the detection of caries.

2.2.6 Applications in endodontics

Finding the minor apical foramen on radiographs by feature extraction stands out among the many uses of this technology. The use of ANNs in this *ex vivo* research showed encouraging findings that need further investigation. Root canal treatments often have better outcomes if they begin with an accurate assessment of the working length. The development of more precise methods for pinpointing the root canal's point of maximum restriction (minor apical foramen) is a major driving force behind the resurgence of interest in clinical endodontics. The purpose of this research was to assess the efficacy of ANNs in the detection of vertical root fracture in a modest set of removed teeth. A bigger sample size of teeth and different dental groups may provide more trustworthy findings, notwithstanding the success of the current study. A look back at earlier investigations that had obtained in vivo data was also performed. CNNs algorithms were tested in these research for their ability to detect vertical root fractures on panoramic radiographs, count the roots on CBCT images of mandibular first molars, and diagnose periapical pathosis. Studies have shown that DL systems can reliably determine whether a patient has one or two roots in the distal roots of their mandibular first molars (89% accuracy) by analyzing low-cost radiography images using CNNs.

2.2.7 Applications in prosthetics, conservative dentistry, and implantology

The use of ML methods has also aided progress in other areas of dentistry, including as prosthodontics, conservative dentistry, and implantology. There has been a push in these areas to use ANN-based algorithms that aid in the prediction of face distortion after implantation of a full prosthesis. The experimental findings validated the method's ability to anticipate the deformation of face soft tissues rapidly and correctly, providing important information for determining next steps in therapy. ANNs have been studied for their potential to improve tooth tone in PC-assisted frameworks, while SVMs have been studied for their use in automating the identification and ordering of dental restorations in panoramic images. A high level of accuracy (93.6%) was found in the SVM analysis. In other contexts, CNNs have been used to foretell the likelihood that computer-aided-design-manufactured composite resin crowns may fall out. In another investigation, the XGBoost algorithm was utilized to create a denture-related tooth extraction treatment prediction clinical decision support model. With a 96.2% success rate, the algorithm demonstrated its efficacy as a robust classifier and regressor, with its best results often being found in structured data. CNNs have shown promising results in two investigations of implantology, both of which focused on the detection of implant systems. Patients who need more aesthetically and functionally sound dental prosthesis rehabilitation have made dental implant insertion a frequent kind of rehabilitation therapy in recent years. Because of the wide variety of implant systems now in use, it may be difficult for clinical dentists to distinguish between them based on standard radiography imaging alone due to the variety of fixation structures and features they use. If these systems were properly identified, patients who needed repairs or repositioning of their implant systems would have fewer instances that required more intrusive procedures. Predicting patients' average peri-implant bone levels is another use in this field; this helps with calculating implant survival and investigating possible therapy avenues that lead to the best results.

2.3 Role of AI in implant dentistry

Clinical preference has always favored dental implants for treating whole, partial, and single-tooth edentulism. Successful implant surgery depends on careful pre-operative planning that allows the implant to be placed in the ideal location while minimizing or avoiding any potential complications. Alveolar bone parameters (bone quality, bone thickness, and bone level) and physical variations in the useable site are surveyed using a variety of radiographic techniques that are embedded in a medical operation (like nasal fossa, mandibular trench, mental foramen, and sinuses). Despite the fact that dental embed medicine relies on straightforward radiographic techniques like all-encompassing and intraoral x-rays to provide an outline of the jaws and promote a vital concept, these approaches are insufficient for comprehensive embed arranging. These techniques are being replaced by computed tomography (CT) and CBCT, which provide cross-sectional tomograms from which experts may access 3D data (CBCT). When comparing CT scanners to redesigned CBCT devices for mark maxillofacial imaging, the latter option is more cost- and space-efficient. Inspecting takes less time now without compromising image quality. CBCT devices may accurately predict the need for useful systems (such as directed tissue recovery, splitting, and sinus rise) and help decide the optimum embed sizes (i.e., length and breadth) prior to surgery in cases when there is insufficient bone at the cautious site. Nevertheless, the doctor's expertise in reading CBCT pictures is crucial for the success of the implant design process.

DL and other recent developments in machine learning are facilitating the recognition, categorization, and quantification of patterns in medical pictures, which aids in the diagnosis and treatment of many illnesses [8].

2.3.1 Use of AI in radiological image analysis for implant placement

X-rays and computerized tomography (CT) scans are just two examples of the medical imaging tools that dentists have been employing for decades to diagnose issues and plan treatment. Today, dental professionals rely heavily on computer tools to aid them in the diagnosis and treatment of such conditions [9].

Using AI has allowed for the development of CAD systems for use in radiology clinics for the purpose of making accurate diagnoses. An effective DL application applied on medical diagnostic pictures is the deep convolutional neural network (DCNN) approach. Tooth numbering, periapical pathosis, and mandibular canal recognition are just a few of the dental diagnoses that have benefited from this technique, which also allows the analysis of more complicated pictures like CBCT imaging. Despite the importance of radiographic image assessment and precise implant design and interpretation of anatomical data, many experts and general practitioners lack the necessary expertise in these areas. This scenario creates difficulties for dentists and has yet to be resolved. The use of AI systems in radiographic interpretation offers several benefits to the doctor and may help with this issue. In dentistry, this might also mean less time wasted on incorrect diagnoses and treatment plans and less work for you [10].

Several DL-based algorithms have also been studied in medical image analysis procedures involving a wide range of organs, including the brain, the pancreatic, breast cancer diagnostics, and the identification and diagnosis of COVID-19. Dental implant identification might benefit from DL's established efficacy in the field of medical imaging. Recognizing dental implants is critical for many areas of dentistry, including forensic identification and reconstructing damaged teeth and jaws. Implants in the context of implant dentistry provide patients enticing prosthetic repair options. The accurate classification of a dental implant placed in a patient's jaw prior to the availability of dental records is a significant challenge in clinical practice. To determine the manufacturer, design, and size of an implant, dentists will commonly examine an X-ray picture of the device. This data is useful for determining the implant's connection type. As soon as the tooth is extracted, the dentist may place an order for a new abutment and a replacement tooth. When the improper abutment or replacement tooth is purchased, it may be quite expensive for dentists. Therefore, it stands to reason that dentists might benefit greatly from an automated system that analyzes X-rays of patients' jaws to determine which category the patient's dental implant best fits into [8]. Using periapical and panoramic radiographs, several AI models have been built for implant image identification. Additionally, dental radiographs have been exploited by AI models to identify periodontal disease and dental cavities. AI has also been used to optimize dental implant designs by integrating FEA calculations with AI models, and prediction models for osteointegration success or implant prognosis have been developed utilizing patient risk variables and ontology criteria [3].

Radiographic images of 10,770 implants of three different kinds were used to train the deep CNN model developed by Lee and Jeong. The authors compared the implant recognition abilities of several examiners (board certified periodontists and the AI model) and different types of radiography images (periapical, panoramic, and both). While there was some variation in accuracy for recognizing implants among the three kinds evaluated, using both periapical and panoramic photos improved the AI model's and the periodontists' specificity and sensitivity [3].

2.3.2 Deep learning in implant classification

DL methods have seen extensive usage in other related situations, such as the categorization of dental implants. Implant identification using transfer learning (TL) and periapical radiographs was found to have a 98% success rate. The same approach yielded similar findings, with the exception of X-ray pictures. In another experiment, radiograph pictures and CNNs were used to make predictions about dental implants from various companies [11].

2.3.3 AI techniques to detect implant bone level and marginal bone loss around implants

There has been significant development in the use of AI in healthcare recently, and this has implications for digital dentistry and telemedicine. When it comes to recognizing and categorizing objects, CNNs thrive. Studies on dental caries, osteoporosis, periodontal bone loss, impacted primary teeth, and dental implants, among others, have all made use of CNNs for counting teeth and collecting data. CNNs could be able to recognize images directly from raw data without any human feature extraction. R-CNNs were developed specifically for use in object identification tasks; they can recognize and label areas of interest that include the targets of a given identification task automatically. An improved version of R-CNN, known as Faster R-CNN, was created later. Constructed on top of Faster R-CNN, the Mask R-CNN method successfully detects targets in images and provides precise segmentation results. Identifying periapical radiographic evidence of marginal bone loss surrounding dental implants has been the focus of a few studies that have used faster R-CNN [12]. For the purpose of forecasting implant bone levels and dental implant failure by bagging, several research have employed clinical and SVMs or trees models [11].

2.3.4 Comparison of the accuracy performance of dental professionals in classification with and without the assistance of the DL algorithm

In a recently published study, a robotic DL model was able to accurately locate and rank dental embed frameworks (DISs) from dental radiography images. Both the location (AUC = 0.984; 95% CI 0.900–1.000) and the order (AUC = 0.869; 95% CI 0.778–0.929) of broken inserts were more accurately predicted by the robotic

DL model employing periapical pictures than by the pre-prepared and modified VGGNet-19 and GoogLeNet models. Not only has the automated DL model for full-mouth and periapical images shown highly accurate performance (AUC = 0.954; 95% CI 0.933–0.970), but its results are on par with or better than those of dental experts like board-certified periodontists, periodontics residents, and dental specialists without training in implantology [13].

2.3.5 AI in fractured dental implant detection

Due to their excellent survival and success rates, dental implants (DIs) have established themselves as a crucial and reliable therapeutic option for replacing lost teeth. A recent comprehensive examination of DI rehabilitation outcomes indicated that the cumulative survival rate after 15 years of follow-up was 82.6%, whereas the survival rate after 10 years was reported to be 96.4% (95% CI 95.2%–97.5%). A broad range of natural concerns (such as peri-embed mucositis and peri-implantitis) and mechanical issues (such as chipping, screw loosening and cracking, and clay and equipment breakage) that need additional medications may thus become more common. One of the most fundamentally difficult-to-fix or-change mechanical faults that may lead to DI disappointment and explantation is a crack. Potentially the most well-known risk factors for DI fracture are biomechanical and physiological strain and stress associated with a non-latent prosthetic fit. The likelihood of DI fracture in late examinations can be affected by a number of clinical factors, including age, sex, DI width, length, situation position, bone join history, apparatus material (CP4 vs. amalgam), cleaned versus unpolished cervical component, butt versus tapered projection association, miniature versus full scale string, and stage exchanging. For 19,006 broken DIs in 5,125 patients, a late analysis with a 12-year follow-up suggested a recurrence of 0.92; however, a rigorous examination of long-term outcomes for over 5 years identified a percentage of 0.18. Early identification of fracture is a tricky undertaking in real clinical practice due to the condition's low frequency and incidence and the fact that it is frequently asymptomatic. If a DI fracture goes undetected or is found too late, substantial bone loss may result in the area surrounding the fracture due to post-traumatic and inflammatory responses. In the recent decade, advances in AI have allowed its widespread use in the medical and dentistry disciplines; in particular, DL and neural network-related technologies.

Researchers found that although VGGNet-19 and GoogLeNet Inception-v3 performed similarly well, the automated DCNN architecture employing solely periapical radiography images performed the best in detecting and classifying fractured DIs. To learn whether or not DCNN architecture can be used in dental practice, further prospective and clinical data is required [14].

2.4 Software initiatives for dental implant

Software such as Digital Smile Design (DSD), 3Shape (3Shape Design Studio and 3Shape Implant Studio), Exocad, and Bellus 3D are just a few of the options available to dentists today that practice digital dentistry. They hoped that by coordinating efforts across disciplines and making more use of digital dentistry, they might better ensure patients received treatment that was both timely and predictable.

3Shape has created a suite of specialized software applications that provide an end-to-end digital workflow, from diagnosis to treatment planning to prosthetic process and implant design and visualization. More importantly, it provides sufficient adaptability for the dental practitioner to make any necessary adjustments. An intraoral digital scanner is required for use with these applications, which process digital stills and moving pictures. In addition to being able to see and alter teeth, the program also allows for the construction of 3D implants using a wide range of pre-existing manufacturers and customization choices. In addition, it works with specialized printers to produce the final output. To create dental implants and other dental applications in a completely digital workflow, Exocad may be used as a CAD (computer-aided design) tool. A custom tooth set may be created from scratch using a variety of methods, one of which is by importing a single tooth or a whole set of teeth from one of numerous dental libraries. When working with a 3D model created using 3Shape, it is simple to make adjustments like moving the teeth around or enlarging them. The program facilitates implant design with an intuitive interface that walks the user through the process's numerous sophisticated possibilities. The whole facial structure may be scanned in 3D with Bellus 3D Dental Pro Integration. The primary goal of this program is to streamline the patient acceptance process and improve the efficacy of dental treatment by integrating the treatment plan with the patient's facial configuration in full 3D [11].

2.5 AI models and implant success predictions

Papantonopoulos *et al.* [15] with the use of demographic, clinical, and radiological data from 72 people with 237 implants, sought to classify prospective implant "phenotypes" and predictors of bone levels surrounding implants. Using AI, scientists mapped implant locations, finding two separate populations of prostheses. The scientists interpreted these groups to stand for two different "phenotypes" of implants: those that are susceptible to peri-implantitis and those that are resistant to it. In order to determine the stress at the implant-bone contact, Li *et al.* used an AI approach, taking into account the implant's length, thread length, and thread pitch rather than a FEA model. The primary goal of the AI model was to determine the values of design factors that would both reduce stress at the implant-bone interface and improve the implant's lifetime. Comparing the FEA model to experimental data, we find that stress at the implant-bone contact is reduced by 36.6% in the former. In lieu of FEA calculations, Roy *et al* used + GA to maximize the porosity, length, and diameter of the implant [16]. To create a neural network architecture, Zaw *et al.* similarly used a reduced-basis approach to modeling the responses of the dental implant-bone system. The suggested AI method proved successful in calculating the implant-bone interface's elastic modulus. While there was consensus across studies that AI models might be used to enhance implant designs, researchers acknowledged that further work was required to refine AI calculations for designing implants and assess their efficacy in in-vitro, animal, and clinical settings [3].

2.6 Discussion

AI systems have been shown to accurately identify a wide variety of dental anomalies, including dental caries, root fractures, root morphologies, jaw pathologies, periodontal bone damages, periapical lesions, and tooth count, according to the dental literature. Before DCNN applications were used in dentistry, research like these analyzed data from several dental radiographic images such periapical, panoramic, bitewing cephalometric, CT, and CBCT. To be sure, there is not a lot of research using CT and CBCT. According to a CBCT study by Johari *et al.*, the probabilistic neural network (PNN) method is effective in identifying vertical root fractures [17]. Hiraiwa *et al.* also found that AI was able to detect impacted teeth in CBCT with acceptable results [17]. Good news for the future of this field was revealed in a study of periapical lesions in CBCT images by Orhan *et al.* (2019), who discovered that volume estimations predicted using the CNN approach are congruent with human measurements [18]. In both dentistry and medicine, treatment planning is a crucial process stage. If the therapy is to be effective, it is necessary to first arrive at the proper diagnosis, which may then be used to develop the most appropriate treatment plan for the individual patient. Planning a course of treatment requires extensive organization and is heavily dependent on a number of variables, including the doctor's level of expertise. Over the last several years, AI systems have been utilized to help doctors with anything from diagnosis to treatment planning. Promising outcomes were achieved using the neural network machine learning system in conjunction with a variety of treatment modalities, including radiation therapy and orthognathic surgery. Dental implant planning relies heavily on radiographic imaging, as is common knowledge. Before a surgery, it is advisable to use 3D imaging technology to inspect the area and make precise preparations by taking a number of measures in accordance with the anatomical differences expected. The key anatomic variables that influence the implant planning are the mandibular canal, sinuses, and nasal fossa that were examined in the present research. In a recent paper, Kwak *et al.* found that the CNN approach worked well for identifying the mandibular canal in CBCT images, suggesting that this might be a future potential for dental planning [19]. To find out where the third mandibular molar should be in relation to the mandibular canal, Fukuda *et al.* examined 600 panoramic radiographs. According to our best estimation, Jaskari *et al.* have successfully used the CNN method to segment the mandibular canal in all CBCT images. AI algorithms, they said, provide sensitive and trustworthy findings in canal determination, suggesting a potential role for AI in implant design in the future [20].

The accuracy of the measures used in implant planning will improve along with the accuracy of AI's ability to identify anatomical components. But there is at least one research that has been done, and the findings of which use AI to successfully identify sinus diseases in panoramic photos [21]. Bone thickness and height were measured for this investigation to see how well the implant planning had gone. This research demonstrates the need for a DL system to enhance AI bone thickness assessments. As a result, doctors will appreciate the use of these technologies in implant design, and the field of implantology will benefit from the added stability they provide [10].

2.7 Final considerations

More and more dentistry subfields are already using ML for usage in their practices. A number of aspects of dental clinical practice might benefit from the use of algorithms like CNNs and SVMs. They provide a wealth of resources for enhancing clinical decision-making and aiding in diagnosis and prognosis. Large volumes of sensitive information need careful consideration of the ethical implications of accessing and using this data. Extraction of meaningful models from raw data requires careful preprocessing. These cutting-edge methods should inform the development of longitudinal research designs and the validation of findings in clinical trials. In order to ensure effective and generalizable model usage, external validation must be expanded. In addition, for a better grasp of such recommendations, the standardization of procedures for presenting these findings in clinical practice should be systematically enhanced. Researchers in the dental sector need to do more work to verify the efficacy of these models before advocating for their widespread use in clinical practice.

References

[1] Alharbi MT and Almutiq MM. Prediction of dental implants using machine learning algorithms. *J Healthc Eng* 2022;2022:7307675. doi:10.1155/2022/7307675. PMID: 35769356; PMCID: PMC9236838.

[2] Reddy S, Fox J, and Purohit MP. Artificial intelligence enabled healthcare delivery. *J R Soc Med* 2019;112(1):22–28.

[3] Revilla-León M, Gómez-Polo M, Vyas S, *et al.* Artificial intelligence applications in implant dentistry: a systematic review. *J Prosthetic Dentistry* 2021;129:293–300. doi:10.1016/j.prosdent.2021.05.008.

[4] Patil S, Albogami S, Hosmani J, *et al.* Artificial intelligence in the diagnosis of oral diseases: applications and pitfalls. *Diagnostics* 2022;12:1029. https://doi.org/10.3390/diagnostics12051029.

[5] Reyes LT, Knorst JK, Ortiz FR, and Ardenghi TM. Scope and challenges of machine learning-based diagnosis and prognosis in clinical dentistry: a literature review. *J Clin Transl Res* 2021;7(4):523–539. PMID: 34541366; PMCID: PMC8445629.

[6] Kunz F, Stellzig-Eisenhauer A, Zeman F, and Boldt J. Artificial intelligence in orthodontics. *J Orofac Orthop Fortschritte Kieferorthop* 2020;81:52–68.

[7] Papantonopoulos G, Takahashi K, Bountis T, and Loos BG. Artificial neural networks for the diagnosis of aggressive periodontitis trained by immunologic parameters. *PLoS One* 2014;9:e89757.

[8] Kohlakala A, Coetzer J, Bertels J, and Vandermeulen D. Deep learning-based dental implant recognition using synthetic X-ray images. *Med Biol Eng Comput* 2022;60(10):2951–2968. doi:10.1007/s11517-022-02642-9. Epub 2022 Aug 18. PMID: 35978215; PMCID: PMC9385426.

[9] Khan, Nag MVA, Mir T, and Dhiman S. Dental image analysis approach integrates dental image diagnosis. *Int J Cur Res Rev* 2020;12:16:47–52.

[10] Kurt Bayrakdar S, Orhan K, Bayrakdar IS, *et al.* A deep learning approach for dental implant planning in cone-beam computed tomography images. *BMC Med Imaging* 2021;21:86. https://doi.org/10.1186/s12880-021-00618-z.

[11] Carrillo-Perez F, Pecho OE, Morales JC, *et al.* Applications of artificial intelligence in dentistry: a comprehensive review. *J Esthet Restor Dent.* 2022;34 (1):259–280. doi:10.1111/jerd.12844. Epub 2021 Nov 29. PMID: 34842324.

[12] Liu M, Wang S, Chen H, and Liu Y. A pilot study of a deep learning approach to detect marginal bone loss around implants. *BMC Oral Health* 2022;22(1):11. doi:10.1186/s12903-021-02035-8. PMID: 35034611; PMCID: PMC8762847.

[13] Lee JH, Kim YT, Lee JB, and Jeong SN. Deep learning improves implant classification by dental professionals: a multi-center evaluation of accuracy and efficiency. *J Periodontal Implant Sci.* 2022;52(3):220–229. doi:10.5051/jpis.2104080204. PMID: 35775697; PMCID: PMC9253278.

[14] Lee D-W, Kim S-Y, Jeong S-N, and Lee J-H. Artificial intelligence in fractured dental implant detection and classification: evaluation using dataset from two dental hospitals. *Diagnostics* 2021;11:233. https://doi.org/10.3390/ diagnostics11020233

[15] Papantonopoulos G, Gogos C, Housos E, Bountis T, and Loos BG. Prediction of individual implant bone levels and the existence of implant "phenotypes". *Clin Oral Implants Res* 2017;28:823–832.

[16] Roy S, Dey S, Khutia N, Roy Chowdhury A, and Datta S. Design of patient specific dental implant using FE analysis and computational intelligence techniques. *Appl Soft Comput* 2018;65:272–279.

[17] Johari M, Esmaeili F, Andalib A, Garjani S, and Saberkari H. Detection of vertical root fractures in intact and endodontically treated premolar teeth by designing a probabilistic neural network: an ex vivo study. *Dentomaxillofac Radiol* 2017;46:20160107.

[18] Orhan K, Bayrakdar I, Ezhov M, Kravtsov A, and Özyürek T. Evaluation of artifcial intelligence for detecting periapical pathosis on cone-beam computed tomography scans. *Int Endod J* 2020;53:680–689.

[19] Kwak GH, Kwak E-J, Song JM, *et al.* Automatic mandibular canal detection using a deep convolutional neural network. *Sci Rep* 2020;10:1–8.

[20] Jaskari J, Sahlsten J, Järnstedt J, *et al.* Deep learning method for mandibular canal segmentation in dental cone beam computed tomography volumes. *Sci Rep* 2020;10:1–8.

[21] Kim Y, Lee KJ, Sunwoo L, *et al.* Deep learning in diagnosis of maxillary sinusitis using conventional radiography. *Investig Radiol* 2019;54:7–15.

Chapter 3

Review of machine learning algorithms for breast and lung cancer detection

Krishna Pai[1], Rakhee Kallimani[2], Sridhar Iyer[3] and Rahul J. Pandya[4]

In the innovative field of medicine, malignant growth has attracted significant attention from the research community due to the fact that genuine treatment of such diseases is currently unavailable. In fact, diseases of such types are so severe that the patient's life can be saved only when the disease is identified in the early stage, i.e., stages I and II. To accomplish this early-stage disease identification, machine learning (ML) and data mining systems are immensely useful. Specifically, using the large available data existing over the web-based repositories, ML techniques and data mining can be implemented to gather valuable information in view of cancer identification or classification. This chapter is oriented towards the aforementioned with an aim to conduct a point-by-point study of the most recent research on various ML techniques such as Artificial Neural Networks (ANNs), k-Nearest Neighbours (KNN), Support Vector Machines (SVMs), and Deep Neural Networks (DNNs). The main contribution of the chapter is the review followed by the decision on the 'best' calculation for an a priori finding of breast and lung malignancy. The crude information from the mammogram or tomography images or the datasets which have been obtained are utilized as the information. The pre-processing of the data and related processes are conducted following which the best prediction model is obtained. Also, the processing time for testing, training, and compliance of all the cases is determined. The results of this study will aid in determining the most appropriate ML technique for the detection of tumours in breast and lung cancer.

[1]Department of Electronics and Communication Engineering, KLE Technological University, Dr. M.S. Sheshgiri College of Engineering & Technology, Belagavi Campus, India
[2]Department of Electrical and Electronics Engineering, KLE Technological University, Dr. M.S. Sheshgiri College of Engineering & Technology, Belagavi Campus, India
[3]Department of CSE(AI), KLE Technological University, Dr. M.S. Sheshgiri College of Engineering & Technology, Belagavi Campus, India
[4]Department of Electrical and Electronics Engineering, Indian Institute of Technology, Dharwad, WALMI Campus, India

3.1 Introduction

The aberrant cell division which leads to the body's cells becoming irregular causes the growth of tumours. These unregulated aberrant cell divisions have the potential to kill healthy body tissues and end up forming a mass, generally termed a tumour, whose growth causes multiple disorders. These tumours can be broadly classified into two types namely, malignant and benign [1]. The tumour with high rates of spreading and influencing capabilities to other healthy parts of the body is called a malignant tumour. On the other hand, benign tumours are known for not spreading or influencing other healthy parts of the body. Hence, not all tumours move from one part of the body to another, and not all types of tumours are necessarily carcinogenic.

Growth of tumours, weight reduction, metabolic syndrome, fertility, lymphoedema, endocrine, peripheral neuropathy, cardiac dysfunction, pulmonary, altered sleep, psychological, fear of recurrence, long haul, fatigue, and irregular bleeding are few of the short-term and long-term side effects faced by the cancer survivors and patients [2,3]. The processes of detection, comprehension, and cure are still in the early stages and are a focused research domain within the therapeutic sector. In the past and currently, the traditional cancer diagnosis process has had many limitations such as high dependence on the patient's pathological reports, multiple clinical trials and courses, and slow indicative procedure [4]. With a prevalence of 13.6% in India, breast cancer is the second most common cause of death for women [5]; Nonetheless, tobacco smoking causes 90% of occurrences of lung cancer, the most common cause of death among males [6]. Unfortunately, lung cancer, along with breast cancer, is a significant cause of mortality in women [7,8]. Considering the year 2020, Figures 3.1 and 3.2 illustrate the top five common cancers which cause the majority of mortality in males and females around the globe [5].

It has been found that early-phased treatment can either postpone or prevent patient mortality when a patient is early diagnosed with cancer in the breast or lung. However, with the current technology, early identification is difficult, either

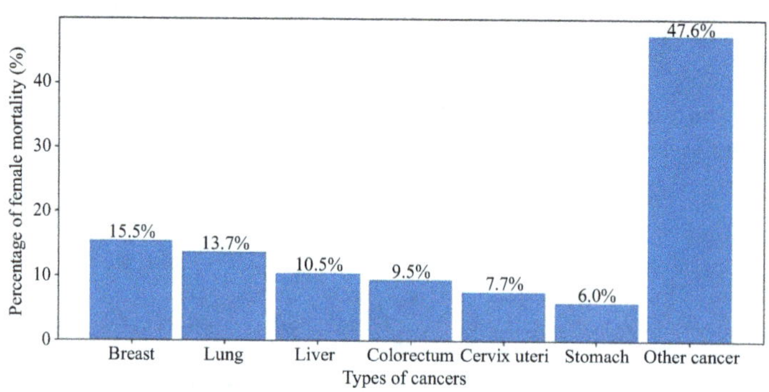

Figure 3.1 For the year 2020, the top five most commonly occurring cancers cause the majority of female mortality worldwide

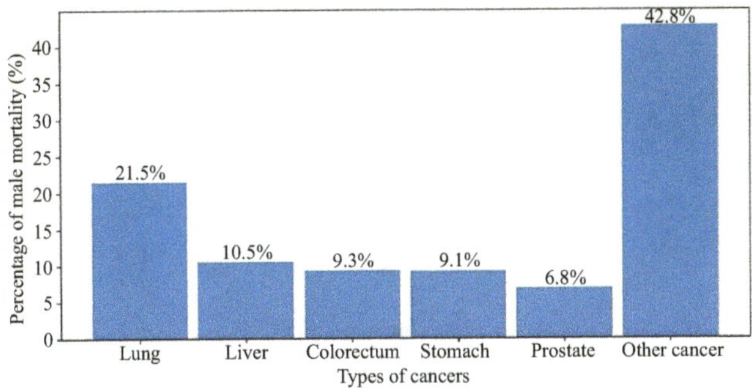

Figure 3.2 For the year 2020, the top five most commonly occurring cancers cause the majority of male mortality worldwide

because medical equipment is unavailable, or due to the negligence of the potential victim. To tackle this issue, potential high-level solutions are needed to be integrated with existing technologies such as tomography and mammography. The implementation of ML and deep learning (DL) algorithms such as KNN, SVM, and convolutional neural network (CNN) can be integrated with the existing technologies to improve the accuracy, precision, and early detection of the potential cancer victims even before the condition moves beyond phase 1. Along these lines, this chapter reviews the various available methods for cancer detection and provides an overview of the proposed model for future implementation.

The taxonomy of this chapter is shown in Figure 3.3. The chapter is organized as follows. In Section 3.2, we provide the literature review followed by Section 3.3 in which we discuss the reviewed methods and accuracies with respect to their performance. Section 3.4 details the overview of the proposed model for cancer detection. Finally, Section 3.5 concludes the chapter.

3.2 Literature review

The authors in [9] stated the existing limitations faced while the detection and classification of breast cancer using traditional methods such as computed tomography (CT), 3D mammography, magnetic resonance imaging (MRI), and histopathological imaging (HI). Overcoming these limitations is the key as invasive breast cancer is the second leading cause of women's mortality considering that every one in eight women in the USA suffers due to this disease throughout their lifetime. The traditional methods are prone to a larger range of errors and highly expensive processes. Analysis of images obtained from the above-mentioned methods can only be possible by an experienced pathologist or radiologist. Hence, advanced methods such as ML algorithms can be used over a larger dataset consisting of MRI, CT, mammography, ultrasound, thermography, and histopathology images. This will help in the early and accurate prediction of breast cancer to help experienced and unseasoned pathologists or radiologists. The scope will be to develop a completely automated and unified framework for accurate classification with minimal effort.

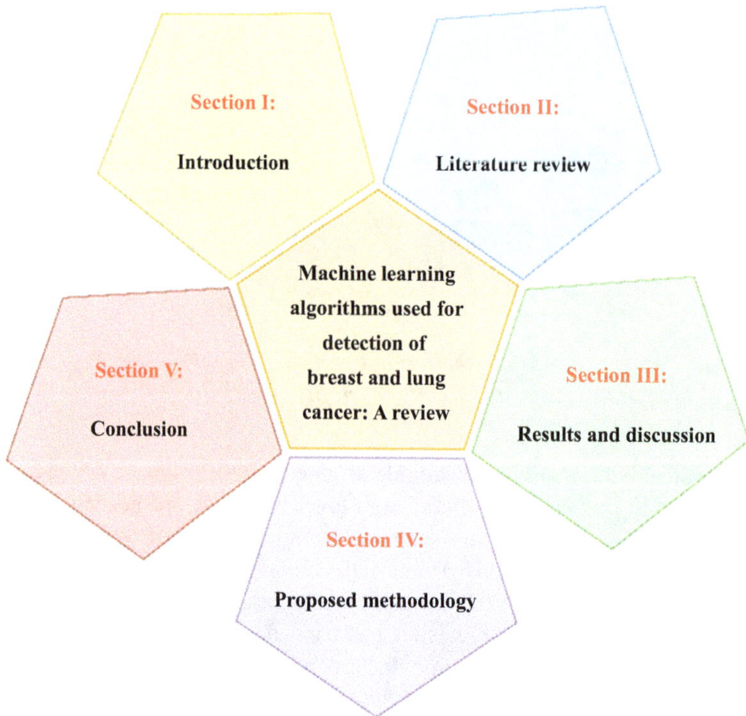

Figure 3.3 Taxonomy of the chapter

The authors in [10] developed an opportunity to improve the mammography-based medical imaging process. The traditional ongoing process was found less efficient in breast cancer detection. With an exponential rise in the ageing population in the regions of Malaysia, the risk of breast cancer development has raised to 19% according to the Malaysia National Cancer Registry Report (MNCRR) 2012–2016 [11]. This implies that everyone in 19 women suffers from life-threatening cancer. The authors conducted a comparative performance study between two DL networks using the image retrieval in medical application (IRMA) dataset [12] consisting of 15,363 images. VGG16, with a total of 16 layers network, was observed to perform better with a difference of 2.3% over ResNet50 with 152 layers.

Authors [13] have researched radiographic images and have developed an efficient automatic system for nodular structures. The proposed system aids early diagnosis for radiologists by detecting the initial stage of cancer. Also, classification using the curvature peak space is demonstrated. The entire system uses three blocks: block 1 used the normalization process and enhanced quality of the image structures; block 2 used the segmentation process to find the suspected nodule areas (SNA); block 3 used the classification of the SNAs. The authors reduced the number of false positives (FP) per image and demonstrated a high degree of sensitivity. Thus, it is established that the problem of early lung cancer detection is connected with a reduction in the number of FP classifications while keeping a high degree of true-positive (TP) diagnoses, i.e., sensitivity.

The detection of lung cancer based on CT images was proposed by the authors in [14]. The authors employed CNN and the developed model was found to be 96% accurate as compared with the previous study [11]. The model was implemented in MATLAB® and the dataset was obtained from the lung image database consortium (LIDC) and image database resource initiative (IDRI). The system was also able to detect the presence of cancerous cells. Authors in [15] provided an overview of the technical aspects of employing radiomics; i.e. analysis of invisible data from the extracted image and the significance of artificial intelligence (AI) in the diagnosis of non-small cell lung cancer. The technical implementation limitations of radiomics such as harmonized datasets, and large extracted data led to the exploration of AI in the diagnosis of cancer. The authors discussed the multiple steps employed in the study which include data acquisition, reconstruction, segmentation, pre-processing, feature extraction, feature selection, modelling, and analysis. A detailed study of the existing dataset, segmentation method, and classifiers for predicting the subtypes of pulmonary nodules was presented.

The authors in [16] conducted a study to predict the risk of cancer on patients' CT volumes. The model exhibited an accuracy of 94.4% and outperformed when compared to the radiologists. The study is unique in nature as the research is conducted in comparison to the previous and current CT images. Trial cases of 6,716 numbers were considered, and the model was validated on 1,139 independent clinical datasets. The data mining technique was used in the study by the authors in [17], and the experimentation aimed at providing a solution to the problem which arises after pre-processing the data during the process of cleaning the data. The authors experimented with applying the filter and resampling the data with three classifiers on two different datasets. The study was conducted over five performance parameters. The results demonstrated the accuracy level of the classifiers to be better after the resampling technique was employed. The dataset considered were Wisconsin Breast Cancer (WBC) and breast cancer dataset. Results proved the performance of the classifiers to be improved for the WBC dataset with the resampling filter applied four times. Whereas, for the breast cancer dataset, seven times the resampling filter was applied. J48 Decision Tree classifier showed 99.24% and 98.20%, Naïve Bayes exhibited 99.12% and 76.61%, and sequential minimal optimization (SMO) showed 99.56% and 95.32% for WBC and breast cancer datasets, respectively. The WBC dataset was used in the study by the authors in [18] who applied visualization and ML techniques to provide a comparative analysis on the same. The predictions were made by visualizing the data and analyzing the correlation of the features. The result demonstrated an accuracy of 98.1% by managing the imbalanced data. The study categorized the original dataset into three datasets. All the independent features were in one dataset, all the highly correlated features were in one dataset, and the features with low correlation were grouped as the last dataset. Logistic regression showed the accuracy results as 98.60%, 95.61%, and 93.85%, KNN demonstrated the scores as 96.49%, 95.32%, and 94.69, SVM obtained 96.49%, 96.49%, and 93.85%, decision tree showed

95.61%, 93.85%, and 92.10%, random forest obtained 95.61%, 94.73%, and 92.98%, and rotation forest algorithm showed 97.4%, 95.89%, and 92.9%.

A systematic review was presented by the authors in [19] who presented the DL and ML techniques for detecting breast cancer based on medical images. The review summarized research databases, algorithms, future trends, and challenges in the research field. It provided a complete overview of the subject and related progress. It proposed that computer-aided detection can be more accurate than the diagnosis of a radiologist. A computer-aided design (CAD) system was developed in [20] to classify mammograms. The study employed feature extraction by discrete wavelet transformation. The principal component analysis is used to extract the discriminating features from the characteristics of the original vector features. A weighted chaotic salp swarm optimization algorithm is proposed for classification. The dataset under study is the Mammographic Image Analysis Society (MIAS), Digital Database for Screening Mammography (DDSM), and Breast Cancer Digital Repository (BCDR). A complete review of the CAD methods is discussed by the authors in [21] for detecting breast cancer. The study was conducted on mammograms, and enhancement of the image and histogram equalization techniques were proposed. Table 3.1 summarizes the entire literature review with limitations, motivation, and aim.

3.3 Review and discussions

This section with Table 3.2 summarizes the key findings from the literature review which is detailed in Section 3.2. Several algorithms have been examined and noted which may be used in the diagnosis of breast and lung cancer. The use of widely accepted datasets such as IRMA, WBC, The Cancer Imaging Archive (TCIA), National Lung Cancer Screening Trial (NLST), and others helped to increase the accuracy. Most algorithms have performed incredibly well, ranging from 89% to 99.93% accuracy, with relatively smaller margins. Additionally, it is found that algorithms such as Sequential Minimal Optimization (SMO), Decision Tree (J48), Naive Bayes, and Salp Swarm Optimization fared better than others.

3.4 Proposed methodology

Following the completion of the literature review, we learned about many algorithms which may be used for early cancer detection. As part of our continued research, in our next study, we will develop and place into practice, an intelligent system model for the diagnosis and prediction of breast and lung cancer. Figure 3.4 illustrates the proposed model which will be put into practice.

The generalized model, as shown in Figure 3.4, starts by receiving input in the form of images or datasets. Real-time tomography (CT scan) and mammography are two examples of classic medical procedures that provide input images. Even well-known medical institutes' or research organizations' datasets may be used as the input. The stage of pre-processing removes interruption and crispiness

Table 3.1 Summary of literature review

Reference	Existing methods	Limitations	Motivation	Aim
[9]	Computed tomography (CT), 3D mammography, magnetic resonance imaging (MRI), and histopathological imaging (HI)	Needed experienced pathologist or radiologist. The process has high error rates and is expensive	Every eighth American woman develops invasive breast cancer over her lifetime, making it the second highest cause of death for females	To critically assess the research on the detection and classification of breast cancer using DL using a dataset made up of images from MRI, CT, mammography, ultrasound, thermography, and histopathology
[10]	Mammography	The traditional screening practice using mammography was less efficient	With the exponential rise in the ageing population in the regions of Malaysia, the risk of breast cancer raised to every 1 in 19 women	To conduct a comparative performance study between two DL networks i.e. VGG16 and ResNet50 using IRMA dataset
[13]	Two-level networks: Level 1: Identifying a suspicious area in a low-resolution picture Level 2: Curvature peaks of the suspicious region	High computational cost and generalization procedure	Even with small nodules and a limited number of FPs, achieve a high level of TP detection rate	Computerized chest radiographs show lesions consistent with lung cancer
[14]	CNN	The hidden neurons need to be improved by employing a 3D CNN	Classify the tumours based on CT images	Detection of malignant tissue in the lung image
[15]	Radiomics and AI	The physiological relevance of radiomics needs attention as reproducibility is affecting the quality score	The performance study of radiomics and AI in predicting the cancerous cell	Add the clinical information into the developed predictive models such that the efficacy could be improved as the models would mimic the human decision

(Continues)

Table 3.1 (Continued)

Reference	Existing methods	Limitations	Motivation	Aim
[17]	Machine algorithm	The number of resampling filters was randomly selected	A resampling technique can be implemented to enhance the performance of the classifier	Aimed to provide a comparison between the classifiers
[18]	Data visualization techniquesML techniques	The comparative analysis focused only on one type of dataset	Visualization of the data and then classify the data as benign or malignant	For detection of breast cancer and diagnosis, aimed at providing a comparative analysis with respect to data visualization and ML
[20]	Classification of the digital mammograms using the CAD system	The abnormalities classification for digital mammograms	Develop CAD systems to attain better accuracy models with a lesser number of features	The goal was to present a kernel extreme learning machine based on the weighted chaotic salp swarm algorithm
[22]	You Only Look Once (YOLO), RetinaNet	Duplication of data leads to memory being extensively used and results in reduced model accuracy for small objects.	Comparison of conventional CNN networks with YOLO and RetinaNET	A comparative study with the recent models proposed for detecting breast cancer

Table 3.2 Collection of various algorithms and their respective accuracies

Reference	Dataset	Algorithm	Accuracy	Cancer type
[10]	IRMA (15,363 IRMA images of 193 categories)	VGG16 with 16 layers	94%	Breast
[10]	IRMA (15,363 IRMA images of 193 categories)	ResNet50 with 152 layers	91.7%	Breast
[13]	90 real nodules and 288 simulated nodules	The classifier of SNA (suspected nodule areas)	89–96%	Lung nodule
[14]	LIDC and IDRI	Deep CNN algorithms	96%	Lung
[15]	TCIA	Radiomics and AI	–	Pulmonary nodule
[16]	NLST, LUng Nodule Analysis (LUNA) and LIDC	3D CNN	94.4%	Lung
[17]	WBC and Breast Cancer dataset	DecisionTree(J48), Naïve Bayes and SMO	J48: 98.20% SMO: 99.56%	Breast
[18]	WBC	Logistic regression, KNN, SVM, Naïve Bayes, decision tree, random forest and rotation forest	Classification accuracy of 98.1%	Breast
[20]	MIAS, DDSM, and BCDR	Salp swarm optimization algorithm	For normal-abnormal category 99.62% (MIAS) and 99.93% (DDSM), For benign-malignant classification 99.28% (MIAS), 99.63% (DDSM), and 99.60%(BCDR)	Breast
[22]	DDSM, Curated Breast Imaging Subset of DDSM (CBIS-DDSM), MIAS, BCDR, and INbreast	YOLO and RetinaNet	YOLO – 93.96% RetinaNet – 97%	Breast
[23]	Influential genes dataset	Random forest	84.375%	Lung
[24]	The University of California, Irvine, online	SVM	98.8%	Lung

(Continues)

Table 3.2 (Continued)

Reference	Dataset	Algorithm	Accuracy	Cancer type
	repository named Lung Cancer (32 instances and 57 characteristics with one class attribute			
[25]	Chest X-ray and CT images (20,000 images) with 247 chest radiographs [26]	VGG19-CNN, ResNet152V2, ResNet152V2 + Gated Recurrent Unit (GRU), and ResNet152V2 + Bidirectional GRU (Bi-GRU)	VGG19+CNN – 98.05% ResNet152V2+GRU – 96.09% ResNet152V2 – 95.31% ResNet152V2+ Bi-GRU – 93.36%	Lung
[27]	The Cancer Genome Atlas (TCGA) dataset contains 533 lung cancer samples and 59 normal samples. The International Cancer Genome Consortium (ICGC) dataset contains 488 lung cancer samples and 55 normal samples	DNN based on Kullback – Leibler divergence gene selection	99%	Lung
[28]	NLST	CNN	99%	Lung

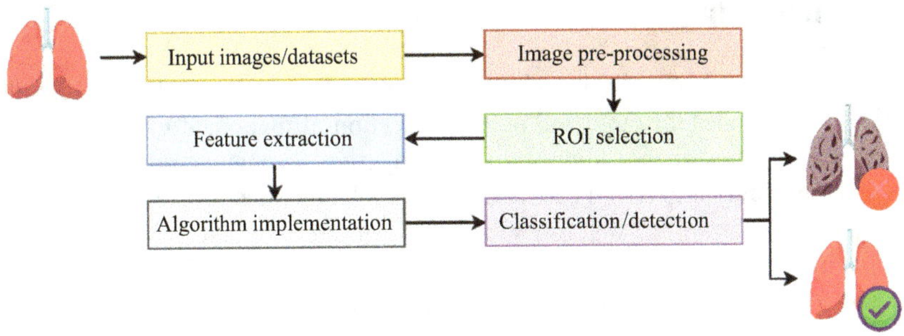

Figure 3.4 Flowchart of proposed methodology

from the image as well as the salt and pepper noises, also known as impulse noises. By using techniques such as the Otsu threshold and statistical threshold, additional disturbance elements such as artefacts, black backgrounds, and labels existing on the mammography or tomography images can be eliminated. Other elements such as contrast and brightness can also be enhanced by utilizing a variety of techniques including intensity-range based partitioned cumulative distribution function (IRPCDF) and background over-transformation controlled (BOTC) [29].

The selection of the region of interest (ROI) and related annotation is one of the best ways to diagnose cancer by concentrating on the tumour-grown areas with a high degree of accuracy and precision. However, the laborious and time-consuming traditional ROI selection procedure is discouraging [30]. Therefore, in our next study, we will propose to develop an automated technique that can speed up the ROI selection process compared to the manual methods. In most of the methods, feature extraction is an optional yet effective process. The features such as wavelet energy values, standard deviation, and mean are derived from the images based on texture analysis methods such as grey level co-occurrence matrix (GLCM). A metaheuristic process based on natural selection known as genetic algorithm (GA) can also be implemented for feature extraction and selection. This method helps in enhancing the quality of the data obtained from the input images.

In the next study, the best algorithms and methodologies will be evaluated and implemented based on the literature review presented in this chapter. This stage will involve performing a thorough analysis of the top five algorithms considering diverse circumstances. The final phase of the suggested procedure will involve the classification-based diagnosis. The lung cancer dataset will be the primary testing ground for this suggested strategy which will then be applied to all other cancer datasets. The proposed methodology will primarily focus on understanding and improving the fault regions in low-resolution pictures, understanding the development patterns of tumour sizes, determining whether or not the tumour is malignant, and ultimately improving diagnosis errors and computational cost.

3.5 Conclusion

The main focus of this study is the identification of numerous implementable algorithms and techniques for rapid and precise detection of malignant tumour growth in the breast and lungs. These algorithms will function in conjunction with the conventional medical methods of a cancer diagnosis. The fact that lung cancer and breast cancer are the two major causes of mortality for both men and women leads us to perform numerous research on the early detection of these malignancies. For the proposed methodology to function, the images from mammography or tomography will be helpful as input images.

Understanding the wide range of algorithms that can be employed is immensely aided by this literature review on various algorithms used to diagnose breast and lung cancer. The proposed methodology will henceforth be expanded and put into practice for lung cancer in future studies considering other cancer datasets. Alternate approaches for image processing will be explored and combined with the same model, including ROI selection and feature extraction. The best-performing algorithms will then be the subject of extensive research.

References

[1] Cancer.Net. What is Cancer? *American Society of Clinical Oncology (ASCO)*, https://www.cancer.net/navigating-cancer-care/cancer-basics/what-cancer (2019, accessed 7 December 2022).

[2] Tonorezos ES, Cohn RJ, Glaser AW, *et al.* Long-term care for people treated for cancer during childhood and adolescence. *Lancet* 2022; 399: 1561–1572.

[3] Emery J, Butow P, Lai-Kwon J, *et al.* Management of common clinical problems experienced by survivors of cancer. *Lancet* 2022; 399: 1537–1550.

[4] Shandilya S and Chandankhede C. Survey on recent cancer classification systems for cancer diagnosis. In *Proceedings of 2017 International Conference on Wireless Communication Signal Process Networking, WiSPNET 2017* 2018; 2018 January, pp. 2590–2594.

[5] Sung H, Ferlay J, Siegel RL, *et al.* Global Cancer Statistics 2020: GLOBOCAN estimates of incidence and mortality worldwide for 36 cancers in 185 countries. *CA Cancer J Clin* 2021; 71: 209–249.

[6] Chaitanya Thandra K, Barsouk A, Saginala K, Sukumar Aluru J, and Barsouk A. Epidemiology of lung cancer. *Współczesna Onkol* 2021; 25: 45–52.

[7] Musial C, Zaucha R, Kuban-Jankowska A, *et al.* Plausible role of estrogens in pathogenesis, progression and therapy of lung cancer. *Int J Environ Res Public Health* 2021; 18: 648.

[8] van der Aalst CM, ten Haaf K, and de Koning HJ. Implementation of lung cancer screening: what are the main issues? *Transl Lung Cancer Res* 2021; 10: 1050–1063.

[9] din NM ud, Dar RA, Rasool M, *et al.* Breast cancer detection using deep learning: datasets, methods, and challenges ahead. *Comput Biol Med* 2022; 149: 106073.

[10] Ismail NS and Sovuthy C. Breast cancer detection based on deep learning technique. In: *2019 International UNIMAS STEM 12th Engineering Conference (EnCon)*. IEEE, pp. 89–92.

[11] Ministry of Health Malaysia. *Malaysia National Cancer Registry Report (MNCRR) 2012–2016*, 2019, http://nci.moh.gov.my.

[12] Deserno T and Ott B. 15,363 IRMA images of 193 categories for ImageCLEFmed 2009. RWTH Publications. Epub ahead of print 2009, doi:10.18154/RWTH-2016-06143.

[13] Penedo MG, Carreira MJ, Mosquera A, *et al.* Computer-aided diagnosis: a neural-network-based approach to lung nodule detection. *IEEE Trans Med Imaging* 1998; 17: 872–880.

[14] Sasikala S, Bharathi M, and Sowmiya BR. Lung cancer detection and classification using deep CNN. *Int J Innov Technol Explor Eng* 2018; 8: 259–262.

[15] Devi VA, Ganesan V, Chowdhury S, Ramya G, and Dutta PK. Diagnosing the severity of covid-19 in lungs using CNN models. In *6th Smart Cities Symposium (SCS 2022), Hybrid Conference*, Bahrain, 2022, pp. 248–252, doi:10.1049/icp.2023.0427.

[16] Ardila D, Kiraly AP, Bharadwaj S, *et al.* End-to-end lung cancer screening with three-dimensional deep learning on low-dose chest computed tomography. *Nat Med* 2019; 25: 954–961.

[17] Mohammed SA, Darrab S, Noaman SA, *et al.* Analysis of breast cancer detection using different machine learning techniques. In: *Data Mining and BigData. DMBD 2020. Communications in Computer and Information Science*, vol. 1234. Springer, Singapore, pp. 108–117.

[18] Dutta PK, Vinayak A, and Kumari S. Asymptotic patients' healthcare monitoring and identification of health ailments in post COVID-19 scenario. In: O Jena, AR Tripathy, AA Elngar, and Z Polkowski (eds.), *Computational Intelligence and Healthcare Informatics*, 2021, https://doi.org/10.1002/9781119818717.ch16.

[19] Houssein EH, Emam MM, Ali AA, *et al.* Deep and machine learning techniques for medical imaging-based breast cancer: a comprehensive review. *Exp Syst Appl*; 167. Epub ahead of print 1 April 2021, doi:10.1016/j.eswa.2020.114161.

[20] Mohanty F, Rup S, Dash B, *et al.* An improved scheme for digital mammogram classification using weighted chaotic salp swarm algorithm-based kernel extreme learning machine. *Appl Soft Comput* 2020; 91: 106266.

[21] Ramadan SZ. Methods used in computer-aided diagnosis for breast cancer detection using mammograms: a review. *J Healthc Eng* 2020; 2020: 1–21.

[22] Hamed G, Marey MAE-R, Amin SE-S, *et al.* Deep learning in breast cancer detection and classification. In: *Proceedings of the International Conference on Artificial Intelligence and Computer Vision (AICV2020). AICV 2020.*

Advances in Intelligent Systems and Computing, vol. 1153. Springer, Cham, 2020, pp. 322–333.

[23] Dutta PK, Ghosh A, De P, and Soltani M. A proposed model of a semi-automated sensor actuator resposcopy analyzer for 'covid-19' patients for respiratory distress detection. In: *Proceedings of 11th International Conference on Cloud Computing, Data Science & Engineering (Confluence)*, Noida, India, 2021, pp. 618–623, doi: 10.1109/Confluence51648.2021.9377180.

[24] Anil Kumar C, Harish S, Ravi P, *et al.* Lung cancer prediction from text datasets using machine learning. *Biomed Res Int* 2022; 2022: 1–10.

[25] Ibrahim DM, Elshennawy NM, and Sarhan AM. Deep-chest: multi-classification deep learning model for diagnosing COVID-19, pneumonia, and lung cancer chest diseases. *Comput Biol Med* 2021; 132: 104348.

[26] Shiraishi J, Katsuragawa S, Ikezoe J, *et al.* Development of a digital image database for chest radiographs with and without a lung nodule. *Am J Roentgenol* 2000; 174: 71–74.

[27] Liu S and Yao W. Prediction of lung cancer using gene expression and deep learning with KL divergence gene selection. *BMC Bioinf* 2022; 23: 175.

[28] Heuvelmans MA, van Ooijen PMA, Ather S, *et al.* Lung cancer prediction by deep learning to identify benign lung nodules. *Lung Cancer* 2021; 154: 1–4.

[29] Senguttuvan D and Pichai S. Mammogram image preprocessing using intensity range based partitioned cumulative distribution function. *J Anal*. Epub ahead of print 26 October 2022, doi:10.1007/s41478-022-00499-7.

[30] Lou Q, Li Y, Qian Y, *et al.* Mammogram classification based on a novel convolutional neural network with efficient channel attention. *Comput Biol Med* 2022; 150: 106082.

Chapter 4

Deep learning for streamlining medical image processing

Sarthak Goel[1], Ayushi Tiwari[1] and B.K Tripathy[1]

According to National Institution for Transforming India (NITI Aayog), India's healthcare sector is taking off with an annual growth rate of around 22% since 2016. With a market close to 372 billion in 2022, the healthcare sector will be one of the largest employment sectors. While the frontline workers are promoting health infrastructure, technology is complementing their efforts. The healthcare industry is witnessing a golden period as advancements in medical imagery, data analysis, computational sciences, and robotics are streamlining complex medical procedures. Integrating technology in healthcare has not only made the entire system efficient but has also reduced dependency on physicians. Even though developed countries have world-class health infrastructure, the fact that doctors are human enough to make mistakes, cannot be ignored. Moreover, different doctors have different intelligence levels and therefore interpret medical records differently. They adopt unique approaches while treating the same disease, which might not always work. For all these challenges, artificial intelligence stands as a one-stop solution. It can learn from past results to produce an unbiased, balanced, and objective report without preconceived notions. Its capability to process large datasets and produce personalized results with high precision makes it the most optimized approach for solving these complex healthcare challenges. While healthcare infrastructure is not up to the standards everywhere, a simple yet rigorously trained artificially intelligent prediction software is efficient enough to diagnose diseases in the initial stages based on the symptoms. Today, deep learning is aiding as a tool to diagnose complex diseases such as diabetic retinopathy, with minimal medical imagery, thereby eradicating the requirements of tedious tests. Advanced image processing algorithms coupled with deep learning analysis techniques have made it possible to re-create low-resolution medical images and automate analysis to produce conclusions in real time. Neural networks are facilitating in performing detailed analysis of medical data produced through magnetic resonance imaging, cardiac tomography, electrocardiography, and other scanning technology, thus making

[1]School of Information Technology & Engineering, Vellore Institute of Technology, India

it significantly convenient to diagnose cancer, cardiovascular diseases, retinal diseases [1], genetic disorders [2], etc.

This chapter is an attempt to highlight the possible use cases of deep learning algorithms and techniques in the healthcare industry. Even though the contribution of other technologies such as the internet of things, robotics, smart medical devices, IT systems, blockchain, surgical equipment, electronic health record management systems, staffing management systems, hybrid operation theatres, kiosks, vending machines, and telehealth tools can never be neglected. Moreover, this chapter tries to focus on how deep learning algorithms are supplementing existing technologies to make them more efficient and widely used. In addition, enhancing the efficiency will not only reduce the burden on existing infrastructure but also reduce the expenditure by eliminating unnecessary biopsies. The World Health Organization's annual report of 2022 determines the devastating impact of COVID-19 worldwide due to poor infrastructure. Scientists believe that deploying digital solutions facilitated by deep learning technologies can prevent such collapses in healthcare facilities in the future.

4.1 Introduction

Throughout the history of humans, medical research has always been a top priority. Be it the discovery of vaccines, anesthesia, micro-surgeries, or radiology, each of them has had a huge impact on the human population. Deep learning can become an indispensable tool for doctors just like a stethoscope. With its exemplary image segmentation capability, deep learning has made significant contributions to biomedical image processing. Using natural language processing and computer vision capabilities, deep learning furnishes diverse solutions that are not only limited to processing the image but also in delivering adequate analysis with regard to results achieved. Non-linear processing units make up a layered architecture which facilitates feature extraction and image transformation [3]. This layered architecture supported by deep learning algorithms allows the system to adjust weights and biases depending on the effect of respective parameters. Each layer is responsible for a specific kind of processing such as gray scaling the biomedical image, noise reduction, color balancing, and ultimately feature detection. This constant "adjust and tune" process makes deep learning algorithms extremely useful with medical data.

Today, the advancements in photographic technologies have enabled physicians to capture high-resolution images. While each image measures as high as 32 MB, processing images using general-purpose processing algorithms is extremely redundant and time-consuming. Deep learning algorithms when put into action, not just analyze these images (a connected database with parameters) but can even diagnose the disease or disorder, eliminating the need for a doctor. It is often reported that the difference in the approach of doctors leads to different paths of treatment. Deep learning predetermines the disease by analyzing symptoms, thus saving a lot of time and effort. Moreover, the physician can now propose treatment in relevant directions. Smart deep learning-enabled software applications can utilize supporting technologies like computer vision, natural language processing, and the Internet of Things to eliminate the need of doctors in at least the initial stage of diseases. Such solutions are economically efficient, scalable, and readily available even in remote locations. Certain diseases have very long incubation periods for which symptoms are blurry during the initial stages. Deep learning algorithms, through their immense data processing abilities, can analyze hundreds of data points such as age, gender, lifestyle, genetics, enzyme analysis, and blood count to consign a comprehensive report. Be it X-ray scans, CT scans, MRI scans, mammograms, ultrasound results, PET scans, or any other medical image, deep learning is versatile enough to adapt to any space.

Big Data is another emerging field that goes hand in hand with deep learning. The three V's of big data such as velocity, volume, and variety complement deep learning, enabling such analytical systems to process more information than humans at any instance. The state-of-the-art deep neural networks (DNNs) are demonstrating exceptional results in the field of image processing, classification, data analytics, and visualizations. They are replacing classical artificial neural networks (ANNS), because of the accessibility of high-dimensional big data sets. Common medical imaging use cases produce datasets as large as 20TB per month, which needs to be collected, stored, and optimized for efficient usage. The capability of deep learning algorithms to process highly unstructured medical data which not only includes images [4] but also signals, sounds, genomic expressions and patterns, text data, and metadata files is one of the major reasons for the adoption of such intelligent systems in the healthcare sector. Intuitive dashboards can abstract complex deep learning algorithms that perform advanced analytics in the background, therefore restricting the required skillset to utilize such systems.

4.2 Deep learning: a general idea

In most simple terms, deep learning is a subset of artificial intelligence which functions on the concept of DNN inspired by biological neural networks. These DNN mimic a human brain by learning from context to perform similar tasks but with immense speed and accuracy. Deep learning and neural networks have become intensely popular in countless applications some of which include classification, prediction, computer vision and image recognition, designing intelligent systems such as smart home and self-driven cars, and data analysis. By impersonating the human brain, neural networks after getting trained adapt quickly using the concept of weights and take precise decisions. With each decision made, the efficiency of the algorithm enhances, thereby producing an optimized model after each iteration.

ANNs are the building blocks of deep learning solutions. They have a computational workflow which is analogous to a biological neural network. Just like a human brain, neurons or nodes combine to form a network. Weights attached to these links help inhibit or enhance the effect of each neuron [5]. Neural networks

Table 4.1 Popular activation functions [6]

Activation function	Function equation
Sigmoid	$f(x) = \frac{1}{1+e^{-x}}$ where e is Euler's number
Hyperbolic Tangent	$f(x) = \tanh(x)$
Soft Sign activation	$f(x) = \frac{x}{1+\|x\|}$
Rectified linear unit (ReLU)	$f(x) = \{0x < 0\ xx \geq 0$
Soft Plus	$f(x) = \ln\ln(1 + e^x)$
Leaky Rectified Linear Unit (Leaky ReLU)	$f(x) = \{\alpha xx < 0\ xx \geq 0$

are broadly classified into two types, i.e., feed-forward neural network (FFNN) or recurrent neural network (RNN) based on the pattern of the association of neurons. An FFNN forms a directed acyclic graph with each layer consisting of nodes, while an RNN generally occurs as a directed cycle. Weights and activation functions are the other two parameters that affect the output of a neural network. Training a neural network is an iterative process where weights are optimized to minimize the loss function. By adjusting these weights, neural networks efficiency can be altered. By using an activation function, the ANN is activated or deactivated. This process is done by comparing the input value to a threshold value. This on-and-off operation throughout the layers of the network, introduces a non-linearity and makes it continuously differentiable. Some activation functions are tabulated in Table 4.1.

Even though deep learning is one of the significant inventions in the field of computational sciences. Unfortunately, there doesn't exist any "one size fits all" solution. Deep learning comes with a lot of dependencies. This trade-off between high dependencies and meticulous results, often expects stakeholders to take rigid decisions. For every AI solution deployed, a general set of preparations need to be followed, which are listed as follows:

- Define the size of the sample from the dataset
- Determining if a previous application domain could be modified to solve the issue at hand. This helps in estimating if the model needs to be trained from scratch or if transfer learning could be applied
- Assess dependent and independent variables to decide the type of algorithms applicable to the given problem statement
- Interpret results based on model logic and behavior

4.3 Deep learning models in medicine

Medical image processing and healthcare systems are some of the highly significant fields when it comes to AI applications. Irrespective of financial recessions, while all other sectors face downtrends, medicine is probably the only sector which is thriving. Moreover, it is the sector where technology can prove to be a game changer as it can automate complex monotonous processes. Artificial Intelligence tools are the exquisite solution to Dr. L. Hood's P4 (predictive, preventive, personalized, and participatory) framework to detect and prevent disease through extensive biomarker testing, close monitoring, deep statistical analysis, and patient health coaching. Deep learning algorithms are extremely adaptable when it comes to use cases. Each algorithm performs a specific operation, however, can fit into contrasting applications. In this section, we discuss some relevant deep learning models that are being utilized in medical image processing use cases.

4.3.1 Convolutional neural networks

Convolutional neural networks (CNNs) are a type of ANNs that use the principle of convolution for data processing [7]. They are apt for use cases involving image analysis such as object detection, edge detection, face recognition, segmentation, and classification tasks. They provide robust and flexible solutions that can work with different modalities. Generally, all CNNs work on a five-layered architecture, which are the input layer, a convolutional layer, a pooling layer, a fully connected layer, a logistic layer, and finally output layer. CNN stands out from its predecessors, as it does not need human intervention for detecting important features. Unlike old solutions, which required input variables (features) to be specified in the initial phase, CNNs work dynamically. This layered architecture is competent in figuring out necessary features on its own. Some popular applications of convolutional neural architecture include AlexNet, VGGNet, GoogleNet, and ResNet. These models aim to classify by concatenating fully-connected layers and a classifier such as Support Vector Machine (SVM). They succeed as an excellent feature extractor and hence find a wide scope of applications in medical image analysis. Some of the most successful implementations of CNNs include AlexNet, VGGNet, ResNet, U-Net, and SegNet. AlexNet was an augmentation to the traditional CNN, AlexNet is 8 layers deep [8]. It contains five convolutional layers, three max-pooling layers, two normalization layers, two fully connected layers, and one soft-max layer. By using a multi-GPU training method, AlexNet is an excellent image classifier. It can work with massive datasets, and classify images into as high as 1,000 object categories. ResNet brings forth a simplified structure of the ever-dense neural network structure. By using a recursive-like approach, the ResNet algorithm is comparatively easier to train. A deep residual network resolves the vanishing gradient problem. It creates a bypass connection that skips certain layers to create an efficient network. A VGGNet, commonly referred to as a very deep convolutional neural network, is an amplification of traditional deep convolutional neural networks. While VGG stands for visual geometry group, VGGNet is appropriate for object recognition models. A U-Net is a generalized full convolutional network (FCN) that is used for quantification tasks [9]. It is deployed for cell detection, the shape measurements in medical image data. A CNN-based U-Net architecture is used for segmentation in medical image analysis [10]. Based on encoder-decoder architecture, SegNet is a semantic segmentation model that aims to achieve an end-to-end pixel-level segmentation. While the encoder uses VGG16 for analyzing object information, the decoder points the parsed information to the final image form. Unlike a FCN SegNet utilizes a large pooling index that is received from the encoder for up-sampling the input non-linearly. Introduced as "You Only Look Once" (YOLO), this algorithm was a substitute for RBCNN. Because of its lucidity and enhanced execution speed, YOLO is becoming extremely popular in the object detection domain. YOLO imparts real-time object detection capabilities by dividing the image into N grids, each with S × S dimension. Each grid is accountable for detecting only a single object. This enables YOLO to a perfect algorithm when detecting large objects [11]. However, when detecting smaller objects like a line of ants, YOLO isn't the best choice here. Nevertheless, YOLO has gone through several upgrades. Today, more than five versions of YOLO are being actively utilized for a variety of use cases.

4.3.2 Recurrent neural networks

Recurrent neural networks (RNNs) are used to process sequential data. Unlike CNNs, RNNs specialize in natural language processing tasks. RNNs emerged as an improvement to feed-forward networks. It actively uses past inputs for making dynamic decisions. Because of its "short-term memory," RNNs can make precise predictions. They are suited for applications such as speech recognition, sentiment analysis, text and language modeling, prediction, and so on. Even though RNNs are slow, complex, and can't be stacked up, it is the only neural network which can map out to many to many, one to many, and many to one inputs and outputs. They are the only neural networks with memory. Long- and short-term memory (LSTM) is one of the most successful RNNs and has been used in a wide range of applications [12].

4.3.3 Auto-encoders (AE)

An Auto-encoders (AE) finds its application in noise reduction use cases. It is best suited for unsupervised learning-based tasks [13]. By encoding the input into a lower dimensional space, the auto-encoder uses a hidden layer for de-noising. An auto-encoder generally follows a three-step process, i.e., encode, decode, and calculate the square error. Most common auto-encoder algorithms include de-noising auto-encoder (DAE), vibrational auto-encoder, and stacked auto-encoder (SAE).

4.4 Deep learning for medical image processing: overview

Computer-aided diagnosis (CAD) is a widely used jargon in the field of medical image processing. The versatile nature of deep learning algorithms helps doctors automate the tedious process of analyzing medical scans and diagnosing diseases [14]. CAD systems streamline the process of disease diagnosis by analyzing hidden patterns in medical scans such as MRI scans [15], CT scans, etc. A CAD system isn't just limited to diagnosing the disease, instead, advanced systems are multi-process and can be easily integrated with smart health management systems (HMS) to provide a one-stop solution for encapsulating patient's data in one place, thus making it readily available. From processing to analyzing followed by archiving results, smart CAD systems come in handy in sharing and preserving the medical history of patients. A CAD system is more accurate and expeditious when compared to a human physician. Though technology can never replace human physicians, it can always supplement their efforts, to smoothen the process. Countries like Japan are heavily relying on such technologies. Healthy Brain Dock is a Japanese brain screening system, which detects high-risk groups for Alzheimer's disease. It uses a traditional MRI system combined with technology to preclude and detect the onset of asymptomatic brain diseases such as dementia and aneurysm.

4.5 Literature review

To deliver a diverse and wholesome view of marvels of deep learning in medical image processing, we referred to plenty of trending research articles from reputed journals and magazines. Our research is not limited to medical image processing, but also to how deep learning is streamlining the complete healthcare and medicinal domain. This chapter points out some recent patents and deep learning-based algorithmic inventions that have excited the interest among researchers. Deep learning is preferred when used as a tool to support other technologies, such as the Internet of Things, data analysis, virtual reality, image processing, and other standalone technologies as it helps amplify the benefits of other technologies.

Throughout the survey, we discovered how deep learning is finding its applications throughout the medical industry. Not just limited to image analysis, the deep learning approach (DLA) is also being utilized actively in other branches such as genomics, transcriptomics, proteomics, and metabolomics. From analyzing DNA structure to disease prediction, RNN- and CNN-based algorithms are used to automate solutions. Some of the use cases include predicting missing values from DNA structure, predicting mutation effect through DNA pattern, reducing data dimensionality and sparsity in RNA structure, classification of RNA components, drug discovery, drug target monitoring, simulating drug effects, optimizing molecular activities, and the list goes on. Algorithms based on CNN, SAE, GAN, and AE are applicable in such use cases. A combination of certain algorithms such as applied LSTM on CNN's output enhances the accuracy of the model in use cases such as protein classification. An interesting application of deep learning algorithms was observed in [16,17] by using these techniques in face mask detection.

Throughout the literature review, deep learning stood out as a cost-effective, highly available, versatile, and robust technology. From CAD to advanced research use cases such as drug discovery, drug effects, and generating medical simulations, deep learning can fit into almost all medical use cases. Table 4.2 mentions some popular solutions provided by deep learning algorithms against medical use cases.

Some interesting use cases of medical image analysis were solved by using Transfer learning in one of the studies we came across [20]. Transfer learning is the process of storing and applying knowledge gained by solving one problem into a similar problem that may arise in the future. In medical imaging, such algorithms are helpful for segmentation and classification-related tasks. A comprehensive list of these methods, mapped against respective disease domains is tabulated as in Table 4.3.

Table 4.2 Deep learning algorithms mapped with their medical use cases

Image processing use case	Domain specific use case	Suitable deep learning algorithm
Segmentation	Cardiovascular image segmentation	CNN
	Tumour segmentation	Deep CNN
	Retinal anatomy segmentation [1]	CNN
	Prostate anatomy segmentation	BOWDA-Net
	Cell-segmentation in microscopy images	U-Net
	Cell segmentation of 2D phase-contrast	U-Net, Multi-Resolution Net
	Dense cell population segmentation	U-Net
	Vessel segmentation	CNN
	Microvasculature segmentation	FCN
	Mast cells segmentation	U-Net, CNN
Detection	Object detection	Marginal space DL
	Lung cancer detection	3D neural network [18]
	Detecting nuclei in breast images	SSAE
	Skin cancer [19]	Softmax classifier
Classification	Lung nodule classification	Artificial CNN, Multi-scale CNN
	Pneumonia	CheXNet DL model
	Skin lesion classification	Multi-layer CNN
	Organ classification	CNN
	Breast cancer detection	CNN
	Cell type classification	CNN
	Mutation prediction and lung cancer classification	CNN
	Red blood cell classification	CNN
	Mitochondrial images classification	CNN
	Fine grained-leukocyte classification	Res-Net
	White blood cell identification	CNN
	Stem cell multi-label classification	CNN
Localization	Prostate localization	SSAE
	Multi-organ disease	Single-layer SSAE
	Localize the fetal	CNN
Registration	Cancer registration	Elastix automated 3D deformable registration software
	Cardiovascular registration	Multi-atlas classifier
	3D image registration	Self-supervised learning model
	Detect motion-free abdominal images	CNN image registration model
Tracking	Cell tracking	U-Net, Faster R-CNN
	Submicron scale particles	CNN, RNN
	Data association in cell tracking	ResCnn
	Cell segmentation, tracking,	U-Net

(Continues)

Table 4.2 (Continued)

Image processing use case	Domain specific use case	Suitable deep learning algorithm
	lineage reconstruction	
	Instance-level microtubule tracking	CNN, LSTM
	Stem cell motion tracking	CNN, TDNNs
	Nuclei detection in time-lapse phase images	Mask R-CNN

Table 4.3 Transfer learning applications in medical image processing

Organ	Application in disease domain	Transfer learning method
Lung	Lung CT	Fine-tuning, feature extractor
	Diffuse lung disease	Fine-tuning
	Lung module classification	Feature extractor
	Lung module detection	Fine-tuning, feature extractor
	Lung cancer	Feature extractor
	Lung lesion	Feature extractor
Breast	Mammographic tumor	AlexNet as feature extractor
	Breast cancer	Fine-tuning, feature extractor
	Breast tomosynthesis	Fine-tuning
	Breast MRIs	Feature extractor
	Mammographic breast lesions	Feature extractor
	Breast mass classification	Feature extractor, fine-tuning
	Mammograms	Fine-tuning
	Breast lesions	Fine-tuning
Brain	Brain tumor	Feature extractor
	Gliomas	Fine-tuning on AlexNet and GoogleNet
	Brian tumor	Feature extractor (VGG-19), Fine-tuning
	Alzheimer	Fine-tuning on Inception-V2
	Brain Lesion	Fine-tuning on U-Net and ResNet
	Glioblastoma multiforme	Feature extractor
	Medulloblastoma tumor	Feature extractor (VGG-16)
Kidney	Kidney segmentation	Feature extractor
	Kidney ultrasound pathology	Feature Extractor (ResNet)
	Renal ultrasound images	Feature extractor
	Glomeruli classification	Fine-tuning, feature extractor (Multi-gaze attention networks, Inception_ResNet_V2, AlexNet)
Heart	Arrhythmia	Feature extractor (DenseNet), Fine-tuning (AlexNet, GoogleNet)
	Cardiopathy	Fine-tuning (CaffeNet)
	Cardiovascular	Fine-tuning (VGG-19, VGG-16, Inception, ResNet)
	Vascular bifurcation	Fine-tuning

4.6 Medical imaging techniques and their use cases

Moving forth, we take up the applications and contributions of DL based on the type of medical disease analyzed by particular images.

4.6.1 X-Ray image

Radiography is one of the most common forms of medical imaging techniques. Chest radiography is being used tremendously, for diagnosing heart- and lung-based diseases [21]. Tuberculosis, pneumothorax, cardiac inflation, and atelectasis are some common use cases where X-ray images are useful [22]. The accessible, affordable, and reliable nature of these scans makes them a recurrent choice compared to other scans. Based on X-ray, deep convolutional neural networks-based screening systems have been designed. Transfer learning is a significant component in such systems. Other similar solutions use modality-specific ensemble learning and class selective mapping of interest for visualizing abnormalities in chest X-rays, popularly abbreviated as CXRs. GAN-based deep transfer training algorithms gained popularity during COVID-19 peak [23,24]. Auxiliary classifier generative adversarial network (ACGAN) is a COVID-GAN model that produces synthetic CXR scans to detect COVID-19 in patients [25]. Apart from cardiovascular diseases, X-rays find wide applications in orthopaedics.

4.6.2 Computerized tomography

Popularly known by the name of computerized tomography (CT) scan, CT utilizes computers and rotary X-rays to produce a cross-sectional view of the body. CT scans are widely accepted because they not only image bones and blood vessels of the body but also the soft tissues. Such scans are used for pulmonary nodule identification, which is fundamental to the early detection of cancer. Deep CNN-based algorithms such as GoogLeNet are beneficial for nodule detection which include semisolid, solid, and ground-glass opacity. Other applications include classification such as liver lesion classification, lung nodule detection and classification, kidney segmentation, COVID-19 detection, and feature extraction. Popular algorithm choices include MRFCN, U-Net, 3DU-Net, etc.

4.6.3 Mammography

Popularly known as mammogram (MG), is the process of using low-energy X-rays for diagnosing and screening breast cancer [26]. The history of mammography begins in 1913. Since then, there have been several advancements in this field. Still, detecting tumors is a deadly task given their small size. Today, MG is a reliable tool, however, the expertise of a physician is a must. Deep learning provides a two-step solution here, which includes detection, segmentation, and classification. CNN-based algorithms are tremendously valuable in such use cases for feature extraction tasks. Innovations over a period of time have enabled these intelligent systems to diagnose and detect cancer at early stages [27]. Classification algorithms permit analysts to quickly determine the type of tumor. This helps start treatment at an early stage. However, this requires human intervention to a significant extent. Creating an end-to-end scalable automated classification and detection system for the masses is still a hot topic of research.

4.6.4 Histopathology

Histopathology is the assessment of illness symptoms under a microscope, using a mounted glass slide from a biopsy or surgical specimen. It is extensively used in the identification of different diseases such as the presence of tumor in the kidney, lungs, and breast [28]. Using dyes, tissue sections are stained to identify lung cancer, Crohn's disease, ulcers, etc. The samples are accumulated through endoscopy, colonoscopy, or adopting surgical procedures such as biopsy. A crucial challenge in existing histopathology infrastructure is identifying disease growth at a species level. Hematoxylin and Eosin (H&E) staining has played a significant role in diagnosing cancer, however, identifying disease patterns by dying technique requires competence [29]. Today, digital pathology has automated this tedious challenge. Using histopathology images, deep learning is automating tasks like cell segmentation, tumor classification, labeling and annotating, nucleus detection, etc. Deep learning models have successfully simulated cell activities, using histopathology images to predict the future conditions of the tissue.

4.6.5 Endoscopy

A long nonsurgical mounted camera is directly inserted through a cavity into the body for visual examination of internal organs of the body. Endoscopy is a pretty mature test that has been in practice for a long. It is best suited for diagnosing ulcers, inflammation, celiac disease, blockages, gastroesophageal reflux disease, and sometimes cancerous linkages. Even though, physicians treat patients with anesthesia before beginning endoscopy, the test can be uncomfortable for most people. A painless non-invasive inspection of the gastrointestinal tract can be done using a recent invention – Wireless capsule endoscopy (WCE). As the name suggests, this capsule can be taken orally. Deep learning comes into the picture after endoscopy images start appearing. Images received from WCE are fed to deep learning algorithms. CNNs make real-time image segmentation, detection, classification, and identification possible. From detecting hookworm through WCE images to analyzing symptoms for predicting disease, endoscopy has evolved a lot. Today, such solutions are used to detect cancer in early stages by performing real-time analysis of tumors. VGG-16 is one of the popular CNN-based algorithms for the diagnostic assessment of the esophageal wall.

4.6.6 Magnetic resonance imaging

An magnetic resonance imaging (MRI) image corrupted by measure noise ε needs to be reconstructed from the k-space signal. The following equation represents MR image reconstruction

$$y = Ax + \varepsilon$$

where x is the image, and A is the linear forward operator [30]. Post-image reconstruction, image denoising, and optimization are performed. Deep learning techniques significantly decrease acquisition times. It is beneficial in imaging of the upper abdomen and in cardiac imaging due to the necessity of breath holding. Convolutional neural networks and stacked auto-encoders trained using MRI images are used for detecting and predicting brain-related diseases [31], segmentation of prostate and left ventricle [32].

4.6.7 Bio-signals

Apart from medical imaging techniques, bio-signaling techniques such as ECG, EEG, PCG, PPG, EMG, SS, NSS, and a lot more are common. Electrocardiography is one of the most common techniques for diagnosing cardiovascular disease. Deep learning algorithms allow early detection of heart-related diseases by analyzing ECG patterns. DNNs detect anomalies from electrocardiography scripts. They are being utilized for electrocardiography interpretation, arrhythmia classification, and systolic dysfunction detection. Such algorithms can work with data involving hundreds of parameters, thus handling complex medical data in the most optimized pattern. Bio-signals being intensely sensitive to body factors, DNNs can be used to eradicate inconsequential labels from the dataset.

4.7 Application of deep learning in medical image processing and analysis

With state-of-the-art, image capturing technologies available today, physicians are able to produce detailed scans. Digital Radiography, mammography, and echocardiography scans are being widely adopted for diagnosing diseases. Though the scans are extremely explicit with each individual projection image measuring as high as 32 MB, a significant composition of the image comprises noisy features. Often such features are blurry and invisible to human eyes. However, Deep Learning is the exemplary solution for eliminating unnecessary features and producing highly visible scans. DL can amplify necessary features and inhibit the visibility of unnecessary features to facilitate diagnosis. In other words, deep learning-based image processing algorithms allow noise reduction, smoothening of images, contrast optimization, elimination of irrelevant artifacts, and much more which are crucial for precise disease diagnosis. Microscopy imaging produces images with a high signal-to-noise ratio, deep learning can deal with use cases involving thousands of variable parameters and thus perform complex calculations with better robustness, higher speed, and precision to yield rich information. Applying deep learning techniques to microscopy, has enabled biologists to reconstruct high-resolution images without relying on sophisticated hardware setups. Deep learning's application in medical image processing, can be broadly classified as segmentation, classification, and tracking. Some other applications are detailed as follows.

4.7.1 Segmentation

It is a crucial step in medical image analysis, as it enables researchers to focus on key areas with relevant information. It is the process of dividing the image into several regions each concentrating on certain features that must be taken care of according to the researcher's interest. While there are dozens of ways of classifying segmentation techniques, some of the most prominent types on the basis of deep learning include semantic level segmentation and instance level segmentation. U-Net, a FCN allows semantic segmentation, while the latter (level segmentation) extends R-CNN. Image segmentation finds its application in almost every image processing workflow which we will discuss some of them in later sections. From the above discussion, we conclude that image segmentation is one of the most significant use cases of deep learning when it comes to medical image processing. At the same time, it is the starting point for most of the medical image analysis workflows.

4.7.2 Classification

Using artificial intelligence for image classification is not a new concept. Since the inception of digital image processing, artificial intelligence algorithms have been rigorously tested against complex object detection and classification tasks. However, with the dawn of machine learning algorithms, the results achieved were remarkable. Deep learning algorithms took it to the whole next level. Today, object detection and classification solutions are in high demand. Coupled with Internet of Things (IoT) technology, deep learning-based image classification systems are actively deployed in industries.

Image classification is the task of annotating input images based on business logic. Certain algorithms such as Bayesian classifiers, neural network-based classifiers, and geometric classifiers are easy to deploy and hence are more commercial. However, CNN-based classifiers, though complex to use, provide high accuracies over traditional machine learning-based classifiers [33]. CNN-based classifiers are at par when it comes to medical image analysis. This is possible because of the layered architecture of neural networks. Most CNN-based neural networks comprise a feature extraction module and a classification module. The input image is initially passed through convolutional and pooling layers to extract features. The output then is passed through the classification module. Deep learning-based methods achieve satisfying performance on low-resolution medical images. They are able to identify different types of cells at multiple stages. Fluorescent images are generally used to train deep learning algorithms in such scenarios.

Classifiers can also help in identifying diseases based on features extracted. By feeding feature parameters, classifiers can differentiate sickle cells from normal cells in case of anemia. It can identify white blood cells, leukemia, autoimmune diseases, lung cancer subtypes, hepatic granuloma, and so on. Deep CNN-based classifiers achieve higher accuracies compared to their competitors. Moreover, they have high execution speeds and low throughput, which make them an ideal choice.

4.7.3 Detection

As discussed in the previous sections of this chapter, object detection is a crucial step in any image analysis. A major challenge in detecting lesions is that multiple false positives arise while performing object detection. In addition, a good proportion of true positive samples are missed. CNN-based algorithms actively solve engrossing use cases such as the identification of enlarged thoracoabdominal lymph nodes, diagnosing lung diseases using CT scans, identification of prostate cancer using biopsy specimens, breast cancer metastasis identification, and so on.

Agglomerative nesting clustering filtering is another remarkable object detection framework that can be used for detecting tumors. Image saliency is another object detection technique that delivers eye-catching results. Saliency maps are a common tool for determining important areas useful in CNN algorithm training.

4.7.4 Deep learning-based tracking

Another fancy application of deep learning is measuring the velocity of cells, to comprehend biological signals. This technique is not just limited to cells alone, but all intracellular targets. A simple example could be tracking nuclei's trajectory. Algorithms facilitating such applications are generally based on RNN. Advanced algorithms such as the attention model and long- and short-term memory (LSTM) solve gradient exploding issues.

4.7.4.1 Object tracking

During diagnosing a disease, monitoring is one of the key activities to be performed. Tests such as endoscopy, laryngoscopy, esophagogastroduodenoscopy (EGD), and so on demand real-time image analysis. Object tracking methods are useful in the dynamic analysis of internal body parts' activities. Using a set of time-lapse images, object-tracking methods are used to monitor sub-cellular structures, drug effects, and cell biology [34]. Deep learning-based algorithms utilize SNR (signal-to-noise ratio) images to detect objects. However, frequent deformations inside the body, overlapping of organs, and disappearance of the target is a major challenge in traditional object detection algorithms. Today, a two-step process involving instance-level localization and data association is developed. Mask R-CNN and RNN are some of the techniques that are used to segment internal organs. RNNs are able to preserve information. Moreover, gating RNNs solve the problem of gradient explosion and disappearance. LSTM is an example of such an algorithm, popular for object tracking in microscopy images.

4.7.4.2 Cell tracking

Some common use cases where cell tracking is of importance while determining a cell's reaction to certain drugs, performing rapid susceptibility tests, tumor analysis, etc. Image segmentation algorithms play a significant role here. U-Nets and other CNN-based models are popular choices for cell tracking. Deep Learning allows monitoring cell reproduction over long periods with high precision.

4.7.4.3 Intracellular particle tracking

For analyzing particle mobility and studying intra-cellular dynamics, deep learning is used to develop software that can predict the trajectory of particles using the dynamics of fluorescent particles, molecules, and cell organelles. This is used to monitor biological processes such as the cell division cycle and other cellular activities. RNN-based trackers track each microtubule activity in real-time and deliver velocity-time graphs.

4.7.5 Using deep learning for image reconstruction

Medical images tend to be noisy and therefore need to be reconstructed. Sometimes medical scans are not very accurate and overlook salient features. While it might be perceived by the human eye, missing data can significantly impact a deep learning model's performance. Hence, deep learning provides some solutions for reconstructing microscopy images. In addition to that, deep learning also optimizes image parameters, de-noises the images, and restores any lost features. Deep learning automatically sets the parameters for imaging which makes the system adaptive. Deep learning systems like content-aware image restoration (CARE) outperform classic digital image de-noising algorithms such as non-local means (NLM). GANs are useful for generating super-resolution images. They find applications in converting diffraction-limited images to super-resolved images. Super-resolution fluorescence microscopy is quite expensive and requires delicate hardware, GANs provide a convenient way of gathering such data [35]. Deep learning algorithms can be used to translate images from one modality to another thereby increasing the significance of medical images. Deep learning-based algorithms can accelerate imaging speed. CNN-based techniques have been successful in providing a robust solution to solve ground truth pairs. Deep-STORM is one such example which was trained using low-resolution images. Conditional GANs (cGAN) are useful in reconstructing super-resolution images using low quality localized and wide-field images. U-Nets can also be utilized for augmenting the performance of SIM imaging limiting the number of high-resolution images required for training purposes. Transfer learning is another technique that can be used to reconstruct microtubule fluorescence images.

Deep learning can be applied throughout the brain disease diagnosis treatment [36]. KNN, SVM, and binary classification algorithms are the most often used algorithms that are useful in processing brain scans. SVM is an exemplary binary linear classifier with an approximate accuracy of 90%. Figure 4.1 outlines major steps carried out in CAD. While the implementation differs based on an adaptation of the algorithm, the general idea remains the same. Pre-processing involves analyzing the input medical images. This collection comprises a mix of both tumor and non-tumor images, fed from the database. The process is succeeded by the acquisition and filtering of noise. The objective is to eradicate all the bugs by reducing artifacts. Image harmonization is applied to stabilize the image quality. Different types of image transformation techniques, namely, discrete wavelet transform (DWT), discrete cosine transform (DCT), and integer wavelet transform are applied to the image. These techniques are suitable for medical scans such as magnetic resonance image (MRI) scans and computed tomography (CT) scans. A quality check sits in the process, which decides if the image qualifies for further steps. Once the image passes quality checks, two significant steps, namely, image

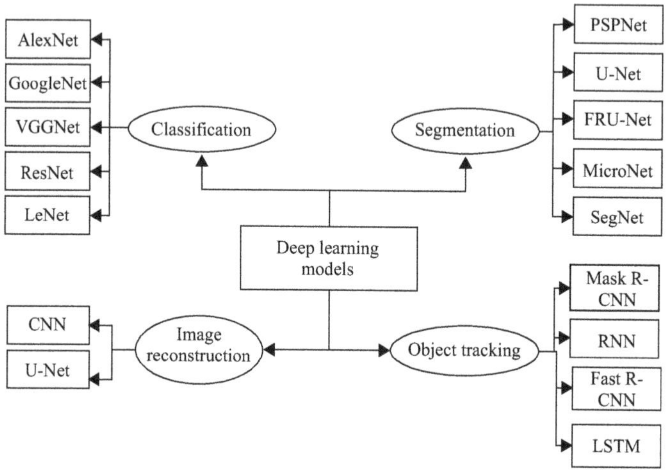

Figure 4.1 Network models for medical image analysis and processing

segmentation and feature extraction, are started. In case the sample contains lots of low-resolution images, deep learning allows rapid and convenient image reconstruction. The process of medical image analysis in case of requirement of image enhancements, is more intensive. Consider a brain tumor image sample if passed through this process, after image segmentation, the image is rigorously scanned by the neural network and passed on to the feature extraction engine. Abnormal tissues here are identified based on the patterns in the image. It is at this point, neurologists can identify certain logics such as the classification of tumor or start the analysis of the entire image sample. Image segmentation is one of the key processes involved in medical image processing and analysis. There existed several methods of image segmentation techniques (Figure 4.2). A widely accepted listing is picturized in Figure 4.3.

An important sub-process throughout medical image analysis using deep learning methodologies is defining region of interest (ROI). Detection and analysis of morphological features, texture variations, shading variations, and gray-level feature analysis are some of the outcomes of this process. It is after establishing ROI, the evaluated image scans are fed to classification algorithms for labeling.

While there exist several classification methodologies, namely, pixel-wise classification, sub-pixel-based classification, and object-based classification, a DNN-based classification model is a go-to classification technique due to their high accuracy along with ease of incorporating query and item features. Other algorithms popular in this space include K-Means, ISODATA, and SOM which rely on unsupervised classification techniques.

The aforementioned approach is common across all medical image analysis use cases. The process can be condensed into three major steps:

1. **Image formation:** This part involves data acquisition and image reconstruction. In the case of real-time image analysis, the algorithm processes images

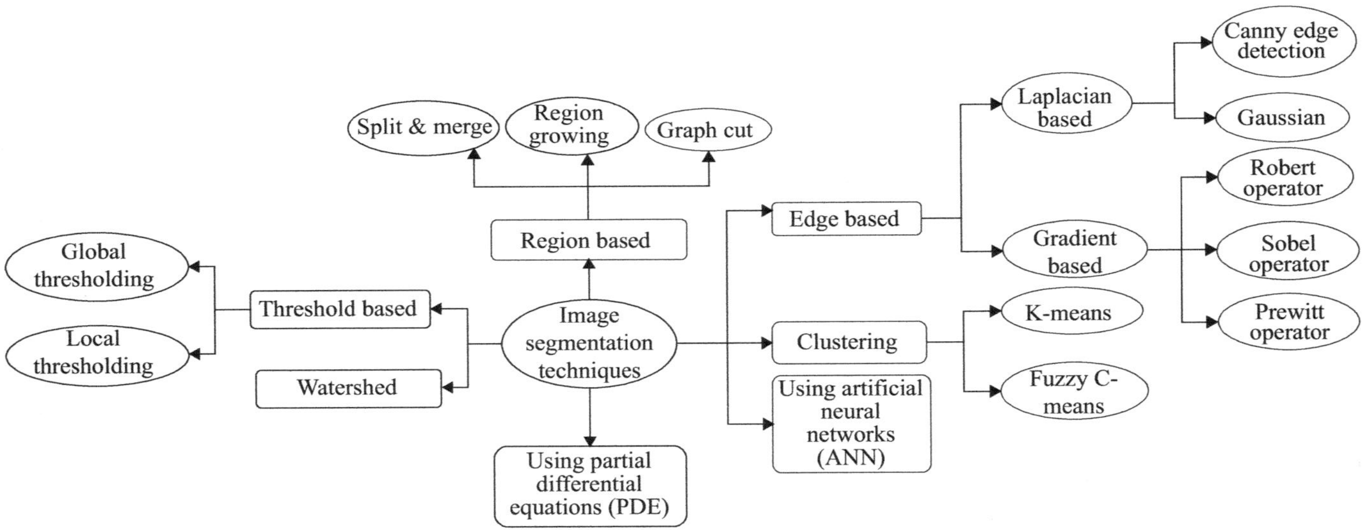

Figure 4.2 Image segmentation techniques

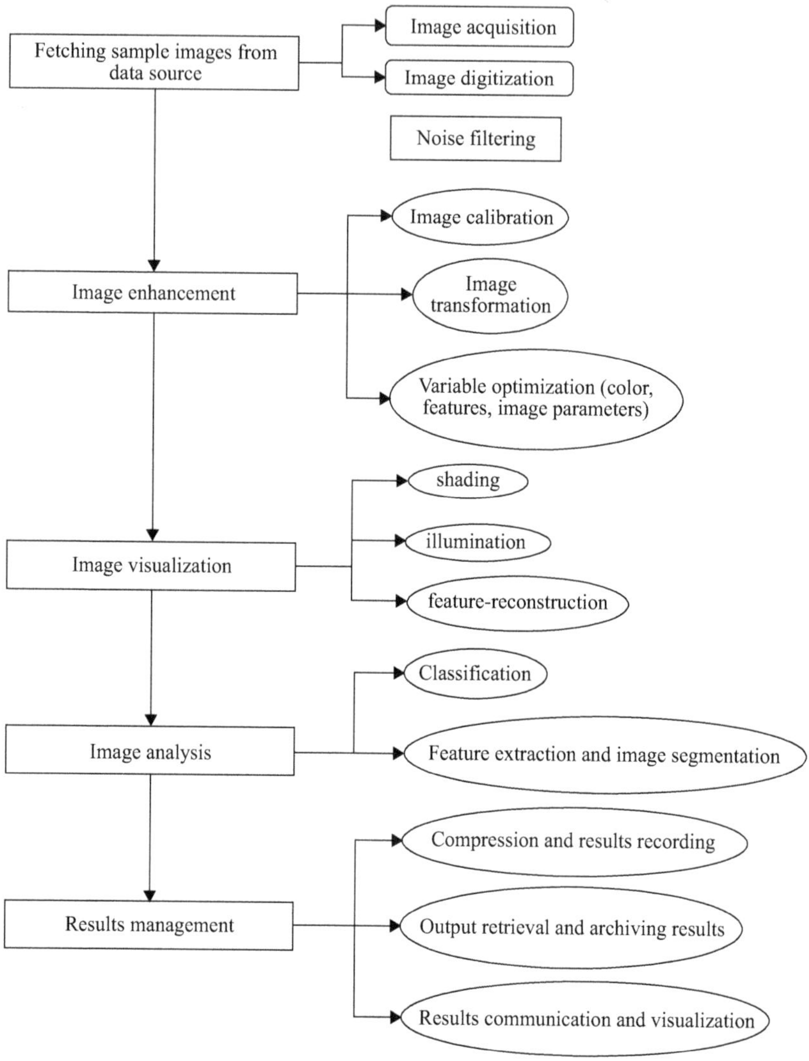

Figure 4.3 Working of a CAD system

coming from an active data source. In the case of batch processing, a collection of medical images sits on a central repository from where it is passed onto the deep learning algorithm. Data acquisition broadly consists of the detection of the image, converting it to a specific format and scale, preconditioning the image, and digitalizing the acquired image signal. The obtained raw image contains original data about captured image parameters, which is the exact description of the internal characteristics of patients' bodies. This is the primary source of image features and must be preserved, as it becomes the subject of all subsequent image processing. Depending on the medical scan, physical quantities of images may differ. For instance, a CT scan energy of incident photons is the primary physical quantity. Similarly, for PET photons' energy is a primary physical quantity, in ultrasonography acoustic echoes forms primary physical quantity and in MRI the radio-frequency signal emitted by excited atoms forms the primary physical quantities. Often medical images tend to be blurry and noisy. Hence, image reconstruction is often preferred to regain the lost features. Using analytical and iterative algorithms, an inverse image reconstruction operation is applied using the acquired raw data to regain lost features and discard unwanted noise. Depending on the medical imaging technology used, the algorithm varies. For tomography scans, filtered back projection is adopted and for MRI scans Fourier transformation is adopted while for ultrasonography delay and sum (DAS) beamforming is what the system relies on [25]. Furthermore, iterative algorithms like maximum-likelihood expectation maximization (MLEM) and algebraic reconstruction (ARC) are used to improve image quality by removing noise and reconstructing optimal images.

2. **Image computing:** Image obtained from the previous step is now passed into the image transformation engine, for enhancement, visualization, and analysis purposes. The transformed image is processed into a digital signal and digital signal transforms are applied to enhance its quality. Transformation techniques are applied to improve image interpretability. This is followed by segmentation and image quantification. Logics are applied, and regions of interest are determined to gather appropriate results. Finally, relevant data are used to formulate clinically relevant results. These results are visualized by rendering image data. Image enhancement refines the image using the spatial approach for contrast optimization and a frequency approach for smoothening and sharpening the image.
3. **Results management:** Once an image is processed, it can be passed for further analysis where business logic are matched against obtained image data. Here, reports are generated, stored, and communicated. For any disease diagnosis, the reports are matched with pre-available data to train neural networks and deliver human-understandable results.

4.8 Training testing validation of outcomes

This chapter tries to highlight some possible ways in which deep learning is streamlining the processing of medical images. Being an extremely sensitive domain, it is essential that each and every innovation is tested thoroughly before it is available to the public. In the following section, we present some training and testing techniques, essential before deploying deep learning solutions.

The training phase refers to the time when the model learns to implement tasks by using available data known as training data. Depending on the data, a supervised or unsupervised learning approach is adopted. The testing phase is

Table 4.4 Popular metrics to evaluate efficiency of deep learning algorithm

Metrics	Formula	Description
Over-segmentation rate	$OSR = \frac{O}{R+O}$ O refers to pixels appearing in actual segmented image but not in theoretical segmented image and R is the reference area of segmented image manually drawn by doctor	Ratio of pixels divided into the reference area of ground truth image
Under-segmentation rate	$USR = \frac{U}{R+O}$ Where U pixels appear in theoretical segmented image but not in actual segmented image.	Ratio of segmentation results to missing pixels in ground truth image
Jaccard Index	Jaccard (A, B) = $\lvert A \cap B\rvert / \lvert A \cup B\rvert$	Used for calculating overlap between 2 sets
Dice Index	Dice (A, B) = 2 * $((\lvert A \cap B\rvert)/(\lvert A\rvert+\lvert B\rvert))$	Used to calculate overlap between two samples. Output ranges from 0 to 1. The closer to 1, the better is the segmentation effect
Segmentation accuracy (SA)	SA = $(1 - (\lvert R_s - T_s\rvert / R_s))$ * 100% Where R_s denotes reference area in ground truth and T_s represents real area of image obtained by the algorithm	The area of SA reflects percentage of real area in ground truth

where the model is exposed to new data known as testing data. This phase is used to verify the accuracy of the model. Testing and training data may come from the same dataset but are expected to be mutually exclusive in order to get accurate results. A validation phase is one that lies between the testing and training phases, where the performance of the model is gauged. Following we tabulate some popular metrics to determine the accuracy of deep learning algorithms (Table 4.4).

4.9 Challenges in deploying deep learning-based solutions

Despite all the aforementioned contributions of deep learning, which prove it to be a boon for medical image processing and analysis, there are certain technical dependencies that come as a prerequisite for deploying deep learning solutions. Not just a suitable infrastructure, dependency on accurate data is a must to yield relevant results. The fascinating results received from medical image analysis come with the heavy cost of gathering consistent data. Image parameters such as resolution, contrast, signal-to-noise ratio, shading, sharpening, tuning, exposure, highlights, vibrancy, saturation, texture, clarity, and all other factors should be in place before the image is sent for analysis.

One of the major impediments in deep learning-based image processing is the requirement of matched training pairs of low-resolution input images and high-resolution ground truth data which can be used for super-resolution reconstruction of images [37]. Even though deep learning delivers satisfactory results for classification and segmentation-based tasks, when it comes to particle tracking and object detection in microscopy images the performance of algorithms deteriorates with increasing density of decorated multiple-target tracking tasks.

Data quality is the most significant issue in biomedical applications of all artificial intelligence-based solutions. The low prevalence of certain features across the dataset creates an imbalance that can make an entire sample obsolete. Imbalance learning caused due to class imbalance is a serious problem that refers to the distribution of sample data across biased or skewed classes. Since negative samples are more abundant than positive features in medical images such sets cannot be used for CNN-based algorithms. Data resampling is a common solution to class imbalance problems. Under-sampling and oversampling approaches can be used to resize the training dataset to achieve a balanced distribution. This can significantly mitigate class imbalance. Synthetic minority over-sampling technique (SMOTE) is a standard technique adopted for learning from imbalanced data. By calculating the difference between the feature vector and the nearest neighbor, the result is multiplied by a random number lying between 0 and 1. Finally, this quantity is added to the feature vector. Some other popular approaches for dealing with class imbalance include borderline SMOTE which is an enhancement to the existing SMOTE. The adaptive synthetic sampling approach (ADASYN) is another improvement over traditional SMOTE. Such approaches are common when dealing with structured data such as demographic-related data. However, when it comes to unstructured data such as biosignals and medical images, playing with the dataset can often lead to unexpected outcomes. The requirement of annotated data is another technical challenge that arises with deep learning solutions in the medical sphere. Labeled and annotated data ease the designing of large DL systems. However, labeling requires domain knowledge of radiology which is highly expensive and time-consuming to achieve.

Medical data privacy is another bottleneck when it comes to gaining access to large datasets [38]. Public datasets cannot be blindly trusted for training enterprise-grade systems when it comes to medical use. Therefore, finding appropriate data for training algorithms is challenging. Moreover, several countries enforce strict laws governing their data confidentiality. Often this lack of data from a specific community hinders innovation among scientists. Though the laws are aimed at protecting the data of European citizens from being misused, it also nullifies the possibility of a breakthrough that a scientific institution working in an Asian country would have given to the world. Y. Ding proposes a stream cipher generator that uses deep learning for medical image encryption and decryption [39].

Even though deep learning is an enhancement over machine learning, there are several use cases, where deep learning fails to deliver expected results. In case the sample size is small or when the outcome class is a continuous variable, machine learning models are preferred over deep learning approaches. According to image biomarker standardization initiative (IBSI), when there are a lot of factors which influence the predicted class such as risk factors (age of patient, genetic history of a disease, consumption of intoxicants, mutations, and infections) which are quite common in biological data, machine learning models stand as a robust and reliable feature selector. Harmonization and de-noising techniques prevent overfitting to enhance relevant features.

High upfront cost in establishing deep learning-based solutions for the consumption of public is a major setback for institutions. AI-based solutions are computationally expensive and therefore require expensive infrastructure. They require highly skilled professionals at least in the initial phases. All these arrangements come with a cost, which institutions are skeptical about. A brain dock system designed by a Japanese company uses an MRI system for detecting Alzheimer's disease in a high-risk group. This revolutionary technology takes just 2 hours for the entire check-up. Results produced are further analyzed by a group of physicians to propose advice concerning enhancement in daily lifestyle that can mitigate such severe disease. While such a solution appears invaluable for society, a healthy brain dock test can cost somewhere around 600,000 JPY.

Dimensionality reduction is another challenge when it comes to processing unstructured data such as images. Medical images tend to contain more noise and redundant features than a casual landscape picture. Moreover, advanced image-capturing devices produce mixed outputs, which if not analyzed using suitable algorithms can lead to a loss of opportunity. Treating a 3D computed tomography image with regular algorithms can neglect a lot of data points which not only limits our scope of analysis but also produces incorrect results. Currently, CNN-based algorithms are unable to deliver promising results when it comes to 3D medical images. Hence, a general approach is to break an image into several components to imitate a 2D image and individually process it. The final result is summed up to produce reports before patients. Certain image analysis algorithms perform a 3D reconstruction of subcellular fabric to produce relevant medical results [31]. Though this approach delivers satisfactory results, advanced algorithms are required to simplify this two-step process. Due to this lengthy two-step process, feeding medical records to a simple machine learning-based prediction system is far more efficient than relying on the aforementioned algorithm.

Deep learning stands as a potential alternative for medical image segmentation tasks. However, constant requirement discovery and solution improvement are expected to fit it into complex use cases. Furthermore, enhancements in deep learning and microscopy techniques will help develop intelligent systems that deliver super-resolution high-content imaging with automatic real-time objective image analysis. Predicting and diagnosing diseases using intelligent systems opens up doors for sustainable medical treatment.

4.10 Conclusion

Among the endless possibilities offered by deep learning in the field of medical image scanning, this chapter outlines some of the major breakthroughs deep learning has caused in medical image processing and analysis. The fast and efficient processing abilities of deep learning algorithms make it a revolutionary technology, which can mitigate slow, error–prone, and labor-intensive image analysis tasks. This chapter was an attempt to highlight how deep learning is streamlining the complex process of medical image analysis to yield exemplary results. Even though medical image analysis requires knowledge of various domains such as mathematics, computer science, pharmacology, physics, biology, physiology, and much more, deep learning systems have the ability to outshine a set of physicians. However, we believe a deep learning system can never be a complete replacement for physical doctors, but it can definitely serve as a 'second set of eyes', thus establishing a healthy coexistence between humans and intelligent machines. Physicians will always be required to act as guides and supervisors. Physicians will be required to exhibit soft skills at all times that will exercise a constructively critical approach utilizing the enormous potential of intelligent systems while reducing the possibility of the scientific dystopian nightmare of the "machines in power".

Deep learning systems are highly relevant and practical in the context of developing nations where medical facilities are limited. In real context, they have high execution speeds, provide significant cost reduction, better diagnostic accuracy with better clinical and operational efficiency, and are scalable along with better availability. Such intelligent algorithms can be easily integrated into mobile software applications which can touch remote locations, thus benefiting the masses who otherwise were isolated because of geographical, economic, or political reasons. These solutions can even be extended towards designing mental health solutions such as scoring sleep health by monitoring EEGs to prevent the onset of possible diseases [40].

Medical image research has a bright future. Deep learning solutions will eventually use transfer learning and then meta-learning [41]. The amalgamation of these technologies along with data augmentation, self-supervising learnings, reinforcement learning, and business domain adaptation will significantly improve the current performance of neural networks and thus solve advanced use cases.

References

[1] Prabhavathy, P., Tripathy, B.K., and Venkatesan, M. Analysis of diabetic retinopathy detection techniques using CNN models. In: S. Mishra, H.K. Tripathy, P. Mallick, and K. Shaalan (eds.), *Augmented Intelligence in Healthcare: A Pragmatic and Integrated Analysis, Studies in Computational Intelligence*, vol. 1024, Springer. https://doi.org/10.1007/978-981-19-1076-0_6

[2] Gupta, P., Bhachawat, S., Dhyani, K., and Tripathy, B.K. A study of gene characteristics and their applications using deep learning, studies in big data (Chapter 4). In: S. S. Roy and Y.-H. Taguchi (eds.), *Handbook of Machine Learning Applications for Genomics*, Vol. 103, 2021. ISBN: 978-981-16-9157-7, 496166_1_En

[3] Tripathy, B.K., Garg, N., and Nikhitha, P. In: L. Perlovsky and G. Kuvich (eds.), Introduction to deep learning, cognitive information processing for intelligent computing and deep learning applications, IGI Publications.

[4] Debgupta, R., Chaudhuri, B.B., and Tripathy B.K. A wide ResNet-based approach for age and gender estimation in face images. In: A. Khanna, D. Gupta, S. Bhattacharyya, V. Snasel, J. Platos, and A. Hassanien (eds.), *International Conference on Innovative Computing and Communications, Advances in Intelligent Systems and Computing*, vol. 1087, Springer, Singapore, 2020, pp. 517–530, https://doi.org/10.1007/978-981-15-1286-5_44.

[5] Ravi Kumar Rungta, P.J. and Tripathy, B.K. A deep learning based approach to measure confidence for virtual interviews. In: A.K. Das *et al.* (eds.), *Proceedings of the 4th International Conference on Computational Intelligence in Pattern Recognition (CIPR)*, CIPR 2022, LNNS480, pp. 278–291, 2022.

[6] Puttagunta, M. and Ravi, S. Medical image analysis based on deep learning approach. *Multimedia Tools and Applications, 2021;80:*24365–24398. https://doi.org/10.1007/s11042-021-10707-4

[7] Karan Maheswari, A.S., Arya, D., Tripathy, B.K. and Rajkumar, R. Convolutional neural networks: a bottom-up approach. In: S. Bhattacharyya, A.E. Hassanian, S. Saha, and B.K. Tripathy (eds.), *Deep Learning Research with Engineering Applications*, De Gruyter Publications, 2020, pp. 21–50. doi:10.1515/9783110670905-002

[8] Yu, H., Yang, L.T., Zhang, Q., Armstrong, D., and Deen, M.J. Convolutional neural networks for medical image analysis: state-of-the-art, comparisons, improvement and perspectives. *Neurocomputing*, 2021;444:92–110. https://doi.org/10.1016/j.neucom.2020.04.157

[9] Kaul, D., Raju, H. and Tripathy, B. K. Deep learning in healthcare. In: D.P. Acharjya, A. Mitra, and N. Zaman (eds.), *Deep Learning in Data Analytics – Recent Techniques, Practices and Applications), Studies in Big Data*, vol. 91. Springer, Cham, 2022, pp. 97–115. doi:10.1007/978-3-030-75855-4_6.

[10] Alalwan, N., Abozeid, A., ElHabshy, A.A., and Alzahrani, A. Efficient 3D deep learning model for medical image semantic segmentation. *Alexandria Engineering Journal*, 2021;60(1):1231–1239. https://doi.org/10.1016/j.aej.2020.10.046.

[11] Liu, Z., Jin, L., Chen, J. *et al.* A survey on applications of deep learning in microscopy image analysis. *Computers in Biology and Medicine*, 2021;134:104523. https://doi.org/10.1016/j.compbiomed.2021.104523

[12] Adate, A. and Tripathy, B.K. S-LSTM-GAN: shared recurrent neural networks with adversarial training. In: A. Kulkarni, S. Satapathy, T. Kang, and A. Kashan (eds.), *Proceedings of the 2nd International Conference on Data Engineering and Communication Technology. Advances in Intelligent Systems and Computing*, vol. 828, Springer, Singapore, 2019, pp. 107–115.

[13] Liu, X., Song, L., Liu, S., and Zhang, Y. A review of deep-learning-based medical image segmentation methods. *Sustainability*, 2021;13(3):1224. https://doi.org/10.3390/su13031224.

[14] Adate, A. and Tripathy, B.K. A survey on deep learning methodologies of recent applications. In D.P. Acharjya, A. Mitra, and N. Zaman (eds.), *Deep Learning in Data Analytics – Recent Techniques, Practices and Applications)*, *Studies in Big Data*, vol. 91. Springer, Cham, 2022, pp. 145–170. doi:10.1007/978-3-030-75855-4_9

[15] Vaidyanathan, A., van der Lubbe, M. F. J. A., Leijenaar, R. T. H., *et al.* Deep learning for the fully automated segmentation of the inner ear on MRI. *Scientific Reports*, 2021;11(1):Article no. 2885. https://doi.org/10.1038/s41598-021-82289-y

[16] Sihare, P., Bardhan, P., A.U.K., and Tripathy, B.K. COVID-19 detection using deep learning: a comparative study of segmentation algorithms. In: K. Das *et al.* (eds.), *Proceedings of the 4th International Conference on Computational Intelligence in Pattern Recognition (CIPR)*, CIPR 2022, LNNS480, 2022, pp. 1–10.

[17] Yagna Sai Surya, K., Geetha Rani, T., and Tripathy, B.K. Social distance monitoring and face mask detection using deep learning. In: J. Nayak, H. Behera, B. Naik, S. Vimal, D. Pelusi (eds.), *Computational Intelligence in Data Mining. Smart Innovation, Systems and Technologies*, vol. 281. Springer, Singapore. https://doi.org/10.1007/978-981-16-9447-9_36

[18] Jungo, A., Scheidegger, O., Reyes, M., and Balsiger, F. pymia: a Python package for data handling and evaluation in deep learning-based medical image analysis. *Computer Methods and Programs in Biomedicine*, 2021;198:105796. https://doi.org/10.1016/j.cmpb.2020.105796

[19] Abdar, M., Samami, M., Dehghani Mahmoodabad, S. *et al.* Uncertainty quantification in skin cancer classification using three-way decision-based Bayesian deep learning. *Computers in Biology and Medicine*, 2021;135:104418. https://doi.org/10.1016/j.compbiomed.2021.104418

[20] Wang, J., Zhu, H., Wang, S.-H., and Zhang, Y.-D. A review of deep learning on medical image analysis. *Mobile Networks and Applications*, 2020;26 (1):351–380. https://doi.org/10.1007/s11036-020-01672-7

[21] Ahmedt-Aristizabal, D., Mohammad Ali Armin, S.D., Fookes, C., and Lars P. Graph-based deep learning for medical diagnosis and analysis: past, present and future. *Sensors, 2021;21*(14):4758. https://doi.org/10.3390/s21144758

[22] Çallı, E., Sogancioglu, E., van Ginneken, B., van Leeuwen, K.G., and Murphy, K. Deep learning for chest X-ray analysis: a survey. *Medical Image Analysis*, 2021;72:102125. https://doi.org/10.1016/j.media.2021.102125

[23] Shorten, C., Khoshgoftaar, T.M., and Furht, B. Deep learning applications for COVID-19. *Journal of Big Data*, 2021;8(1):Article no. 18. https://doi.org/10.1186/s40537-020-00392-9

[24] Gaur, L., Bhatia, U., Jhanjhi, N. Z., Muhammad, G., and Masud, M. Medical image-based detection of COVID-19 using deep convolution neural

networks. *Multimedia Systems*. 2021;29:1729–1738. https://doi.org/10.1007/s00530-021-00794-6

[25] Skandarani, Y., Jodoin, P.-M., and Lalande, A. GANs for medical image synthesis: an empirical study, n.d. https://arxiv.org/pdf/2105.05318.pdf

[26] Bhattacharyya, S., Snasel, V., Hassanian, A.E., Saha, S., and Tripathy, B.K. *Deep Learning Research with Engineering Applications*, De Gruyter Publications, 2020. ISBN: 3110670909, 9783110670905. DOI: 10.1515/9783110670905

[27] Jain, S., Singhania, U., Tripathy, B.K., Nasr, E. A., Aboudaif, M.K., and Kamrani, A.K. Deep learning based transfer learning for classification of skin cancer. *Sensors (Basel)*, 2021;21(23):8142. doi:10.3390/s21238142

[28] Papadimitroulas, P., Brocki, L., Christopher Chung, N. *et al.* Artificial intelligence: deep learning in oncological radiomics and challenges of interpretability and data harmonization. *Physica Medica*, 2021;83:108–121. https://doi.org/10.1016/j.ejmp.2021.03.009

[29] Salvi, M., Acharya, U.R., Molinari, F., and Meiburger, K.M. The impact of pre- and post-image processing techniques on deep learning frameworks: a comprehensive review for digital pathology image analysis. *Computers in Biology and Medicine*, 2021;128:104129. https://doi.org/10.1016/j.compbiomed.2020.104129

[30] Ranjbarzadeh, R., Bagherian Kasgari, A., Jafarzadeh Ghoushchi, S. *et al.* Brain tumour segmentation based on deep learning and an attention mechanism using MRI multi-modalities brain images. *Sci Rep*, 2021;11:10930. https://doi.org/10.1038/s41598-021-90428-8

[31] Tripathy, B.K., Parikh, S., Ajay, P., and Magapu, C. Brain MRI segmentation techniques based on CNN and its variants (Chapter 10). In: J. Chaki (ed.), *Brain Tumor MRI Image Segmentation Using Deep Learning Techniques*, Elsevier Publications, 2022, pp. 161–182. doi:10.1016/B978-0-323-91171-9.00001-6

[32] Gassenmaier, S., Küstner, T., Nickel, D., *et al.* Deep learning applications in magnetic resonance imaging: has the future become present? *Diagnostics (Basel, Switzerland)*, 2021;11(12):2181. https://doi.org/10.3390/diagnostics11122181

[33] Castiglioni, I., Rundo, L., Codari, M., *et al.* AI applications to medical images: from machine learning to deep learning. *Physica Medica*, 2021;83:9–24. https://doi.org/10.1016/j.ejmp.2021.02.006

[34] Bhardwaj, P., Guhan, T., and Tripathy, B.K. Computational biology in the lens of CNN, studies in big data (*Chapter 5*). In: S. S. Roy and Y.-H. Taguchi (eds.), *Handbook of Machine Learning Applications for Genomics*, Vol. 103, 2021. ISBN: 978-981-16-9157-7 496166_1_En

[35] Bhandari, A., Tripathy, B., Adate, A., Saxena, R., and Thippa Reddy, G. From beginning to BEGANing: role of adversarial learning in reshaping generative models. *Electronics, Special Issue Artificial Intelligence Technologies and Applications*, 2023;12(1):155. https://doi.org/10.3390/electronics12010155

[36] Magadza, T. and Viriri, S. Deep learning for brain tumour segmentation: a survey of state-of-the-art. *Journal of Imaging, 2021;7*(2):19. https://doi.org/10.3390/jimaging702001

[37] Pramod, A., Naicker, H.S., and Tyagi, A.K. Machine learning and deep learning: open issues and future research directions for the next 10 years. In *Computational Analysis and Deep Learning for Medical Care*, Wiley, 2021, pp. 463–490. https://doi.org/10.1002/9781119785750.ch18

[38] Ma, X., Niu, Y., Gu, L., *et al.* Understanding adversarial attacks on deep learning based medical image analysis systems. *Pattern Recognition*, 2021;110:107332. https://doi.org/10.1016/j.patcog.2020.107332

[39] Ding, Y., Tan, F., Qin, Z., Cao, M., Choo, K.-K.R., and Qin, Z. DeepKeyGen: a deep learning-based stream cipher generator for medical image encryption and decryption. *IEEE Transactions on Neural Networks and Learning Systems*, 2022;33(9):4915–4929. doi:10.1109/TNNLS.2021.3062754

[40] Nogales, A., García-Tejedor, Á.J., Monge, D., Vara, J.S., and Antón, C. A survey of deep learning models in medical therapeutic areas. *Artificial Intelligence in Medicine*, 2021;112:102020. https://doi.org/10.1016/j.artmed.2021.102020

[41] Singh, R., Bharti, V., Purohit, V., Kumar, A., Singh, A.K., and Singh, S.K. MetaMed: few-shot medical image classification using gradient-based meta-learning. *Pattern Recognition*, 2021;120:108111. https://doi.org/10.1016/j.patcog.2021.108111

Chapter 5
Comparative analysis of lumpy skin disease detection using deep learning models

Shikhar Katiyar[1], *Krishna Kumar*[1], *E. Ramanujam*[1], *K. Suganya Devi*[1] *and Vadagana Nagendra Naidu*[1]

Lumpy skin disease (LSD) is an infectious disease caused by a Poxviridae family of viruses to the cattle, and it is a transboundary disease affecting the cattle industry worldwide. Asia has the highest number of LSD reports from the year 2019 till today, as per the LSD outbreak report data generated by the World Organization of Animal Health. In India, it started in 2022 and resulted in the death of over 97,000 cattle in three months between July and September 2022. LSD transmission is mainly due to blood-feeding insects. The other water feed troughs and contaminated environments are minor cases. According to Cattle India Foundation (CIF) analysis, more than 16.42 lakh cattle have been infected by LSD, and 75,000 have died since July 2022. Thus, the reduction of the livestock mortality rate of cattle is significant today either by analyzing skin diseases or through early detection mechanisms. The LSD research is evolving and attracts various Artificial Intelligence (AI) experts to analyze the problem through image processing, machine learning, and deep learning. This chapter compares the performance of deep and hybrid deep learning models for detecting LSD. The challenges and limitations of this study have been extended into future scope for enhancements in LSD detection.

5.1 Introduction

Cattle are the most widespread species that provide milk, meat, and draft power to humans and remain sizeable ruminant livestock [1]. The term "livestock" is indistinct and may be defined broadly as any population of animals kept by humans for a functional, commercial purpose [2]. Horse, Donkey, Cattle, Zebu, Bali cattle, Yak, Water buffalo, Gayal, Sheep, Goat, Reindeer, Bactrian camel, Arabian camel, Llama, Alpaca, Domestic Pig, Chicken, Rabbit, Guinea pig, etc., are the varieties of livestock raised by the people [3]. Around 21 million people in India depend on

[1]Department of Computer Science and Engineering, National Institute of Technology Silchar, India

Table 5.1 Twentieth Indian livestock census report in a million population

Livestock	# in million population
Cattle	192.49
Goat	148.88
Buffalo	109.85
Sheep	74.26
Pig	9.06
Mithun	0.38
Yak	0.08
Horses and Ponies	0.34
Mule	0.08
Donkey	0.12
Camel	0.25

livestock for their livelihood. Specifically, India has ranked first in cattle inventory per the statistics of 2021 [4]. Brazil and China have the second and third-largest cattle inventory rates. In addition, India ranks first in milk production per the 20th livestock census of India report as shown in Table 5.1 [5]. In India, livestock employed 8.8% of the total population and contributes around 5% of the GDP and 25.6% of the total Agriculture GDP [6].

Cattle majorly serve social and financial roles in societies. Over the world, more than 1.5 billion cattle were present as per the report [7]. It has been raised primarily for their family's subsistence in local sales. Most cattle farmers supply these cattle to the international markets in large quantities [8]. Around 40% of the world's agricultural output has been contributed by livestock which secures food production to almost a billion people [9]. It has also been growing fast in the world, driven by income growth and supported by structural and technical advances, particularly in the agriculture sector. This growth and transformation have provided multiple opportunities to the agricultural sector regarding poverty alleviation and food security.

Livestock is also considered a valuable asset to livestock owners for their wealth, collateral credits, and security during financial needs. It has also been a center for mixed farming systems as it consumes waste products during agricultural and food processing, helps control insects and weeds, produces manure for fertilization and conditioning fields, and provides draught power for plowing and transportation [10]. In some places, livestock has been used as a public sanitation facility to consume waste products; otherwise, it may pose severe pollution and public health problems. In the world, livestock contributes 15% towards food energy and 25% towards dietary protein. Almost 80% of the illiterate and undernourished people have a primary dependency on agriculture and the raising of livestock for their daily needs [11]. Data from the Food and Agriculture Organization (FAO) database on rural income generating activities (RIGA) shows that, in a sample of 14 countries, 60% of rural people raise livestock. A significant proportion of the rural household's livestock is to contribute to household income [12].

5.1.1 Health issues of cattle

All over the world, cattle provide many opportunities for the wealth and welfare of the people. In other views, they have health issues like humans. Health issues are mainly categorized as infectious diseases, deficiency diseases, genetic and non-genetic diseases, etc. [13]. Infectious diseases are the greatest threat to cattle. The majorly affected infectious diseases are Anthrax, a Black quarter (blackleg), blue-tongue, ringworm, foot and mouth disease (FMD), psoroptic disease, bovine viral diarrhea (BVD), transmissible spongiform encephalopathies (TSE), lumpy skin disease (LSD), etc. [14]. Figure 5.1 shows the sample images of the diseases in the

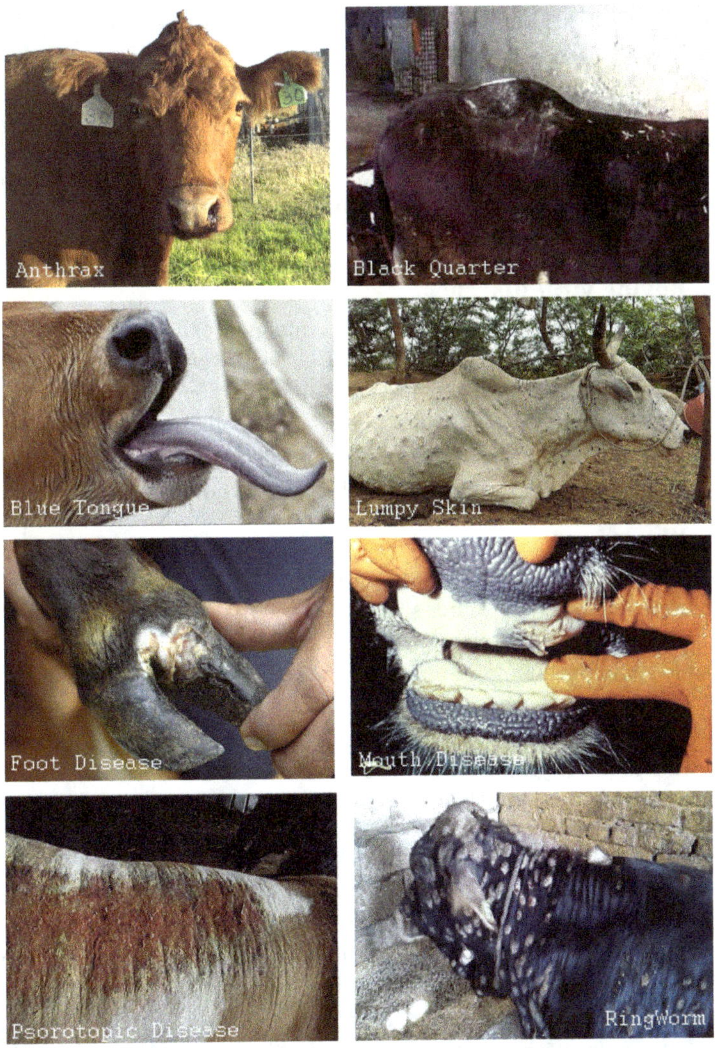

Figure 5.1 Sample images of infectious disease in infected cows

infected cows. Direct causes of diseases are chemical poisons, parasites, fungi, viruses, bacteria, nutritional deficiencies, and unknown causes. Additionally, the well-being of cattle can also be influenced indirectly by elements like food, water, and the surrounding environment. [15]. The detailed description and infection of the diseases are as follows.

5.1.1.1 Foot and mouth disease

Foot and mouth disease (FMD) is a highly serious and infectious viral disease that causes illness and is very expensive to cure [16]. This may affect the cattle, sheep, swine, and goats with cloven hooves. FMD is a transboundary animal disease (TAD) that significantly impacts livestock growth and output. A, O, C, SAT1, SAT2, SAT3, and Asia1 are the seven strains which causes FMD worldwide. Each strain needs a unique vaccination to provide inoculated animals protection.

5.1.1.2 Ringworm

Ringworm is a hideous skin lesion, round and hairless caused by a fungal infection of the hair follicle and the skin's outer layer [17]. *Trichophyton verrucosum* is the most prevalent agent infecting cattle, with other fungi being less common. Ringworm is a zoonotic infection. Ringworm is uncommon in sheep and goats raised for meat.

5.1.1.3 Bovine viral diarrhea

Bovine viral diarrhea (BVD) virus (BVDV) infection can affect livestock of any age [18]. BVDV is a single linear positive-stranded RNA virus belonging to the family Flaviviridae's Pestivirus genus. Wide-ranging clinical symptoms of BVDV infection include intestinal and respiratory disease in all classes of cattle as well as a reproductive and fetal disease after infection of a breeding female who is vulnerable to it. BVDV viruses also depress the immune system.

5.1.1.4 Bluetongue

Bluetongue (BT) is a viral infection spread by vectors. It affects both wild and domestic ruminants such as cattle, sheep, goats, buffaloes, deer, African antelope species, and camels [19]. Although the Bluetongue virus (BTV) does not typically cause visible symptoms in most animals, it can lead to a potentially fatal illness in a subset of infected sheep, deer, and wild ruminants. It transmits primarily by a few species of the genus *Culicoides*, insects that act as vectors. These vectors become infected with the BTV when they feed on viraemic animals and subsequently spread the infection to vulnerable ruminants.

5.1.1.5 Transmissible spongiform encephalopathies

Transmissible spongiform encephalopathies (TSEs) are a cluster of debilitating and fatal disorders that affects the brain and nervous system of various animals [20]. These disorders are caused by the presence of prions in the body, which are abnormal proteins that cause damage to the brain. While the most commonly accepted explanation for the spread of TSEs is through prions, some research suggests that a Spiroplasma infection may also play a role. Upon examination of brain tissue taken after death, it can be observed that there is a loss of mental and physical abilities and numerous microscopic holes in the brain's cortex, giving it a spongy appearance. These illnesses lead to a decline in brain function, including memory loss, personality changes, and mobility issues that worsen over time.

5.1.1.6 Psoroptic disease

Psoroptic mange is a disease that affects sheep caused by the non-burrowing mite Psoropteovis (also known as the Scab mite) [21]. Other Psoroptes mite species infect many animals, including cattle, goats, horses, rabbits, and camelids; however, all mites are host specific. The mite lives in the skin's keratin layer and possesses abrasive mouthparts. It feeds on the exudate of lymph, skin cells, and bacteria induced by the host's hypersensitivity reaction to antigenic mite feces. This results in severe pruritus, self-trauma, crust and scale development, and inflammation.

5.1.1.7 Anthrax

The spore-forming bacteria *Bacillus anthracis* causes anthrax [22]. Anthrax spores in soil are highly resistant and can cause illness even years after an outbreak. Wet weather or deep tilling bring the spores to the surface, and when consumed by ruminants, the sickness emerges. Anthrax is found on all continents which is a major cause of mortality in domestic and wild herbivores, most animals, and some bird species.

5.1.1.8 Black quarter

Clostridium chauvoei, a type of bacteria easily visible under a microscope and known for its gram-positive characteristics, is the most prevalent cause of various livestock diseases such as blackleg, black quarter, quarter bad, or quarter ill [23]. This species of bacteria has a global presence, primarily impacting cattle, sheep, and goats. While it is most commonly observed in these animals, the disease has also been reported in farmed bison and deer. The severity of the symptoms caused by this bacteria makes treatment difficult, and the effectiveness of commonly used vaccinations is often called into question.

5.1.1.9 Lumpy skin disease

An outbreak of Lumpy skin disease (LSD) has resulted in the deaths of approximately 75,000 cattle in India, with Rajasthan most affected. LSD is a viral illness caused by the poxviridae family and belongs to the genus Capripox virus [24]. Like smallpox and monkeypox, it is not a zoonotic virus and, therefore, cannot be spread to humans. The disease is primarily spread through the bites of ticks, mosquitoes, and other flying insects, with cows and water buffaloes being the most susceptible host animals. This virus can also spread through contaminated feed and water and through animal sperm used for artificial insemination reported by FAO of the United Nations. It can spread from oral and nasal secretions of the infected animals and to sick animals, which can also contaminate shared feeding and drinking troughs.

The term "LSD" derives from how the virus affects an animal's lymph nodes, causing them to enlarge and appear as lumps on the skin. Cutaneous nodules of

2–5 cm in diameter appear on various areas of the infected animal's body, including the neck, head, limbs, udder, genitalia, and perineum. These nodules may eventually turn into ulcers and scabs. Other disease symptoms include sudden decrease in milk production, high fever, eyes and nose discharge, saliva secretion, appetite loss, depression, emaciation, miscarriages, infertility, damaged hides, etc. The incubation period for the virus is typically 28 days, though some estimates put it between 4 and 14 days reported by FAO.

The current outbreak of LSD in India has a morbidity rate ranging from 2% to 45%, with a mortality rate of less than 10%. However, the reported mortality rate for the present epidemic in India is as high as 15%, particularly in the country's western region, Rajasthan. The FAO and the World Organization for Animal Health (WOAH) have warned that the spreading of the disease could result in significant and severe economic losses. This is due to reduced milk production as the animal becomes weak and loses appetite due to oral ulcers, poor growth, decreased draught power capability, and reproductive issues such as infertility, abortions, and a shortage of sperm for artificial insemination. Additionally, the movement and trade restrictions imposed due to the infection can significantly impact the entire value chain.

India being the world's largest milk producer, the current outbreak of LSD has a serious and significant threat to the dairy industry. Additionally, India is home to the world's most significant number of cattle and buffalo. The outbreak has had the greatest impact in Rajasthan, where milk production has decreased by three to six lakh liters/day. Reports also indicate that milk output has decreased in Punjab due to the spread of the disease. By early August, the outbreak had spread to Punjab, Himachal Pradesh, Andaman and Nicobar Islands, and Uttarakhand, originating in Gujarat and Rajasthan in July. It subsequently spread to Haryana, Uttar Pradesh, and Jammu & Kashmir.

Recently, LSD has been reported in the Indian states of Maharashtra, Madhya Pradesh, Delhi, and Jharkhand. As of September 2022, the virus has infected roughly 1.6 million animals across 200 districts. Out of the approximately 75,000 animals lost to the infection, over 50,000 bovine fatalities, primarily cows, have been reported in Rajasthan. Currently, there is no direct treatment for LSD. The FAO has suggested a set of control measures to control the spread of the disease, including vaccination of susceptible populations with more than 80% coverage, quarantining and restricting the movement of bovine animals, implementing biosecurity measures through vector control, strengthening active and passive surveillance, raising awareness of risk mitigation among stakeholders, and creating large protection and surveillance zones and vaccinating.

The "Union Ministry of Fisheries, Animal Husbandry, and Dairying" have announced that the "Goat Pox Vaccine" has been proven to be highly effective in controlling the spread of LSD in affected states. As of the first week of September, 97 lakh vaccine doses have been administered. The government has implemented movement bans in the affected states to control the spread of the disease. They also isolate the infected cattle and buffaloes using insecticides to kill infected germs. The Western and North-Western states of India have also established the control roles and helpline numbers to assist farmers whose cattle have been infected.

Since the LSD outbreaks have been heavily in India, this chapter has a deep insight into LSD detection using Artificial intelligence techniques, especially with deep learning models. The details of the research work that deal already with other skin disease detection, hybrid deep learning models proposed for LSD detection, experimental analysis with the models proposed, and discussion are as follows.

5.2 Related works

In the last decade, these infectious diseases were identified and removed from the herd to reduce the spread and mortality of cattle. Many Artificial Intelligence techniques have supported researchers in this area to identify diseases without any medical examination by scanning the images of an infected cow. This section briefly discusses the machine, deep and hybrid deep learning models that are used to detect infectious skin diseases.

Very recently, in the year 2022, LSD research started in India and attracted some researchers to depict a model for the early diagnosis and identification of LSD. The models are categorized here into LSD diagnosis and prognosis and other skin disease detection techniques in cattle. Mostly, the models proposed are machine/deep learning and hybrid model. In the machine learning (ML) based models, the image processing concepts have been used for feature extraction and classified using traditional or improved ML models. In the hybrid models, the deep learning and image processing concepts are integrated for improved performance in the recognition of LSD.

5.2.1 LSD diagnosis and prognosis

The research work [25] has utilized a Random Forest (RF) algorithm to predict lumpy infection cases using Mendeley's data. The dataset has 12.25% lumpy and 87.75% of non-lumpy cow images. The researcher handled the class imbalance problem using Random Under-sampling (RUS) technique and Synthetic Minority Oversampling Technique (SMOTE). RF performs well on RUS and SMOTE data. Experimentation shows an improvement of 1–2% higher performance by SMOTE than RUS. The proposed work focuses only on the class imbalance problem, and there are no chances of early disease diagnosis or detection. Shivaanivarsha *et al.* [26] have introduced ConvNets to predict and analyze diseases such as Bovine Mastitis, Photosensitisation, Papillomatosis, and LSD to the welfare of cattle. The proposed study uses an effective smart mobile application and detects bovine diseases with an accuracy of 98.58%. However, the proposed work utilizes fewer data for model creation, which makes the system infeasible in real-time implementation.

The research work by Lake *et al.* [27] integrates an expert system with deep learning and image-processing concepts for skin disease detection and prognosis. The system collects the image through a cell phone camera, and symptoms are recorded as a text message and sent to the server. The server analyzes the image using the CNN algorithm, uses NLP concepts for the text message, and produces a diagnosis report. The proposed system is very cost-effective as it involves deep learning, an expert system, and an NLP process.

5.2.2 Other skin disease detection technique in cows

Allugunti *et al.* [28] have proposed a Dense Convolutional Network, a two-stage learning model known to detect Melanoma skin disease. The proposed system offered an accuracy of 86.6% accuracy on the melanoma dataset from the research website [29]. CNN's modular and hierarchical structure improves its performance, outperforming traditional machine learning techniques. However, the system is computationally cost-ineffective. Ahsan *et al.* [30] proposed a modified VCG16 model that includes a transfer learning approach and Local interpretable model-agnostic explanations (LIME) for the skin disease dataset collected in real time. The system achieves an accuracy of 97.2% and has been deployed in a smartphone for early diagnosis.

The research work by Karthik *et al.* [31] developed Eff2Net built on the base of EfficientNetV2 and integrated with Efficient Channel Attention (ECA) block. The proposed model replaces the standard Squeeze and Excitation (SE) block in the EfficieintNetV2 in the ECA block. Experimentation has been conducted to classify the skin diseases such as melanoma, acne, psoriasis, and actinic keratosis (AK). The system achieved an accuracy of 84.70% on the dataset collected from [29]. Upadya *et al.* [32] have used Gray-Level Co-Occurrence Matrix (GLCM) method for feature extraction to classify the maculopapular and vesicular rashes in cattle. Otsu, *k*-Means clustering, and the Image Segmenter app have been utilized for image segmentation and classifying using traditional machine learning classifiers. The final model was performed with an accuracy of 83.43% on a real-time dataset collected from various internet sources [29]. Rony *et al.* [33] have utilized conventional deep CNN and pre-trained models such as Inception V3, and VGG-16 for the early detection of external diseases in cattle. Inception-v3 has achieved the maximum specificity of 96% for the data collected from several places like cattle farms, veterinary hospitals, web resources, etc. The research work [34] proposed a system that aims for real-time cow disease diagnosis and therapeutic measures for cow disease. The system has been implemented using image processing and classification by traditional classifiers.

Rathod *et al.* [35] have introduced an automated image-based system for skin disease detection in cows using machine learning techniques. The system utilizes image processing for the extraction of complex features and classification using CNN. The system achieved an accuracy of 70% on the dataset collected from [29]. Tahari *et al.* [36] have evaluated the skin disease classification technique using five pre-trained CNN models. ResNet152V2 model achieved an accuracy of 95.84%, the precision of 96.3%, the recall of 96.1%, and a F1-score of 95.6% for the collected dataset [29].

5.3 Proposed model

The proposed model has two phases for the detection of LSD in cows. The phases are data collection and classification using hybrid deep learning models, and the detailed structure of the proposed system is shown in Figure 5.2.

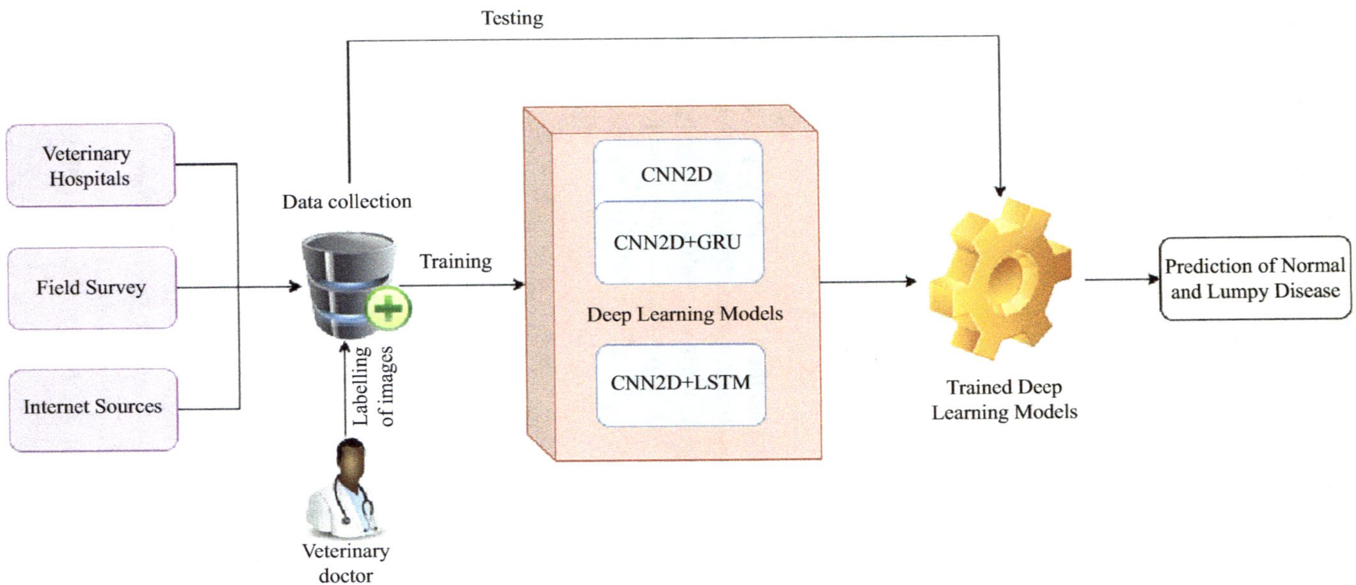

Figure 5.2 Architecture of the proposed model

5.3.1 Data collection

Data collection is essential for any deep or machine learning to react over. Sufficient data makes the model perform well in real-time disease detection and diagnosis implementation. Since the LSD started recently in India, we have collected certain images from various districts of Rajasthan and Punjab by visiting veterinary hospitals and field surveys. The 500 images of LSD-infected cows have been collected. In another view, 800 healthy cow images for binary classification are collected from the same districts and web sources.

5.3.2 Deep learning models

Convolutional neural network (CNN) model

A special type of feed-forward neural network in which the connectivity pattern between its neurons is inspired by visual context. Also termed as ConvNets, nothing but the neural networks share their parameters. In general, CNN uses various operations in the form of layers, as follows:

- **Input layer**—Receives the input image encompassing three channels (mostly for 2D image classification) and resizes the input image if required for classification purposes.
- **Convolution layer**—Computes the neuron's output by receiving an input from the previous layer neurons. It is associated with the input's local regions, so each neuron performs convolution to produce the output. Specifically, the neuron will do a dot product between the weights and a small region of the input volume for the number of filters or kernels assigned to the layer.
- **ReLU layer**—An activation function that removes the input volume weights less than zero (thresholding at zero) and that never unchanged size of the volume.
- **Pooling layer**—To perform a downsampling operation along the spatial dimension of the image from one layer to another layer.

Recurrent neural network (RNN)

Artificial neural networks are primarily used in semantic analysis and language processing. Unlike a feed-forward neural network, it has recurrent connections with the previous state of the network. Thus RNN can process sequential input of arbitrary length, especially with NLP and time series process. RNN has a major challenge of exploding and vanishing gradient problems. Thus this article integrates its variants (alternate architectures)-gated recurrent unit (GRU) and long short-term memory (LSTM) with CNN to improve the performance.

- LSTM—Avoids error backflow problems through special gates with low computational complexity. LSTM has forgotten gate, an input gate, and an output gate with an activation function. LSTM has a memory operation (memory cell) to remember the information, requiring input and the previous memory cell state for the process.
- GRU—A similar approach to LSTM with an adaptive reset gate to update the memory content of the previous cell state. It uses a reset and update gate in

addition to the gates of the LSTM unit and an activation function. GRU does not have an exclusive memory cell rather it exposes the memory at each time step.

To improve the performance of the CNN, the variants of RNN, such as GRU and LSTM, are appended to the feature extracted from the CNN for classification.

5.4 Experimental results and discussions

The experimentation has been carried out using the data collected for the lumpy and healthy (normal) cow images. As reported earlier, the dataset has 800 healthy cows and 500 cows with lumpy infections. The dataset has been split into 80% for training the models and 20% for testing the generated models multilayer perceptron (MLP), convolutional neural network for 2D image (CNN2D), convolutional neural network for 2D image with LSTM model (CNN2D+LSTM), and convolutional neural network for 2D image with LSTM model (CNN2D+GRU). The architecture of each of the models CNN2D, CNN2D+LSTM, and CNN2D+GRU is shown in Figures 5.3, 5.4, 5.5, and 5.6, respectively.

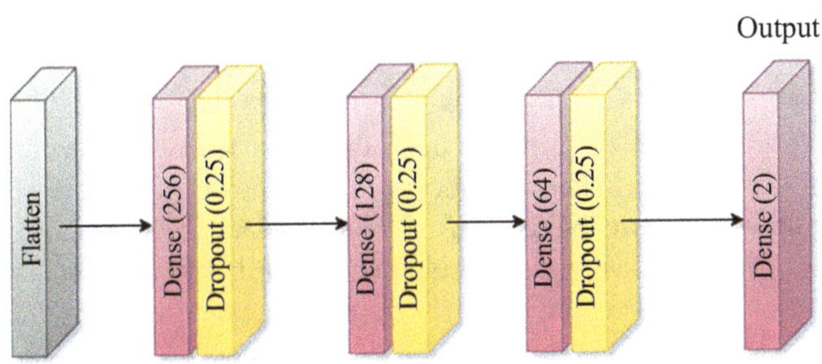

Figure 5.3 Multilayer perceptron model

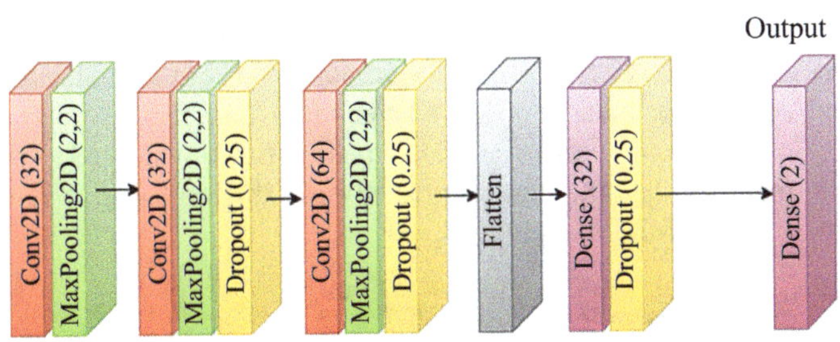

Figure 5.4 Convolutional neural network (2D) model

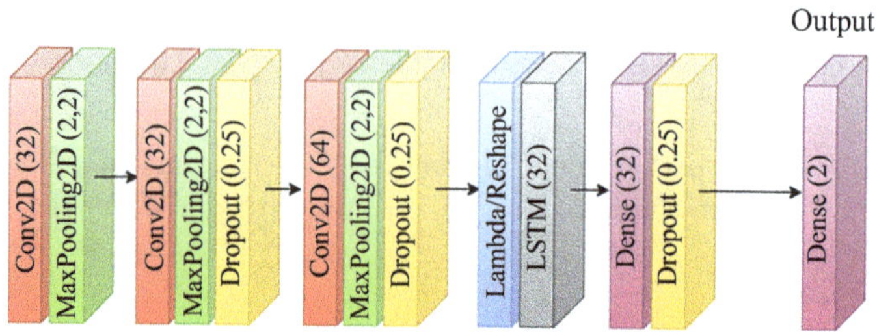

Figure 5.5 A hybrid convolutional neural network (2D)+LSTM model

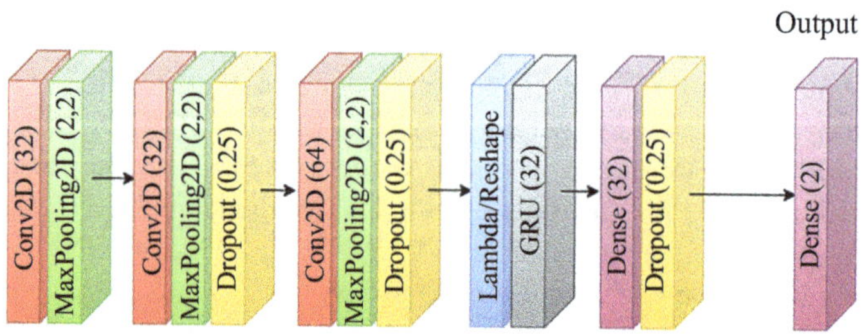

Figure 5.6 A hybrid convolutional neural network (2D)+GRU model

5.4.1 MLP model

The MLP model receives the input through the input layer with an image of size (200×200) followed by a flattened layer and a dense layer of 256 units, dropout of 0.25, and followed by two block sets of dense and dropout, where dense units change from 128 to 64 with same dropout ratio. Finally, a dense layer with a sigmoid function is used to classify the images as normal (healthy), and lumpy-infected cows. The detailed architecture of the MLP model is shown in Figure 5.3.

5.4.2 CNN model

The CNN or CNN2D model receives the input through the input layer with an image of size (200×200) followed by a Conv2D layer of 32 filters and a Max pooling layer of filter size (2,2). Followed by two block sets of Conv2D with 32 filters, a Max pooling layer of (2,2), and a dropout layer of 0.25. Then the outputs are flattened and fed as input to a dense layer of 32 units and a dropout layer of 0.25. Finally, a dense layer with a sigmoid activation function is used to classify the images as normal (healthy) and lumpy-infected cows. Figure 5.4 shows the detailed architecture of the CNN2D model.

5.4.3 CNN+LSTM model

To add up the advantage of sequence modeling in the spatial process, the LSTM layer is integrated with the CNN process. In the proposed system, the CNN model has been used as mentioned in Figure 5.5 up to the second block of Conv2D, Max pooling, and dropout layer. After that, a lambda layer is used to reshape the structure of the tensor. Then it is fed as input to the LSTM of 32 units. Then the sequenced output is fed as input to a dense layer of 32 units, dropout of 0.25, and a final dense layer with sigmoid activation function to classify the images as Normal (healthy) and Lumpy infected cows. Figure 5.5 shows the architecture of the hybrid CNN+LSTM model.

5.4.4 CNN+GRU model

The CNN+GRU model has also been included in the proposed system to compare the performance of the CNN2D and CNN2D+LSTM models. The GRU layer is replaced instead of the LSTM layer of 32 units as mentioned in Figure 5.5 to form a hybrid model of the CNN+GRU model. The detailed architecture of the hybrid CNN+GRU model is shown in Figure 5.6.

5.4.5 Hyperparameters

The hyperparameters utilized to train and test the hybrid deep learning models are shown in Table 5.2.

5.4.6 Performance evaluation

The hybrid deep learning models performance is validated using the familiar metrics such as accuracy, precision, recall, and F-measure as represented in (5.1), (5.2), (5.3), and (5.4), respectively, based on the confusion matrix. The confusion matrix holds the value of

- True positive (TP) is a classification outcome where the system predicts the healthy cows correctly
- True negative (TN) is a classification outcome where the system predicts the Lumpy-infected cows correctly
- False positive (FP) is a classification outcome where the system predicts the healthy cows as Lumpy-infected cows
- False negative (FN) is a classification outcome where the system predicts the Lumpy infected cows as healthy cows

Table 5.2 Hyperparameters utilized for the performance evaluation of proposed system

S. no.	Hyperparameters	Value
1	Epochs	100
2	Loss	Adam
3	Entropy	Categorical cross-entropy
4	Callbacks	Early stopping on monitoring the loss

$$Accuracy = \frac{TP + TN}{TP + TN + FP + FN} \tag{5.1}$$

$$Precision = \frac{TP}{TP + FP} \tag{5.2}$$

$$Recall = \frac{TP}{TP + FN} \tag{5.3}$$

$$F - Measure = \frac{2{*}Precision * Recall}{Precision + Recall} \tag{5.4}$$

The performance of the proposed hybrid deep learning models is shown in Table 5.3. The Conv2D has the highest performance of 99.61% and outperforms the other models. However, accuracy alone does not contribute more to disease classification. Thus, the precision and recall values are analyzed for performance comparison. On comparing the precision and recall, the Conv2D+GRU scores higher value and outperforms the other two models. This results in a higher recognition rate of lumpy-infected disease and healthy cows than Conv2D and Conv2D+LSTM models. The F-measure also evident that the Conv2D+GRU performs better than the other two models.

The experimentation on training and testing with its loss and accuracy values for the MLP, CNN2D, CNN2D+LSTM, and CNN2D+GRU has also been analyzed and shown in Figures 5.7, 5.8, 5.9, and 5.10, respectively. Figures shows no chance of overfitting and underfitting (concept of high bias and high variance). This proves the efficiency of the proposed system in diagnosing lumpy skin disease detection using AI concepts.

Table 5.3 Performance comparison of proposed hybrid deep learning models

S. no.	Model	Accuracy	Precision	Recall	F-measure
1	MLP	96.12	80.71	97.56	88.17
2	Conv2D	99.61	99.70	68.07	80.56
3	Conv2D+GRU	99.42	99.43	99.53	99.48
4	Conv2D+LSTM	98.65	98.42	98.86	98.64

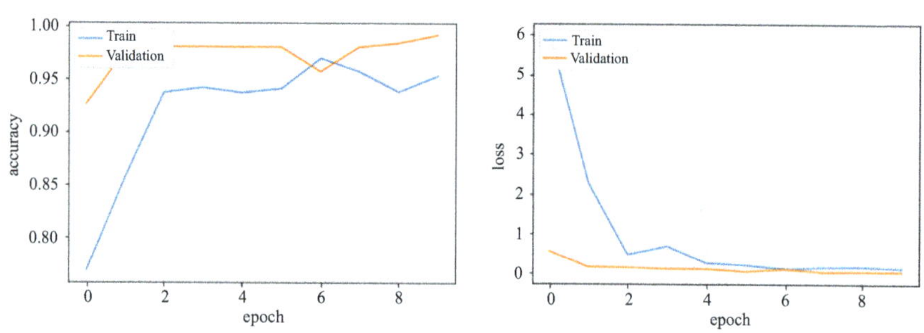

Figure 5.7 Accuracy and loss function of MLP model

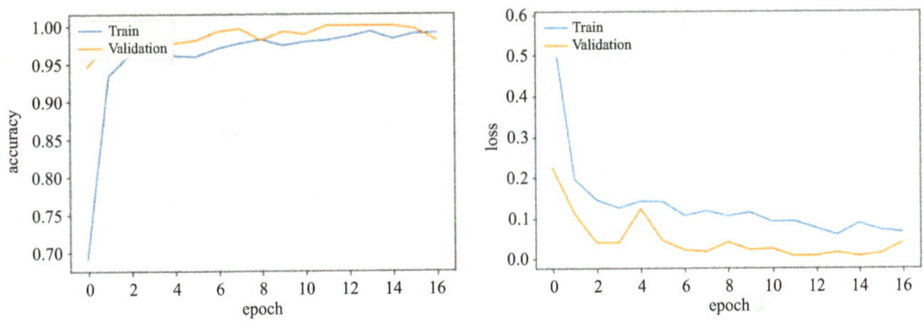

Figure 5.8 Accuracy and loss function of CNN2D model

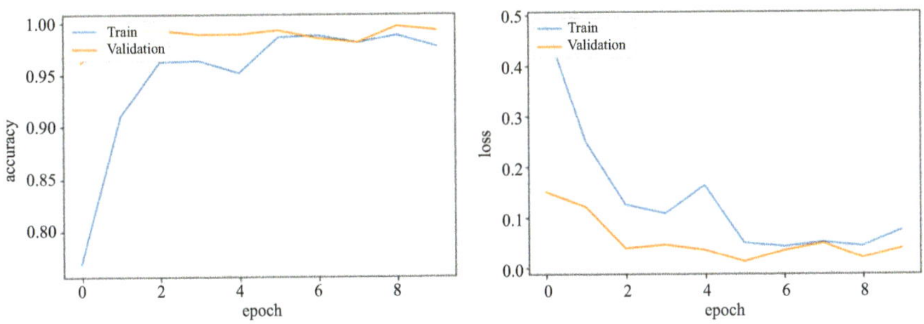

Figure 5.9 Accuracy and loss function of CNN2D+LSTM model

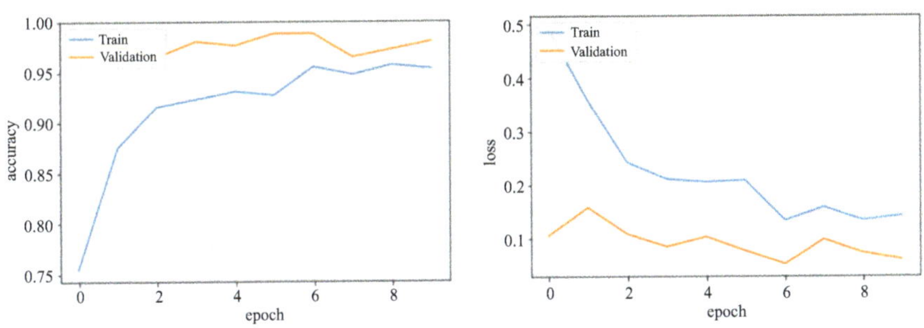

Figure 5.10 Accuracy and loss function of CNN2D+GRU model

5.5 Conclusion

The foundation of every civilization or society is its health care system, which ensures that every living thing receives an accurate diagnosis and effective treatment. Today's world is becoming technologically advanced and automated. Therefore, this industry's use of modern technology, machines, robotics, etc., is both necessary and unavoidable. Thanks to technological advancements, procedures, including diagnosis, treatment, and prescription of medications, have become quicker and more effective. So in this article, we have explained how AI can be utilized in this research in an excellent and surprising approach for treating diseases like lumpy skin disease and not limited to this particular disease as this a generalized model. With a slight tweak in the training dataset, we can also use it to identify more skin diseases from images. The strategy of this study is based on CNN. Regarding LSD classification accuracy, our CNN classifier's modular and hierarchical structure, along with LSTM and GRU, has performed better than traditional machine learning techniques and significantly minimizes the computational effort required. The future scope of the article is to detect the disease's severity level and develop a smart application for the early diagnosis and detection of skin diseases.

References

[1] Gerber PJ, Mottet A, Opio CI, *et al.* Environmental impacts of beef production: review of challenges and perspectives for durability. *Meat Science.* 2015;109:2–12.

[2] Johnston J, Weiler A, and Baumann S. The cultural imaginary of ethical meat: a study of producer perceptions. *Journal of Rural Studies.* 2022;89:186–198.

[3] Porter V. Mason's world dictionary of livestock breeds, types and varieties. *CABI*; 2020.

[4] Martha TR, Roy P, Jain N, *et al.* Geospatial landslide inventory of India—an insight into occurrence and exposure on a national scale. *Landslides.* 2021;18(6):2125–2141.

[5] 20th Livestock Census Report; 2019 [updated 2019 Oct 18; cited 2022 Nov 30]. Department of Animal Husbandry and Dairying. Available from: https://pib.gov.in/PressReleasePage.aspx?PRID=1588304.

[6] Neeraj A and Kumar P. Problems perceived by livestock farmers in utilization of livestock extension services of animal husbandry department in Jammu District of Jammu and Kashmir. *International Journal of Current Microbiology and Applied Sciences.* 2018;7(2):1106–1113.

[7] USDA U. of A. Livestock and Poultry: World Markets and Trade; 2021.

[8] Lu CD and Miller BA. Current status, challenges and prospects for dairy goat production in the Americas. *Asian-Australasian Journal of Animal Sciences.* 2019;32(8_spc):1244–1255.

[9] Crist E, Mora C, and Engelman R. The interaction of human population, food production, and biodiversity protection. *Science.* 2017;356(6335):260–264.

[10] Meissner H, Scholtz M, and Palmer A. Sustainability of the South African livestock sector towards 2050. Part 1: worth and impact of the sector. *South African Journal of Animal Science.* 2013;43(3):282–297.

[11] Hu Y, Cheng H, and Tao S. Environmental and human health challenges of industrial livestock and poultry farming in China and their mitigation. *Environment International.* 2017;107:111–130.

[12] FAO F. Food and Agriculture Organization of the United Nations, Rome, 2018. http://faostat fao org.

[13] Bradhurst R, Garner G, Hóvári M, *et al.* Development of a transboundary model of livestock disease in Europe. *Transboundary and Emerging Diseases.* 2022;69(4):1963–1982.

[14] Brooks DR, Hoberg EP, Boeger WA, *et al.* Emerging infectious disease: an underappreciated area of strategic concern for food security. *Transboundary and Emerging Diseases.* 2022;69(2):254–267.

[15] Libera K, Konieczny K, Grabska J, *et al.* Selected livestock-associated zoonoses as a growing challenge for public health. *Infectious Disease Reports.* 2022;14(1):63–81.

[16] Grubman MJ and Baxt B. Foot-and-mouth disease. *Clinical Microbiology Reviews.* 2004;17(2):465–493.

[17] Lauder I and O'sullivan J. Ringworm in cattle. Prevention and treatment with griseofulvin. *Veterinary Record.* 1958;70(47):949.

[18] Bachofen C, Stalder H, Vogt HR, *et al.* Bovine viral diarrhea (BVD): from biology to control. *Berliner und Munchener tierarztliche Wochenschrift.* 2013;126(11–12):452–461.

[19] Maclachlan NJ. Bluetongue: history, global epidemiology, and pathogenesis. *Preventive Veterinary Medicine.* 2011;102(2):107–111.

[20] Collins SJ, Lawson VA, and Masters CL. Transmissible spongiform encephalopathies. *The Lancet.* 2004;363(9402):51–61.

[21] O'Brien DJ. Treatment of psoroptic mange with reference to epidemiology and history. *Veterinary Parasitology.* 1999;83(3–4):177–185.

[22] Cieslak TJ and Eitzen Jr EM. Clinical and epidemiologic principles of anthrax. *Emerging Infectious Diseases.* 1999;5(4):552.

[23] Sultana M, Ahad A, Biswas PK, *et al.* Black quarter (BQ) disease in cattle and diagnosis of BQ septicaemia based on gross lesions and microscopic examination. *Bangladesh Journal of Microbiology.* 2008;25(1):13–16.

[24] Coetzer J and Tuppurainen E. Lumpy skin disease. *Infectious Diseases of Livestock.* 2004;2:1268–1276.

[25] Suparyati S, Utami E, Muhammad AH, *et al.* Applying different resampling strategies in random forest algorithm to predict lumpy skin disease. *Jurnal RESTI (Rekayasa Sistem dan Teknologi Informasi).* 2022;6(4):555–562.

[26] Shivaanivarsha N, Lakshmidevi PB, and Josy JT. A ConvNet based real-time detection and interpretation of bovine disorders. In: *2022 International Conference on Communication, Computing and Internet of Things (IC3IoT).* IEEE; 2022. p. 1–6.

[27] Bhatt R, Sharma G, Dhall A, *et al.* Categorization and reorientation of images based on low level features. *Journal of Intelligent Learning Systems and Applications.* 2011;3(01):1.

[28] Allugunti VR. A machine learning model for skin disease classification using convolution neural network. *International Journal of Computing, Programming and Database Management.* 2022;3(1):141–147.

[29] Skin Disease Dataset; 2017 [cited 2022 Nov 30]. Dermatology Resource. Available from: https://dermetnz.org.

[30] Ahsan MM, Uddin MR, Farjana M, *et al. Image Data collection and implementation of deep learning-based model in detecting Monkeypox disease using modified VGG16,* 2022. *arXiv preprint arXiv*:220601862.

[31] Karthik R, Vaichole TS, Kulkarni SK, *et al.* Eff2Net: an efficient channel attention-based convolutional neural network for skin disease classification. *Biomedical Signal Processing and Control.* 2022;73:103406.

[32] Upadya P S, Sampathila N, Hebbar H, *et al.* Machine learning approach for classification of maculopapular and vesicular rashes using the textural features of the skin images. *Cogent Engineering.* 2022;9(1):2009093.

[33] Rony M, Barai D, Hasan Z, *et al.* Cattle external disease classification using deep learning techniques. In: *2021 12th International Conference on Computing Communication and Networking Technologies (ICCCNT)*. IEEE, 2021. p. 1–7.

[34] Saranya P, Krishneswari K, and Kavipriya K. Identification of diseases in dairy cow based on image texture feature and suggestion of therapeutical measures. *International Journal of Internet, Broadcasting and Communication.* 14(4):173–180.

[35] Rathod J, Waghmode V, Sodha A, *et al.* Diagnosis of skin diseases using convolutional neural networks. In: *2018 Second International Conference on Electronics, Communication and Aerospace Technology (ICECA)*. IEEE, 2018. p. 1048–1051.

[36] Thohari ANA, Triyono L, Hestiningsih I, *et al.* Performance evaluation of pre-trained convolutional neural network model for skin disease classification. *JUITA: Jurnal Informatika.* 2022;10(1):9–18.

Chapter 6

Can AI-powered imaging be a replacement for radiologists?

Riddhi Paul[1], Shreejita Karmakar[1] and Prabuddha Gupta[1]

Artificial Intelligence (AI) has a wide range of potential uses in medical imaging, despite many clinical implementation challenges. AI can enhance a radiologist's productivity by prioritizing work lists, for example, AI can automatically examine chest X-rays for pneumothorax and evidence of intracranial hemorrhage, Alzheimer's disease, and urinary stones. AI may be used to automatically quantify skeletal maturity on pediatric hand radiographs, coronary calcium scoring, prostate categorization through MRI, breast density via mammography, and ventricle segmentation via cardiac MRI. The usage of AI covers almost the full spectrum of medical imaging. AI is gaining traction not as a replacement for a radiologist but as an essential companion or tool. The possible applications of AI in medical imaging are numerous and include the full medical imaging life cycle, from picture production to diagnosis to prediction of outcome. The availability of sufficiently vast, curated, and representative training data to train, evaluate, and test algorithms optimally are some of the most significant barriers to AI algorithm development and clinical adoption, but they can be resolved in upcoming years through the creation of data libraries. Therefore, AI is not a competitor, but a friend in need of radiologists who can use it to deal with day-to-day jobs and concentrate on more challenging cases. All these aspects of interactions between AI and human resources in the field of medical imaging are discussed in this chapter.

6.1 Artificial Intelligence (AI) and its present footprints in radiology

Radiology is a medical specialty that diagnoses and treats illnesses using imaging technology. Self-learning computer software called artificial intelligence (AI) can aid radiology practices in finding anomalies and tumors, among other things. AI-based systems have gained widespread adoption in radiology departments all over the world due to their ability to detect and diagnose diseases more accurately than human radiologists. Artificial intelligence and medical imaging have experienced

[1]Amity Institute of Biotechnology, Amity University Kolkata, India

rapid technological advancements over the past ten years, leading to a recent convergence of the two fields. Radiology-related AI research has advanced thanks to significant improvements in computing power and improved data access [1]. The ability to be more effective thanks to AI systems is one of their biggest benefits. AI can be used to accomplish much larger, more complex tasks as well as smaller, repetitive ones more quickly. Whatever their use, AI systems are not constrained by human limitations and never get old. The neural networks that operate in the brain served as the inspiration for the deep learning technique, which has large networks in layers that can learn over time. In imaging data, deep learning can uncover intricate patterns. In a variety of tasks, AI performance has advanced from being subhuman to being comparable to humans, and in the coming years, AI performance alongside humans will greatly increase human performance. For diagnosis, staging, planning radiation oncology treatments, and assessing patient responses, cancer 3D imaging can be recorded over time and space multiple times. Clinical work already shows this to be true. A recent study found that there is a severe lack of radiologists in the workforce, with 1.9 radiologists per million people in low-income countries and 97.9 in high-income nations, respectively. An expert clinician was tasked by UK researchers with categorizing more than 3,600 images of hip fractures. According to the study, clinicians correctly identified only 77.5% of the images, whereas the machine learning system did so with 92% accuracy [2]. In a nutshell, AI is a savior for global healthcare due to the constantly rising demand for radiology and the development of increasingly precise AI-based radiology systems.

6.2 Brief history of AI in radiology

Although AI was first applied in radiology to detect microcalcifications in mammography in 1992, it has gained much more attention recently [3].

Over the years there has been a tremendous increase in the number of radiological examinations taken per day. There have also been technological improvements in the machines used, the radiation doses required have decreased, and the recording of image interpretation has improved A radiologist interprets an image based on visual acuity, search patterns, pattern recognition, training, and experience. As the amount of data to be examined has increased in recent years error rates of radiographic studies have spiked to 30% [4] as not all the information present in the image is viewed resulting in misdiagnosis or overdiagnosis. The earliest form of AI usage was a computerized clinical decision support system in 1972 UK called the AAPhelp which computed the likely cause of acute abdominal pain based on patient symptoms [5], over time the system became more accurate. There has been a leap in the progress of AI in the last 10 years with advancements in machine learning, development of deep learning, and development of computer hardware and interface software which improved the accessibility of this technology. In the 1970s, scientists started to get interested in AI for the biological sciences. Earlier attempts, like the Dendral project, had a stronger chemical than medical focus. The goal of contemporary AI is to address real-world healthcare issues. The benefits of technology in healthcare, notably the application of AI in radiology, have been enhanced by cutting-edge methods like deep learning [6]. Machine learning is an approach to AI. The goal of a machine learning algorithm is to develop a mathematical model that fits the data. As such, the five basic components of AI include learning, reasoning, problem-solving, perception, and language understanding.

6.3 AI aided medical imaging

Radiography is a fundamental technology used in clinical medicine and dentistry for regular diagnostic purposes. A radiograph is a representation of a three-dimensional object in two dimensions. This is referred to as projection imaging. As a result, it is necessary to investigate the elements that impact the interpretation of structures in radiographic images. A brief description of the atomic structural issues connected with the creation and absorption of X-rays is followed by an account of the practical techniques of producing X-radiation and the kind of spectrum produced. Numerous new findings and advancements are being made, the majority of which can be categorized into the following four groups: reactive machines, limited memory, theory of mind, and self-aware AI. They will undoubtedly use AI in their daily work to help with repetitive tasks and basic case diagnoses. Radiologists can benefit from AI by quickly analyzing images and data registries, improving patient understanding, expanding their clinical role, and joining the core management team. A 30% usage rate for AI among radiologists was estimated. Overall, it correctly diagnosed patients 92% of the time compared to doctors who did so 77.5% of the time, giving the machines a 19% advantage over doctors. As AI can be used to aid diagnosis and assessments over great distances, it helps reduce waiting times for emergency patients who must be transported from rural and remote areas. AI in teleradiology can be used to support radiologists and facilitate analysis. It is highly likely that in the future, radiologists' innovative work will be required to oversee diagnostic procedures and tackle difficult problems. Radiologists cannot be replaced by AI. On the other hand, it can make radiologists' routine tasks easier. Early adopters of AI will therefore probably lead the radiology industry in the future [7].

There are 10 major benefits of AI in radiology: [2,8]

1. **Early detection**—AI has a greater ability to identify diseases in their earliest stages, avoiding complications and significantly enhancing patient outcomes.
2. **Better prioritization**—AI-based radiology tools can automatically rank scans according to the seriousness of the case, saving time for clinicians and guaranteeing that patients receive timely care.
3. **Greater accuracy**—The majority of radiology AI tools can identify abnormalities more precisely than human radiologists, improving the prognosis for patients.
4. **Optimized radiology dosing**—By minimizing the radiation dose, AI dose optimization systems can help lower the radiation level to which patients are exposed during a scan.

5. **Lessened radiation exposure**—AI can help lessen radiation exposure by producing more precise images with fewer imaging repetitions.
6. **Improved image quality**—AI can enhance the image quality of medical scans, making it easier to find and diagnose anomalies.
7. **Greater satisfaction**—By delivering quicker and more precise diagnoses, AI-powered radiation tools can contribute to greater patient satisfaction.
8. **Quicker diagnosis**—By accelerating the diagnosis process, AI can help patients receive treatment more quickly.
9. **Better access to care**—AI can democratize access to radiology globally by increasing patient throughput and making decisions without human involvement.
10. **Better reporting**—The majority of AI-powered radiology tools generate error-free, standardized reports automatically, which saves time and streamlines workflow [7].

6.4 AI imaging pathway

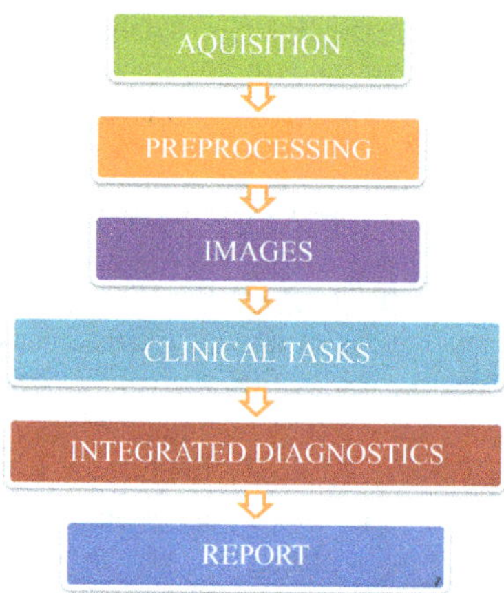

Figure 6.1 Flowchart representing the generalized AI pathway in the medical field like radiology [9]

ACQUISITION—Image acquisition is the action of obtaining an image for subsequent processing from an external source. Since no operation can be started without first getting a picture, it is always the first stage in the workflow. Without related data, such as patient identity [10], study identification, additional photos, and pertinent clinical information, a biological image is worthless (i.e., image acquisition context). For example, with CT scans AI imaging modalities used involve high-throughput extraction of data from CT images [11] (Figure 6.1).

PREPROCESSING—To prepare picture data for use in a deep-learning model, preprocessing is a crucial step. Preprocessing is necessary for both technical and performance reasons [12]. The process of converting an image into a digital format and carrying out specific procedures to extract some useful information from it is known as image processing. When implementing certain specified signal processing techniques, the image processing system typically interprets all pictures as 2D signals [13].

The preprocessing steps include:

1. Converting all the images into the same format.
2. Cropping the unnecessary regions on images.
3. Transforming them into numbers for algorithms to learn from them (array of numbers) [14].

Through preprocessing, we may get rid of undesired distortions and enhance certain properties that are crucial for the application we are developing. Those qualities could alter based on the application. For software to work properly and deliver the required results, a picture must be preprocessed (Figure 6.1).

IMAGES—Following pre-processing and acquisition, we receive a clear pixel of the picture, which the AI and deep learning utilize to compare with the patients' radiographs and perform clinical duties and processes [15,16] (Figure 6.1).

CLINICAL TASKS—AI approaches are also the most effective in recognizing the diagnosis of many sorts of disorders. The presence of computerized reasoning (AI) as a means for better medical services provides new opportunities to recover patient and clinical group outcomes, reduce expenses, and so on [17]. Individual care providers and care teams must have access to at least three key forms of clinical information to successfully diagnose and treat individual patients: the patient's health record, the quickly changing medical-evidence base, and provider instructions directing the patient care process.

The clinical tasks are further sub-divided into the following:

1. **Detection**—This includes automated detection of abnormalities like tumors and metastasis in images. Examples can be detecting a lung nodule, brain metastasis, or calcification in the heart.
2. **Characterization**—After detection, we look forward to characterizing the result obtained. Characterization is done in the following steps:
   (a) **Segmentation**: Detecting the boundaries of normal tissue and abnormality
   (b) **Diagnosis:** Identifying the abnormalities whether they are benign or malignant.
   (c) **Staging:** The observed abnormalities are assigned to different predefined categories.
3. **Monitoring**—Detecting the change in the tissue over time by tracking multiple scans (Figure 6.1).

INTEGRATED DIAGNOSTICS—The usage and scope of advanced practitioner radiographers, who with the use of AI technologies can offer an instantaneous result to the patient and referring doctor at the time of examination, may be expanded if AI is demonstrated to be accurate in picture interpretation. The use of AI in medical imaging allows doctors to diagnose problems considerably more quickly, encouraging early intervention. Researchers found that by evaluating tissue scans equally well or better than pathologists, AI can reliably detect and diagnose colorectal cancer [18] (Figure 6.1).

REPORT—AI has learned to identify ailments in those scans as precisely as a human radiologist after processing thousands of chest X-rays and the clinical records that go with them. The bulk of diagnostic AI models now in use are trained on scans that have been annotated by people, however, annotating scans by humans takes time. It is quite possible that in the future, radiologists' innovative work would be required to monitor diagnostic procedures and tackle difficulties (Figure 6.1).

6.5 Prediction of disease

Medical diagnosis necessitates the employment of clinicians and medical laboratories for testing, whereas AI-based predictive algorithms can be utilized for disease prediction at preliminary stages. Based on accessible patient data, AI can be trained and then utilized to anticipate illnesses [19].

6.5.1 Progression without deep learning

Predefined designed characteristics are used in conjunction with traditional machine learning [20]. A skilled radiologist uses AI to determine the extent of the tumor. The tumor is characterized using a variety of methods, including texture, intratumor homogeneity, form, density, and histogram [21]. These characteristics are then sent into a pipeline for future extraction, followed by future selection and categorization based on data.

6.5.2 Progress prediction with deep learning

Deep learning automates the whole process, from the input to the location, definition of the feature set, selection, and classification of a tumor [22]. During training, features are created that have been optimized for a particular result, such as cancer patient survival prediction. The benefit of this method may be demonstrated when predicting a patient's reaction to therapy or the status of a mutation. By restricting the number of expert inputs, the process is optimized, and the performance is excellent [1,12].

6.6 Recent implementation of AI in radiology

The field of medicine has been transformed by AI. It is the area of computer science that deals with computing that is intelligent. Radiology is the branch of medicine that creates medical imaging, such as X-ray, CT, ultrasound, and MRI pictures, to find cancers and abnormalities. AI systems are capable of automatically identifying intricate abnormal patterns in visual data to help doctors diagnose patients. According to the American Department of Radiology, from 2015 to 2020, there was a 30% increase in radiology's usage of AI, a gradual yet consistent increase. Here are a few applications in the context of radiology.

6.6.1 Imaging of the thorax

One of the most prevalent and dangerous tumors is lung cancer. Pulmonary nodules can be found by lung cancer screening, and for many individuals, early discovery can save their lives. These nodules can be automatically identified and classified as benign or cancerous with the aid of AI [9] (Figure 6.2).

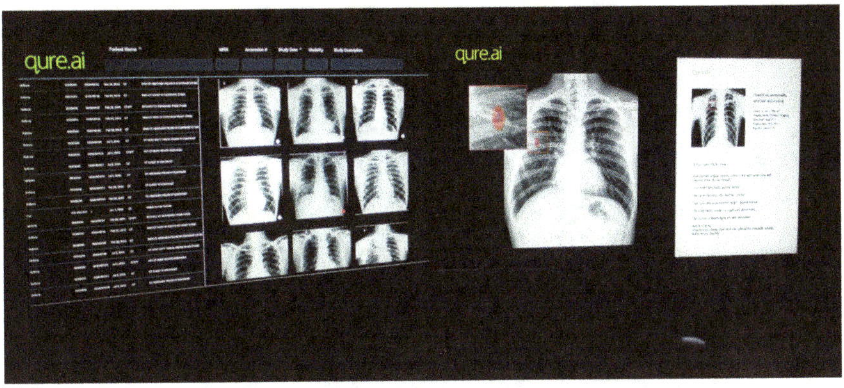

Figure 6.2 Radio imaging of the thorax using an Indian healthcare start-up, Qure.ai [23]

AI model can aid in chest X-ray collapsed lung detection

December 19, 2022—A recent study found that an AI model can correctly identify simple and stress pneumothorax on chest radiographs. The study was published last week in JAMA Network Open. A pneumothorax is a collapsed lung that happens when air seeps into the area between the lung and chest wall, according to Mayo Clinic. This air exerts pressure on the lung's outside, causing partial or complete collapse. According to Johns Hopkins Medicine, a pneumothorax can be brought on by chest trauma, too much pressure on the lungs, or a lung condition such as whooping cough, cystic fibrosis, chronic obstructive pulmonary disease (COPD), asthma, or cystic fibrosis [24]. The study concluded that early pneumothorax identification is essential since the condition's severity will decide whether or not emergency treatment is required. The conventional method for detecting and diagnosing a pneumothorax involves a chest X-ray and radiologist interpretation; however, the scientists proposed that AI may facilitate this procedure.

6.6.2 Pelvic and abdominal imaging

More incidental discoveries, such as liver lesions, are being discovered as a result of the rapid expansion of medical imaging, particularly computed tomography (CT) and magnetic resonance imaging (MRI) [9] (Figure 6.3). By classifying these lesions as benign or malignant, AI may make it easier to prioritize patient follow-up evaluation for those who have these lesions.

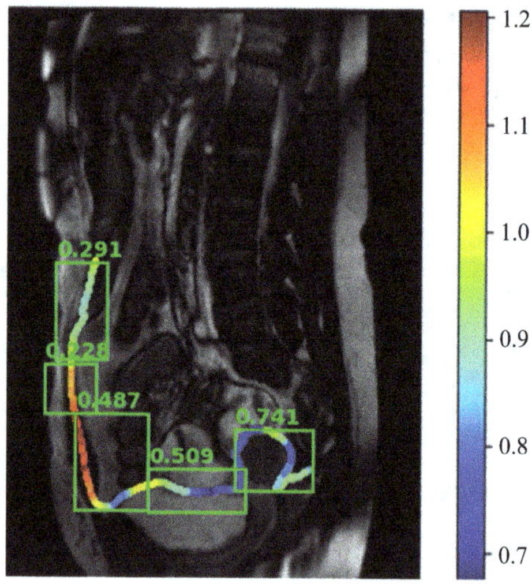

Figure 6.3 AI-assisted detection of adhesions on cine MRI [25]

A three-dimensional pelvic model utilizing artificial intelligence technologies for preoperative MRI simulation of rectal cancer surgery

Automatic segmentation of the pelvic organs from the 3D MRI using an AI-based system, including the arteries, nerves, and bone. This algorithm may be used for urological or gynecological operations as well as pre-operative simulation of rectal cancer surgery. This method can give effective information for the surgeons to understand the anatomical configuration prior to surgery, which we feel to be connected to the execution of safe and curative surgery, especially in challenging instances like locally advanced rectal cancer. This is the first time an automated technology for preoperative simulation has been used in clinical practice. This sort of algorithm can be constructed because of recent advancements in AI technology [26]. This may be used for surgical simulation in the treatment of advanced rectal cancer to autonomously segment intrapelvic anatomies. As a result of its greater usability, this system has the potential for success.

6.6.3 Colonoscopy

Unidentified or incorrectly categorized colonic polyps may increase the risk of colorectal cancer. Even while the majority of polyps start benign, they can eventually turn cancerous. Early identification and consistent use of powerful AI-based solutions for monitoring are essential [27] (Figure 6.4).

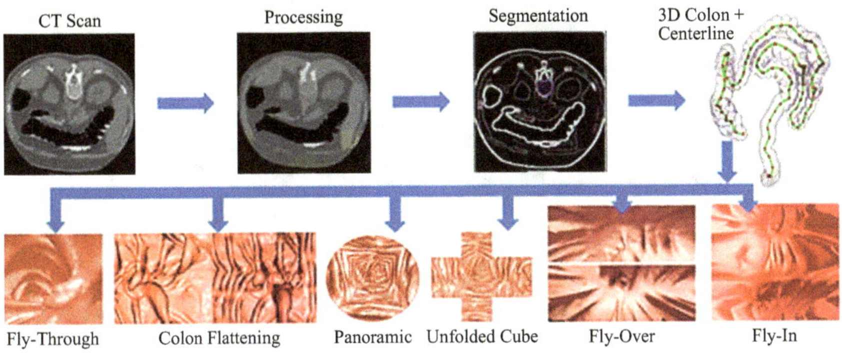

Figure 6.4 *The CTC pipeline: first, DICOM images are cleaned, and then colon regions are segmented. Second, the 3D colon is reconstructed from segmented regions, then the centerline may be extracted. Finally, the internal surface of the colon can be visualized using different visualization methods [28].*

Colonoscopy using artificial intelligence assistance: a survey

All endoscopists should get familiar with Computer Aided Design (CAD) technology and feel at ease utilizing AI-assisted devices in colonoscopy as AI models have been found to compete with and outperform endoscopists in performance. The use of AI in colonoscopy is limited by the absence of solutions to assist endoscopists with quality control, video annotation, design ideas, and polypectomy completion [29]. It appears conceivable that using the most current advances in computer science in colonoscopy practice may improve the quality of patient diagnosis, treatment, and screening. AI technologies still need a lot of study and development before they can be used in healthcare settings. They must be trusted by patients, regulatory authorities, and all medical professionals. The AI-assisted colonoscopy is heavily reliant on the endoscopist, who must endeavor to deliver the clearest picture or video to the AI model for analysis while also taking into consideration other contemporaneous patient characteristics such as a family history of CRC or the outcomes of prior colonoscopies.

6.6.4 Brain scanning

AI could be used to create diagnostic predictions for brain tumors, which are defined by aberrant tissue growth and can be benign, malignant, primary, or metastatic [27] (Figure 6.5).

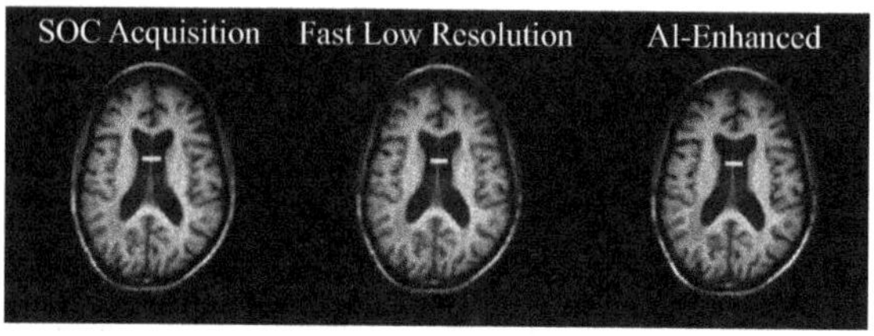

Figure 6.5 Using AI-based image enhancement to reduce brain MRI scan times and improve signal to noise ratio [31]

Current neuroimaging applications in the age of AI

AI has the potential to increase the quality of neuroimaging while reducing the clinical and systemic burdens of other imaging modalities. Patient wait times for computed tomography (CT) [30], magnetic resonance imaging (MRI), ultrasound, and X-ray imaging may be predicted using AI. A machine learning-based AI identified the factors that most influenced patient waits times, such as closeness to federal holidays and the severity of the patient's ailment and projected how long patients would be detained after their planned appointment time. This AI method may enable more effective patient scheduling and expose areas of patient processing that might be altered, thereby enhancing patient results and patient satisfaction for neurological disorders that require prompt treatment. These technologies are most immediately beneficial for neuroimaging of acute situations because MRIs with their high resolution and SNR start to approach CT imaging time scales. It has cost reduction and neurologic care enhancement in the contemporary radiology age is CS-MRI optimization.

6.6.5 Mammography

The interpretation of screening mammography is technically difficult. AI can help with interpretation by recognizing and classifying microcalcifications [9] (Figure 6.6).

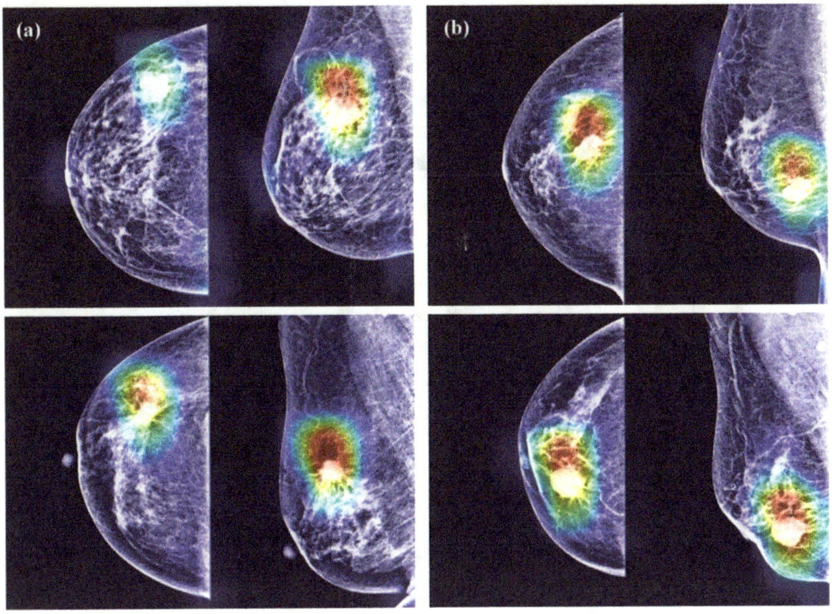

Figure 6.6 Automated breast cancer detection in digital mammograms of various densities via deep learning gradient-weighed class activation mapping for mammograms having breast cancer by (a) DenseNet-169 and (b) EfficientNet-B5 [32]

Thermal imaging and AI technology

Breast cancer is becoming more prevalent in low- and middle-income nations, yet early detection screening and treatment remain uncommon. A start-up in India has created a less expensive, non-invasive test that makes use of AI and thermal imaging. The method has generally been hailed for being a painless, cost-effective way to detect breast cancer in its earliest stages. Patients were pleased with the operation since they did not need to take off their clothes, radiography only takes about 10 min, and AI technology produces quick results. Therefore, it is a very privacy-conscious strategy [33]. The process is free of radiation, convenient to use, and portable. With specially qualified personnel, the exam may even be completed in the convenience of our own homes. Further tests to rule out breast cancer are indicated and prescribed for patients if any unusual detectable mapping is seen in the thermal imaging. The approach necessitates the use of thermal imaging as well as AI formulation (AI technology).

6.7 How does AI help in the automated localization and segmentation of tumors?

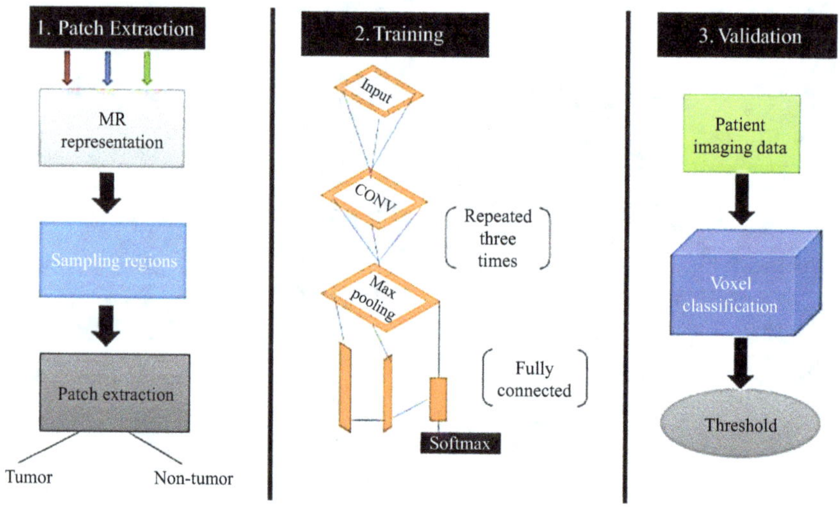

Figure 6.7 AI aided tumour detection.

A schematic representation of the basic approach of the segmentation and localization of tumors with the help of AI is depicted here (Fig 6.7). AI uses three classified steps: (a) patch extraction and locating the tumor from the X-rays, (b) training of the algorithm, and (c) validation of the result and clinical diagnosis [34].

6.7.1 Multi-parametric MR rectal cancer segmentation

Multiparametric MRI (mpMRI) has emerged as a powerful tool in the field of rectal cancer diagnosis and treatment planning, enabling advanced segmentation techniques for precise clinical insights. Several studies have shown that – as an addition to standard morphological MRI – DWI (diffusion-weighted imaging) can aid in assessing response to chemoradiotherapy. For this reason use of DWI is now even recommended in international clinical practice guidelines for rectal cancer imaging. Most of the volumetric measurements and histogram features are calculated from regions of interest (ROI) of the tumour which are typically obtained after manual tumour segmentation by experienced readers. The main problem with manual segmentation approaches, is that these are highly time consuming, and as such unlikely to be implemented into daily clinical practice. Various studies have explored ways to automatically perform segmentations using deep learning. These approaches work best on diffusion-weighted images, as these highlight the tumour and suppress background tissues. The data obtained from a multi-parametric MR were combined to develop multiple modalities available for a convolutional neural

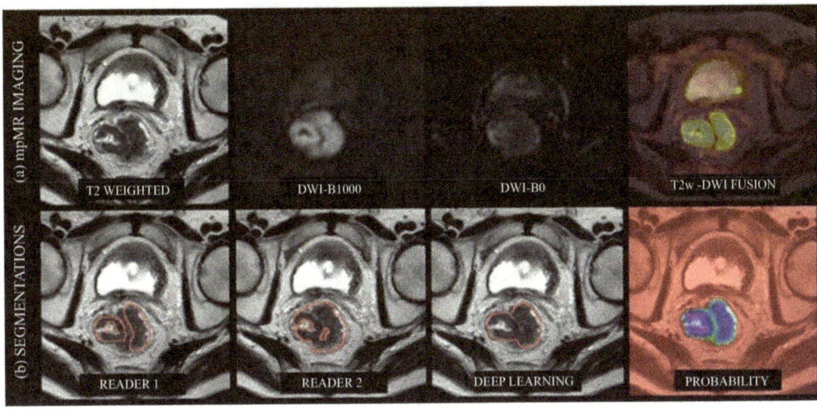

Figure 6.8 (a) mpMR images obtained and (b) tumor segmentation performed by a deep learning algorithm to create the probability map (from right to left) [36]

network (CNN), a deep learning tool which uses it to locate the tumour and its extent within the image [35]. An expert reader is used for training which is followed by the independent reader to generate the algorithm result, and the related probability map created by the algorithm (Figure 6.8). The model application is trained with hundreds of cases of rectal cancer and the performance obtained was comparable to human performance in the validation data set. Therefore, deep learning tools can accelerate accurate identification and segmentation of tumours from patient data [37].

6.7.2 Automated tumor characterization

AI can capture the radiographic phenotypic characteristics like homogeneity, heterogeneity, isolation, or infiltration in between different CT images of cancer [38]. For example, according to a 2014 study published in nature communications, a prognostic radiomics signature quantifying intra-tumor heterogeneity was developed and validated by radiomics analysis on CT imaging of about 1,000 patients with lung cancer. The model was trained on lung cancer data and was validated on head and neck cancer, the performance improved with the head and neck cancer cohorts indicating that the specific radiomics signature can be applied to distinct cancer types.

6.8 The Felix Project

The goal of this project is to develop and apply deep learning algorithms to screen for pancreatic neoplasms using CT and MR imaging techniques [22]. The strategy comprises using normal images of the pancreas as training data and abnormal images to identify pancreatic tumors at an early stage through a background

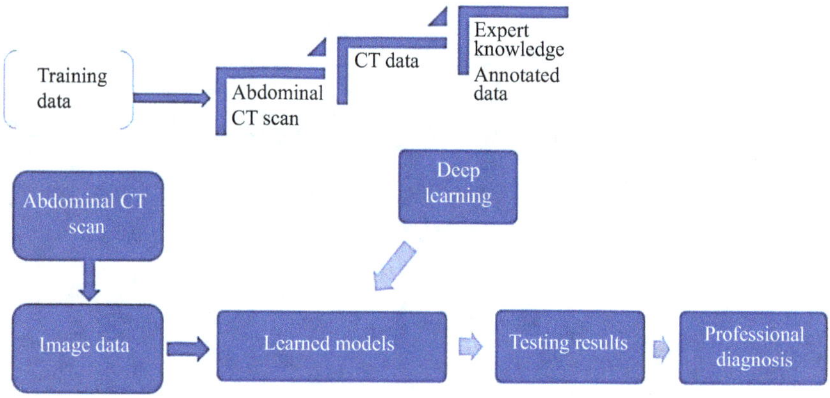

Figure 6.9 The schematic representation of the professional diagnosis done during the Felix Project with deep learning

program running on all abdominal CT images which will alert the radiologist on the abnormal pancreas (Figure 6.9).

6.9 Challenges faced due to AI technology

In the healthcare sector, AI has a wide range of uses and is still expanding as technology develops. But there are also significant drawbacks in this area that prevent AI from being fully incorporated into the existing healthcare systems. The main challenges encountered in radiomics are mentioned below [39].

1. **Matching AI results with medical recommendations**
   The majority of modern AI applications in radiology offer assessments of a patient's propensity for problems. As an illustration, an AI system determines that a patient's breast lesion has a 10% chance of being cancerous. A radiologist could decide to do a biopsy, but the AI system might not recognize the seriousness of the issue and judge a 10% probability of cancer is irrelevant. Working closely together is essential for developers and medical experts. The effectiveness of AI-based solutions can be increased with the help of medical experts' insights.
2. **Feature extraction**
   Model creation is made incredibly simple by the deep learning tools now available, thus many models are starting to appear. Anyone with access to enough properly labeled data may begin creating models. Shape features give information about the volume, maximum diameter along various orthogonal directions, maximum surface, tumor compactness, and sphericity of the traced region of interest (ROI) [40]. For instance, a speculated tumor will have a higher surface-to-volume ratio than a round tumor with the same volume. Selecting which and how many parameters to extract from the images presents

some challenges for the user. Each tool determines a varied quantity of features from various categories,

3. **Effect of acquisition and reconstruction**
   Each institution has its own set of reconstruction parameters and methods, with potential variances among individual patients. All these factors have an impact on image noise and texture, which in turn affects image characteristics. Therefore, rather than reflecting different biological properties of tissues, the features derived from images acquired at a single institution using a variety of acquisition protocols or acquired at various institutions using a variation of scanners in a wide range of patient populations can be affected by a combination of parameters [40]. Certain settings for acquisition and reconstruction may yield unstable features, resulting in different values being derived from successive measurements made under the same circumstances.
4. **Human reluctancy**
   Both developing precise AI algorithms and comprehending how to incorporate AI technologies into routine healthcare operations are difficult. Radiologists' duties and responsibilities are subject to change. Despite the indicated precision and efficacy of algorithms, it is doubtful that they will ever be entirely independent [41].
5. **Inadequate IT infrastructure**
   Despite several AI applications in radiology, many healthcare organizations have yet to begin the digital revolution. Their systems lack interoperability, hardware has to be upgraded, and their security methods are out-of-date. The use of AI in this situation may provide extra challenges [42].
6. **Data integrity**
   The shortage of high-quality labeled datasets is a problem that affects all fields and businesses, including radiology. It is difficult to get access to clean, labeled data for training medical AI [42].

6.10 Solutions to improve the technology

Healthcare providers should make sure that human experts continue to take the lead in decision-making and that human–machine collaboration is effective. In order to combat these issues, IT infrastructure must be gradually changed, ideally with assistance from a consortium of experts. Many healthcare organizations are already undergoing digital transformations and there is an increasing need for high-quality information, it is only a matter of time before most datasets meet these criteria.

6.11 Conclusion

Since the 1890s, when X-ray imaging first gained popularity, medical imaging has been a cornerstone of healthcare. This trend has continued with more recent advancements in CT, MRI, and PET scanning. It is now feasible to identify incredibly minute differences in tissue densities thanks to advancements in imaging equipment quality, sensitivity, and resolution. These alterations can oftentimes be hard to see, even with trained eyes and even traditional AI methods used in the clinic. As a result, these approaches lack the sophistication of imaging tools, but they nonetheless offer another incentive to investigate this paradigm. Furthermore, the deep learning algorithms scale with data, which means that as more data are gathered every day and as research efforts continue, it is anticipated that relative performance will increase [43]. By handling laborious tasks like structure segmentation, AI may considerably reduce the burden. Future possibilities in the next 10 years will incorporate background AI models which will already have reviewed the patient's EMR and images as well as specify probable findings when a radiologist opens a CT image. It will classify normal and abnormal features as a result radiologists will be focused on tackling abnormal results [44]. AI will not only assist in interpreting images but healthcare will be upgraded with intelligent equipment handling acquisition and reconstructions, segmentation, and 3D rendering of imaging data. Finally, it can spot information in photos that people miss, including molecular markers in tumors. It is also important to note that AI varies from human intelligence in a number of areas, and brilliance in one area does not always translate into greatness in another. The potential of new AI techniques should thus not be overstated [45]. Furthermore, it is evident that AI will not take the role of radiologists in the near or far future. Radiologists' jobs will develop as they grow more reliant on technology and have access to advanced equipment. They will provide knowledge and keep an eye on effectiveness while developing AI training models. Therefore, the various forms of AI will eventually be valuable assets in radiography.

References

[1] Oren O, Gersh B, and Bhatt D. Artificial intelligence in medical imaging: switching from radiographic pathological data to clinically meaningful endpoints. *The Lancet Digital Health* 2020;2:E486–E488.

[2] Sandra VBJ. The electronic health record and its contribution to healthcare information systems interoperability. *Procedia Technology* 2013;9: 940–948.

[3] Driver C, Bowles B, and Greenberg-Worisek A. Artificial intelligence in radiology: a call for thoughtful application. *Clinical and Translational Science* 2020;13:216–218.

[4] Berlin L. Radiologic errors, past, present and future. *Diagnosis (Berlin)* 2014;1(1):79–84. doi:10.1515/dx-2013-0012. PMID: 29539959.

[5] Farooq K, Khan BS, Niazi MA, Leslie SJ, and Hussain A. Clinical Decision Support Systems: A Visual Survey, 2017. *ArXiv*.

[6] Wainberg M, Merico D, Delong A, and Frey BJ. Deep learning in biomedicine. *Nature Biotechnology* 2018;36(9):829–838. doi:10.1038/nbt.4233. Epub 2018 Sep 6. PMID: 30188539.

[7] Pianykh O, Langs G, Dewey M, *et al.* Continuous learning AI in radiology: implementation principles and early applications. *Radiology* 2020;297:6–14.

[8] Strohm L, Hehakaya C, Ranschaert ER, Boon WPC, and Moors EHM. Implementation of artificial intelligence (AI) applications in radiology: hindering and facilitating factors. *European Radiology* 2020;30(10):5525.

[9] Hosny A, Parmar C, Quackenbush J, Schwartz LH, and Aerts HJWL. Artificial intelligence in radiology. *Nature Reviews Cancer* 2018;18(8):500–510. doi:10.1038/s41568-018-0016-5. PMID: 29777175; PMCID: PMC6268174.

[10] Benchamardimath B. A study on the importance of image processing and its applications. *International Journal of Research in Engineering and Technology* 2014;03:15.

[11] Zhang X and Dahu W. Application of artificial intelligence algorithms in image processing. *Journal of Visual Communication and Image Representation* 2019;61:42–49.

[12] Yang M, Hu J, Chong L, *et al.* An in-depth survey of underwater image enhancement and restoration. *IEEE Access* 2019;7:123638–123657.

[13] Huisman M, Ranschaert E, Parker W, *et al.* An international survey on AI in radiology in 1041 radiologists and radiology residents, part 2: expectations, hurdles to implementation, and education. *European Radiology* 2021;31 (11): 8797–8806.

[14] Jiao L and Zhao J. A survey on the new generation of deep learning in image processing. *IEEE Access* 2019;7:172231–172263.

[15] Sadek RA. SVD based image processing applications: state of the art. *Contributions and Research Challenges: A R International Journal of Advanced Computer Science and Applications* 2012;3(7).

[16] Wang H, Zhang Y, and Yu X. An overview of image caption generation methods. *Computational Intelligence and Neuroscience* 2020;2020.

[17] Boykov Y, Porikli F, Plaza A, Kehtarnavaz N, and Terzopoulos D. Image segmentation using deep learning: a survey. *IEEE Transactions on Pattern Analysis and Machine Intelligence* 2022;44(7):3523–3542.

[18] Filice R and Kahn C. Biomedical ontologies to guide AI development in radiology. *Journal of Digital Imaging* 2021;34(6):1331–1341.

[19] Dikici E, Bigelow M, Prevedello LM, White RD, and Erdal BS. Integrating AI into radiology workflow: levels of research, production, and feedback maturity. *Journal of Medical Imaging* 2020;7(01):016502.

[20] Rezazade Mehrizi M, van Ooijen P, and Homan M. Applications of artificial intelligence (AI) in diagnostic radiology: a technography study. *European Radiology* 2021;31(4):1805–1811.

[21] Mamoshina P, Vieira A, Putin E, and Zhavoronkov A. Applications of deep learning in biomedicine. *Molecular Pharmaceutics* 2016;13(5):1445–1454. doi:10.1021/acs.molpharmaceut.5b00982. Epub 2016 Mar 29. PMID: 27007977.

[22] Chu LC, Park S, Kawamoto S, *et al.* Application of deep learning to pancreatic cancer detection: lessons learned from our initial experience. *Journal of the American College of Radiology* 2019;16(9 Pt B):1338–1342. doi:10.1016/j.jacr.2019.05.034. PMID: 31492412.

[23] Engle E, Gabrielian A, Long A, Hurt DE, and Rosenthal A. Figure 2: performance of Qure.ai automatic classifiers against a large annotated database of patients with diverse forms of tuberculosis. *PLoS One* 2020;15(1):e0224445.
[24] Kennedy S. AI model can help detect collapsed lung using chest X-rays. The artificial intelligence model accurately detected pneumothorax, or a collapsed lung, and exceeded FDA guidelines for computer-assisted triage devices. News Blog:https://healthitanalytics.com/news/ai-model-can-help-detect-collapsed-lung-using-chest-x-rays.
[25] Artificial Intelligence-Assisted Detection of Adhesions on Cine-MRI. Master Thesis Evgeniia Martynova S1038931.
[26] Hamabe A, Ishii M, Kamoda R, *et al.* Artificial intelligence-based technology to make a three-dimensional pelvic model for preoperative simulation of rectal cancer surgery using MRI. *Ann Gastroenterol Surg.* 2022 May 11;6 (6):788–794. doi: 10.1002/ags3.12574.
[27] Tang X. The role of artificial intelligence in medical imaging research. *BJR Open* 2019;2(1):20190031. doi: 10.1259/bjro.20190031. PMID: 33178962; PMCID: PMC7594889.
[28] Alkabbany I, Ali AM, Mohamed M, Elshazly SM, and Farag A. An AI-based colonic polyp classifier for colorectal cancer screening using low-dose abdominal CT. *Sensors* 2022;22:9761.
[29] Kudo SE, Mori Y, Misawa M, *et al.* Artificial intelligence and colonoscopy: current status and future perspectives. *Digestive Endoscopy* 2018;30:52–53.
[30] Ramasubbu R, Brown EC, Marcil LD, Talai AS, and Forkert ND. Automatic classification of major depression disorder using arterial spin labeling MRI perfusion measurements. *Psychiatry and Clinical Neurosciences* 2019;73: 486–493.
[31] Rudie JD, Gleason T, and Barkovich MJ. Clinical assessment of deep learning-based super-resolution for 3D volumetric brain MRI. *Radiology: Artificial Intelligence* 2022;4(2):e210059.
[32] Suh Y, Jung J, and Cho B. Automated breast cancer detection in digital mammograms of various densities via deep learning. *Journal of Personalized Medicine* 2020;10(4):E211.
[33] Hasan AS, Sagheer A, and Veisi H. Breast cancer classification using machine learning techniques: a review. *IJRAR* 2021;9:590–594.
[34] Swathikan C, Viknesh S, Nick M, and Markar SR. Diagnostic performance of artificial intelligence-centred systems in the diagnosis and postoperative surveillance of upper gastrointestinal malignancies using computed tomography imaging: a systematic review and meta-analysis of diagnostic accuracy. *Annals of Surgical Oncology* 2021;29(3):1977.
[35] Wang PP, Deng CL, and Wu B. Magnetic resonance imaging-based artificial intelligence model in rectal cancer. *World Journal of Gastroenterology* 2021;27(18):2122–2130. doi: 10.3748/wjg.v27.i18.2122. PMID: 34025068; PMCID: PMC8117733.

[36] Trebeschi S, Van Griethuysen JJM, Lambregts DMJ, *et al.* Deep learning for fully-automated localization and segmentation of rectal cancer on multi-parametric. *Scientific Report* 2017;7(1):5301.

[37] Trebeschi S, van Griethuysen JJM, Lambregts DMJ, *et al.* Deep learning for fully-automated localization and segmentation of rectal cancer on multi-parametric MR. *Scientific Reports* 2017;7(1):5301. doi: 10.1038/s41598-017-05728-9. Erratum in: *Sci Rep.* 2018 Feb 2;8(1):2589. PMID: 28706185; PMCID: PMC5509680.

[38] Joy Mathew C, David AM, and Joy Mathew CM. Artificial Intelligence and its future potential in lung cancer screening. *EXCLI J.* 2020;19:1552–1562. doi: 10.17179/excli2020-3095. PMID: 33408594; PMCID: PMC7783473.

[39] Dilmegani C. Top 6 Challenges of AI in Healthcare and Overcoming them in 2023. Updated on December 26, 2022 | Published on March 1, 2022.

[40] Rizzo S, Botta F, Raimondi S, *et al.* Radiomics: the facts and the challenges of image analysis. *European Radiology Experimental* 2018;2(1):36. doi: 10.1186/s41747-018-0068-z. PMID: 30426318; PMCID: PMC6234198.

[41] Lebovitz S, Lifshitz-Assaf H, and Levina N. To incorporate or not to incorporate AI for critical judgments: the importance of ambiguity in professionals' judgment process. *Collective Intelligence, The Association for Computing Machinery* 2020.

[42] Waller J, O'connor A, Eleeza Raafat, *et al.* Applications and challenges of artificial intelligence in diagnostic and interventional radiology. *Polish Journal of Radiology* 2022;87: e113–e117.

[43] Mun SK, Wong KH, Lo S, Li Y, and Bayarsaikhan S. Artificial intelligence for the future radiology diagnostic service. *Frontiers in Molecular Biosciences* 2021;7:Article 614258.

[44] Wagner M, Namdar K, Biswas A, *et al.* Radiomics, machine learning, and artificial intelligence—what the neuroradiologist needs to know. *Neuroradiology* 2021;63:1957–1967.

[45] Koçak B, Durmaz EŞ, Ateş E, and Kılıçkesmez Ö. Radiomics with artificial intelligence: a practical guide for beginners. *Diagnostic and Interventional Radiology* 2019;25(6):485–495. doi: 10.5152/dir.2019.19321. PMID: 31650960; PMCID: PMC6837295.

Chapter 7

Healthcare multimedia data analysis algorithms tools and techniques

Sathya Raja[1], V. Vijey Nathan[1] and Deva Priya Sethuraj[1]

In the domain of Information Retrieval (IR), there exists a number of models which is used for different sorts of applications. The extraction of multimedia is one of the types which specifically deals with the handling of multimedia data with different types of tools and techniques. There are various techniques for handling multimedia data such as feature handling, extraction, and selection. The features selected by these techniques have been classified using machine learning and deep learning techniques. This chapter provides complete insights into the audio, video, and text semantic descriptions of the multimedia data with the following objectives:

(i) Methods
(ii) Data summarization
(iii) Data categorization and its media descriptions

Upon considering this organization, the entire chapter has dealt with as a case study depicting feature extraction, merging, filtering, and data validation.

7.1 Introduction

The information retrieval (IR) domain is considered an essential paradigm in different real-time applications. The advancement in data retrieval techniques was established more than five thousand years ago. In practice, the intent of the data retrieval to that of information retrieval has been raised with the accordance of model development, process analysis, and data interpretation and evaluation. One of the primary forms of data that have multiple supportable formats is multimedia data. This data utilizes different information retrieval models to establish a particular decision support system. In a specific context, feature-based analysis plays a significant role in data prediction and validation. The only advent is that it must adapt to that of the particular database community and the modular applications in which it deals with the formats.

[1]Department of Computer Science and Engineering, SRM TRP Engineering College, India

Until March 2012, this multimedia information retrieval (also known as MMIR) was just a buzzword, just like the metaverse today [1]. Well, that is not a scenario anymore. Nowadays, researchers, industries, and end-users require organized data to feed our machine learning (ML) algorithms [2]. While statistic-based ML algorithms only need a comma separated value (CSV, although it requires correction of data) file, media-based ML algorithms struggle for competent datasets. This struggle evolves the need for MMIR into the current scenario. The MMIR is a blooming research discipline that focuses on extracting text and text-based information (*semantic*, to be more accurate) from various multimedia sources. It may extract explicit media such as audio, video, and image. It can also extract implicit media such as written text and textual descriptions. Moreover, it can extract data from totally indirect multimedia sources such as bio-information, and stock prices. The MMIR data extraction methodology spans three significant steps [3,4]:

1. Feature extraction
2. Filtering
3. Categorization

The first step in MMIR, which is pretty simple and obvious, is feature extraction. The general goal of this particular step can be achieved by completing one but two processes, namely, summarization and pattern detection. Before going over anything, we need a not-accurate summary of what we are onto. That is the summarization process. It takes whatever media it has and summarizes it. The next one is pattern detection. Here we use either auto-correlation or cross-correlation to detect the patterns.

The second step in MIMIR is merging and filtering. As we are to feed multimedia datasets, the pool will likely be a cluster of all available media formats. This step ensures that every relevant data gets into the algorithm by properly merging and filtering them. It sets multiple media channels, and each channel has a label on the supposed data going in. Then it uses a simple filtering method, such as factor analysis, to the more complex one, such as Kalman filter, to effectively filter and merge the descriptions.

The last step in MIMIR is categorization. In this step, we can choose any ML form, as one always performs better than another, respective to the given dataset. As we have an abundance of ML classifiers, we can choose the one that will likely give us acceptable results. We can also let the algorithm choose the classifier using tools such as Weka, data miner, R, and Python.

The process of research practice and its supportive culture has become blooming with the process of handling different types of data. The supporting types are having different issues with the data processing platforms which are suited for analysis. Also, the utilization of data-driven models is increasing daily with its available metrics. Metric-based data validation and extraction is one of the tedious tasks, which certainly makes the data suitable for analysis. The algorithmic models may vary, but the aspect that must be considered is easy. In the present stages of study, the designers choose their way of representing and handling the data to a certain extent, especially [5]:

- Design of decision support systems (DSS) to provide a complete service.
- To utilize the system effectively to communicate with the professionals, this states the expectations behind the system.
- To enhance the researchers to effectively utilize the model in data integration, analysis, and spotting relevant multimedia data.

The extraction of multimedia data sources is analyzed with efficient forms of data analysis and linguistic processes. These methods can be efficiently organized into three such groups:

1. Methods suitably used for summarizing the media data are precisely the result of the feature extraction process.
2. Methods and techniques for filtering out the media content and its sub-processes.
3. Methods that are suitable for categorizing the media into different classes and functions.

7.1.1 Techniques for summarizing media data

Feature extraction is motivated by an innumerably large multimedia object, its redundancy, and possibly nosiness. By feature extraction, two goals can be achieved.

1. Data summary generation
2. Data correlation analysis with specific autocorrelation and comparison

7.1.2 Techniques for filtering out media data

The process of MMIR emphasizes the locally visible and executable channels for the different forms of IR models suitably supported. The results are merged into one description per media content. Descriptions are classified into two based on their size and they are as follows:

- Fixed size
- Variable size

The process of merging takes place through simple concatenation, especially in the form of fixed and variable size lengths. Size has to be normalized to a fixed size before merging. They most often occur as motion descriptions. The most commonly used filtering mechanisms are the following:

- Factor analysis
- Single value decomposition (SVD)
- Kalman filter

7.1.3 Techniques for media description categorization—classes

The central concept of ML is applied to categorizing multimedia descriptions. The list of applicable classifiers is as follows:

- Metric-based model development
- Nearest neighbor-based model development
- Minimization of risk in estimation models
- Density-based evaluation/approaches
- Feed-forward neural network
- Simple heuristic analysis

The main objective of this model is to minimize the overhead of user-needed information. Some major application areas include bio-information analysis, face recognition, speech recognition, video browsing, and bio-signal processing.

7.2 Literature survey

The research work [6] stated that it is better to search a video based on its metadata description to reduce the time complexity during video retrieval. According to the journal, the video is first searched based on its content. The video's two main features are considered: one is the visual screen which is nothing but text, and the other is the audio tracks which are nothing but speech. This process started with video segmentation, in which one has to classify moving objects in a lecture video sequence. Followed by the second step, in which the transition of the actual slide is captured. This process is repeated to reduce the content's redundancy ultimately.

The next step is to create a video OCR for the characters in the text. An OCR is a system that can process an image. The image is recognized based on the similarity of the loaded image and the image model, the image is recognized. In the next step, automatic speech recognition (ASR) technology helps to identify the words spoken by a person, and then it is converted to text. Its ultimate goal is to recognize speech in real time with total accuracy. Finally, after applying OCR and ASR algorithms on those keyframes, the results are stored in the database with a unique identifier and timestamp. When a user searches for content, the video search is successful if it matches the content in the database. So this is called content-based video retrieval.

The work [7] proposed a mechanism for the analysis of retrieval based on multimodal fusion, which includes the component of textual and visual data. They have used data clustering and association rule mining techniques for evaluation to retrieve the content modality and analysis explicitly. They have utilized the possible way of three-mode data segregation and analysis. The proposed model involves a multimodal analysis of a three-way combination of data retrieval platforms. Here the relevant image which is supposed to be retrieved is taken with subsequent forms of model extraction. The fusion subsystem and the LBP pattern are used for the next level of data analysis and retrieval.

Experimental results justify that when visual data is fed into the system based on which textual data is entered into the system, after searching, a relevant image comparison is made between the two data using the LBP. Finally, the matched images are retrieved after the images are mapped with the model to extract the suitable patterns from the test data set.

XaroBenavent and Ana Garcia-Serrano [8] proposed retrieving textual and visual features through multimedia fusion. This method increases the accuracy of the retrieved results based on the existing fusion levels, such as early fusion, which is based on the extracted features from different information sources. Late fusion or hybrid combines the individual decision by the mono-modal feature extraction process and the model development metrics.

In the developed environment, a system involves steps like the Extraction of textual information and textual pre-processing, which includes three steps: elimination of ascents and stop words and stemming. Then indexation is done using the White Space Analyser. Finally, searching is done to obtain the textual results.

The content-based information retrieval (CBIR) subsystem involves two steps: Feature extraction and a similarity module specifically allotted to extract the data similarity content from the contextual part of the data.

The process of the late fusion algorithm is segregated into two types of categories:

1. Relevant score
2. Re-ranking the score normalization process

The work by the authors Lew *et al.* [9] proposed a novel idea for the mechanism of a content-based retrieval process in extracting multimedia image contents. They have also analyzed the phenomenon of text annotations and incomplete data transformations. Media items, including text, annotated content, multimedia, and browsing content, are also analyzed (Table 7.1).

Table 7.1 Summary of literature review

Research work	Methods used	Dataset used	Improvements/accuracy
[11]	Naive Bayes, Random Forest, SVM and Logistic Regression Classifiers, Ensemble classifier	Stanford—Sentiment140 corpus, Health Care Reform (HCR)	Ensemble models perform well for sentiment analysis than other classification algorithms.
[12]	Sentiment classification	Twitter data	Statistical significance proves the validation of the text analysis process for all classification schemes.
[13]	Sentiment analysis and text analysis	Tweet Dataset	The authors conclude this article by comparing papers and providing an overview for challenges in sentiment analysis related to sentiment analysis approaches and techniques.
[14]	Ensemble method	Twitter data (Arabic)	The performance has been tested and evaluated with an F1 score of 64.46%.

(Continues)

Table 7.1 (*Continued*)

Research work	Methods used	Dataset used	Improvements/accuracy
[15]	Sentiment analysis	SauDiSenti, AraSenTi	The proposed model using AraSenTi performs well than the other models and the same can be used for analyzing different categories of sentiments.
[16]	SVM, NB, MNB	Arabic tweets	SVM and NB classifier works well for binary classification with the highest performance in accuracy, precision, and recall and multinomial Naive Bayes works for multi-way classification.
[17]	Sentiment analysis	Twitter data	Results conclude that only 30% of people are unhappy with the demonetization policy introduced by the Indian Government.
[18]	Lexicon-based sentiment analysis	Movie reviews	Built-in lexicons can be used well for categorical sentiments.
[19]	Google's algorithm Word2Vec	Movie reviews	Comparing different types of clustering algorithms and types of clusters.
[20]	Sentiment classification	Movie reviews	Precision: 92.02%

7.3 Methodology

Now, let us discuss in detail the methods available to perform retrieval based on multimedia information retrieval.

7.3.1 Techniques for data summarization

The feature extraction and analysis domain lie at the extent of data processing and analysis. In order to remove the noise and redundancy in the given dataset nature of the data with its available transformation and the consistency levels has to be checked. This can be achieved by the set of derived values and procedures suitable for facilitating the learning and analysis process. In healthcare and big data platform, the intent of analyzing the data lies in the possible states of complete data security and redundancy. This comes under the platform of dimensionality reduction and process generalization. From the observed data in various formats supporting the healthcare data analysis, it should be noted that the features should be

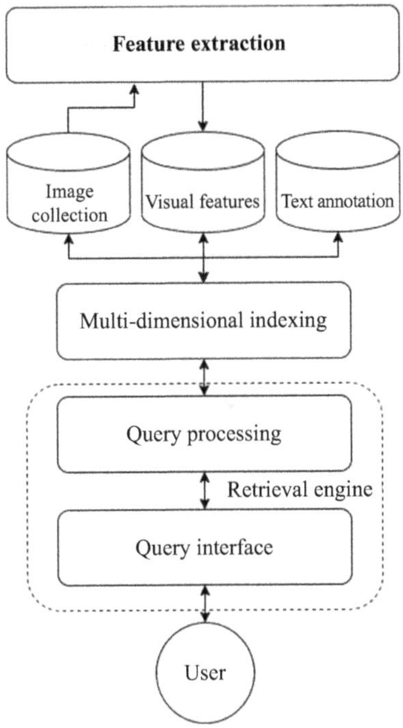

Figure 7.1 Process of summarizing the media content

explicitly classified under one constraint and correction phenomenon, which describes the original dataset. An example of this method is depicted in Figure 7.1.

Suppose the input algorithm is found to be more robust. In that case, the data can be analyzed into different variations with a reduced feature set for subsequent feature extraction and quantization. Image analysis and image-based data segmentation lie in the data quality, which explicitly relies on the pixels generated with the video stream. The shapes may vary, but the same realm lies in analyzing image data segmentation with robust machine-learning algorithms. It specifically involves:

- The low detection rate analysis
- Edge-based data segmentation with a reduced level of noise in digital images
- Facilitating automated recovery of data segments

The rate of low-level edge detection involves specific sub-tasks for analysis as follows:

- Edge detection with mathematical methods determining the brightness and the level of point of segmentation during analysis with mild and high effects.
- Corner analysis to determine the missed feature contents at sure edges with panorama effects during 3D editing, modeling, and object recognition.

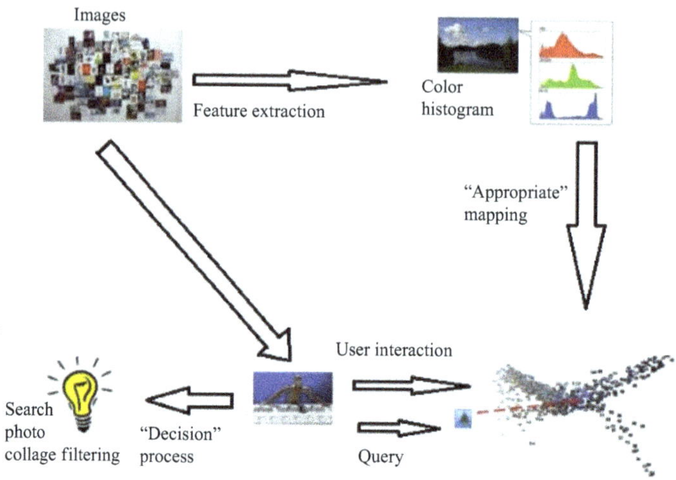

Figure 7.2 Multimedia content extraction process and analysis

- Blob segment with different imaging properties at the curvature points determine similar cases with image analysis properties.
- Ridge analysis makes the functions of two variables to determine the set of curvature points at least at one single dimension.

7.3.2 Merging and filtering method

According to this model, in order to understand the content of the media, multiple channels are employed. Media-specific feature transformation describes these channels. At last, these annotations have to be merged with a single description per object. As already explained, merging is of two types:

- Fixed-size merging
- Variable-sized merging

If the description is of a fixed size, then simple concatenation is done to merge two or more descriptions. The descriptions have to be normalized to a fixed size for variable-sized descriptions, and they most commonly occur in motion descriptions.

Commonly used filtering methods are as follows:

- Factor analysis
- Singular value decomposition
- Extraction and testing of statistical moments
- Kalman filter

7.3.2.1 Factor analysis

This technique is adopted to reduce a more significant number of variables into fewer factors. These techniques extract the common variances from all the variables and place them into a standard score. The types of factoring are the following:

- Principal component analysis extracts a large number of variables and puts them into a single first factor.
- Common factor analysis extracts standard variables and puts them into a single factor.
- Image factoring is based on a correlation matrix, which determines the exact correlation.
- Maximum likelihood method based on a correlation matrix.

7.3.2.2 Singular value decomposition

The process of singular value decomposition involves a set matrix A which has a factorization value with A=UDV T in which the column values are assigned to be ortho-normal. The matrix D is said to be diagonal with real positive values. The process of singular value decomposition is used in different sorts of applications, including explicitly ranking data esteeming the low and high-rank levels as in Figure 7.2.

In medical data analysis, multimedia content is of many forms of representation. A simple example includes the analysis of doctors' textual descriptions needs a text-to-audio conversion and then into structured forms of representation. In this context, variations and complicated structures exist for analyzing the data into different manipulations.

Some of the practical examples include:

- Nearest orthogonal matrix
- The kabsch algorithm
- Signal processing
- Total least square minimization

7.3.2.3 Extraction and testing of statistical moments

In this process, there are unscented transformations, with the process following moment-generating functions. This is more popular in electromagnetic computation for large-scale data analysis. At certain stages, the Monte Carlo approach is used to evaluate and analyze data models. The list of applicable classifiers includes the following:

7.3.3 Evaluating approaches

- Cluster Analysis—the most similar objects are processed using k-means and k-medoids or Self Organizing maps.
- Vector Space Model—the algebraic model concentrates on term frequency and inverse document frequency (tf-idf).
- Support Vector Machine—supervised learning multi-models used for regression processes.
- Linear Discriminate Analysis—a generalization of Fisher's linear discriminate and also uses pattern recognition for learning the objects.
- Markov Process—stochastic model relies on the sequence of events with a probabilistic measure of occurrence.
- Perceptron Neural Networks—one of the significant algorithms primarily used for supervised learning and skilled working of specific classes.

- Decision Tree—a decision support tool that uses a more tree-like model of decisions, their conditions, and their possible consequences, combining event outcomes and utility. It is one of the ways to display an algorithm that only contains conditional control statements.

7.4 Sample illustration: case study

The illustration can be considered with the analysis of predictive maintenance using medical data. This involves the analysis of benchmark data from the UCI ML repository [10], which concerns the heart disease of signified patients. Figure 7.3 describes the data description.

At each stage of the disease, the influencing factors are observed, and the analysis is made accordingly. In order to overcome these factors, the weight level can be introduced and can be increased or decreased at indeed levels. Figure 7.4 provides the incorporation of weight with the influencing rates.

The detection curve as shown in Figure 7.5 when applied to observed situations of the disease in the context of predictive maintenance using medical data, is

Age	Sex	cpType	trestbps	chol	fbs	restecg	thalach	exang	oldpeak	slope	ca	thal	classlabel
63	1	1	145	233	1	2	150	0	2.3	3	0	6	Zero
67	1	4	160	286	0	2	108	1	1.5	2	3	3	One
67	1	4	120	233	0	2	129	1	2.6	2	2	7	One
38	1	3	130	250	0	0	187	0	3.5	3	0	3	Zero
41	0	2	130	240	0	2	165	0	1.4	1	0	3	Zero

Figure 7.3 Dataset description

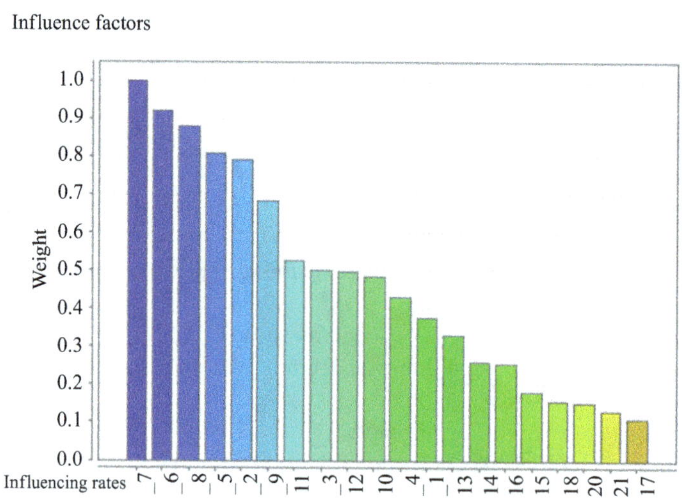

Figure 7.4 Factors influencing the rate of analysis

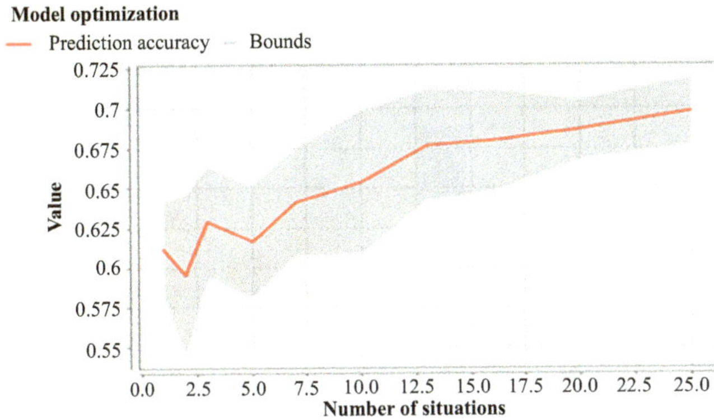

Figure 7.5 Detection curve over the observed situations of the disease

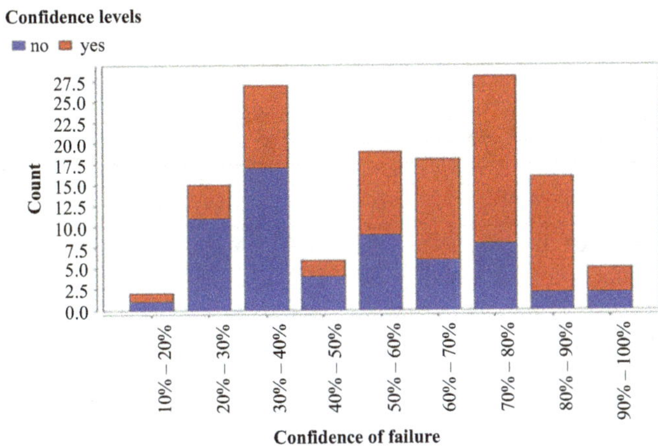

Figure 7.6 Confidence level and failure rate

illustrated through an analysis of benchmark data from the UCI ML repository focusing on heart disease in diagnosed patients in value vs number of situations.

The patients (as in Figure 7.6) at certain stages can vary according to the number of situations leveraging the factors to be considered. This provides a scenario for data curation, and it should be overcome by introducing confidence levels for the situation-based analysis.

Finally, in each situation, the risk level can be determined by the count of patients who have utterly suffered and recovered from illness. Figure 7.7 provides the risk level estimations with different conditions and scenarios of understanding levels.

7.5 Applications

The variants of text-based models are significantly used in different sectors for conceptual analysis of text design and extraction. A significant analysis must be

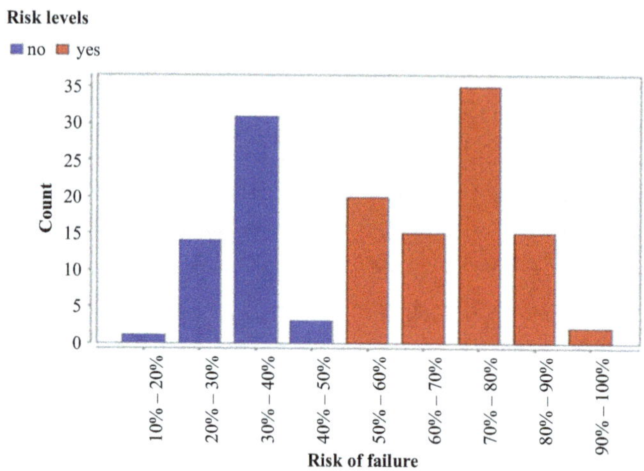

Figure 7.7 Risk levels and failure rate

made to make the data exact and confirm a decision at the best level [9]. Applications include bioinformatics, signal processing, content-based retrieval, and speech recognition platforms.

- Bio-informatics is concerned with biological data analysis with complete model extraction and analysis. The data may be in a semi or unstructured format.
- Bio-signal processing concerns the signals concerning living beings in a given environment.
- Content-based image retrieval deals with the search of digital images for the given environment of extensive data collection.
- Facial recognition system concerned with activity recognition for the given platform in the sequence of data frames.
- Speech recognition system transforms speech to text as recognized by computers.
- Technical chart analysis a market data analysis usually falls under this category of concern. This can be of type chart and visual perception analysis.

7.6 Conclusion

Information analysis from different formats of data is one of the tedious tasks. Analyzing and collecting those variants of data need to be considered a challenging task. Most of the modeling in multimedia data follows significant IR-based modeling to bring out the essential facts and truths behind it. In this chapter, we have discussed the different forms of IR models, tools, and applications with an example case study illustrating the flow of analysis of medical data during the stages of the modeling process. In the future, the aspects of different strategies can be discussed in accordance with the level of data that can be monitored with various tools and applications.

References

[1] Hanjalic, A., Lienhart, R., Ma, W. Y., and Smith, J. R. (2008). The holy grail of multimedia information retrieval: so close or yet so far away? *Proceedings of the IEEE*, 96(4), 541–547.

[2] Kolhe, H. J. and Manekar, A. (2014). A review paper on multimedia information retrieval based on late semantic fusion approaches. *International Journal of Computer Applications*, 975, 8887.

[3] Raieli, R. (2013, January). Multimedia digital libraries handling: the organic MMIR perspective. In *Italian Research Conference on Digital Libraries* (pp. 171–186). Springer, Berlin, Heidelberg.

[4] Rüger, S. (2009). Multimedia information retrieval. *Synthesis Lectures on Information Concepts, Retrieval, and Services*, 1(1), 1–171.

[5] Khobragade, M. V. B., Patil, M. L. H., and Patel, M. U. (2015). Image retrieval by information fusion of multimedia resources. *International Journal of Advanced Research in Computer Engineering & Technology (IJARCET)*, 4(5), 1721–1727.

[6] Sangale, A. P. and Durugkar, S. R. (2014). A review on circumscribe based video retrieval. *International Journal*, 4(11), 34–44.

[7] Aslam, J. A. and Montague, M. (2001, September). Models for metasearch. In *Proceedings of the 24th Annual International ACM SIGIR Conference on Research and Development in Information Retrieval* (pp. 276–284).

[8] Benavent, X., Garcia-Serrano, A., Granados, R., Benavent, J., and de Ves, E. (2013). Multimedia information retrieval based on late semantic fusion approaches: experiments on a wikipedia image collection. *IEEE Transactions on Multimedia*, 15(8), 2009–2021.

[9] Lew, M. S., Sebe, N., Djeraba, C., and Jain, R. (2006). Content-based multimedia information retrieval: state of the art and challenges. *ACM Transactions on Multimedia Computing, Communications, and Applications (TOMM)*, 2(1), 1–19.

[10] Asuncion, A. and Newman, D. (2007). UCI Machine Learning Repository.

[11] Saleena, N. (2018). An ensemble classification system for twitter sentiment analysis. *Procedia Computer Science*, 132, 937–946.

[12] Araque, O., Corcuera-Platas, I., Sánchez-Rada, J. F., and Iglesias, C. A. (2017). Enhancing deep learning sentiment analysis with ensemble techniques in social applications. *Expert Systems with Applications*, 77, 236–246.

[13] Hussein, D. M. E. D. M. (2018). A survey on sentiment analysis challenges. *Journal of King Saud University—Engineering Sciences*, 30(4), 330–338.

[14] Heikal, M., Torki, M., and El-Makky, N. (2018). Sentiment analysis of Arabic tweets using deep learning. *Procedia Computer Science*, 142, 114–122.

[15] Al-Thubaity, A., Alqahtani, Q., and Aljandal, A. (2018). Sentiment lexicon for sentiment analysis of Saudi dialect tweets. *Procedia Computer Science*, 142, 301–307.
[16] Boudad, N., Faizi, R., Thami, R. O. H., and Chiheb, R. (2018). Sentiment analysis in Arabic: a review of the literature. *Ain Shams Engineering Journal*, 9(4), 2479–2490.
[17] Singh, P., Dwivedi, Y. K., Kahlon, K. S., Sawhney, R. S., Alalwan, A. A., and Rana, N. P. (2020). Smart monitoring and controlling of government policies using social media and cloud computing. *Information Systems Frontiers*, 22(2), 315–337.
[18] Anandarajan, M., Hill, C., and Nolan, T. (2019). Practical text analytics *Maximizing the Value of Text Data. (Advances in Analytics and Data Science* (vol. 2, pp. 45–59). Springer.
[19] Chakraborty, K., Bhattacharyya, S., Bag, R., and Hassanien, A. E. (2018, February). Comparative sentiment analysis on a set of movie reviews using deep learning approach. In *International Conference on Advanced Machine Learning Technologies and Applications* (pp. 311–318). Springer, Cham.
[20] Pandey, S., Sagnika, S., and Mishra, B. S. P. (2018, April). A technique to handle negation in sentiment analysis on movie reviews. In *2018 International Conference on Communication and Signal Processing (ICCSP)* (pp. 0737–0743). IEEE.

Chapter 8

Empirical mode fusion of MRI-PET images using deep convolutional neural networks

N.V. Maheswar Reddy[1], *G. Suryanarayana*[1], *J. Premavani*[1] *and B. Tejaswi*[1]

In this chapter, we develop an image fusion method for magnetic resonance imaging (MRI) and positron emission tomography (PET) images. This method employs empirical mode decomposition (EMD) based on morphological filtering (MF) in a deep learning environment. By applying our resolution enhancement neural network (RENN) on PET source images, we obtain the lost high-frequency information. The PET-RENN recovered HR images and MRI source images are then subjected to bi-dimensional EMD to generate multiple intrinsic mode functions (IMFs) and a residual component. Morphological operations are applied to the intrinsic mode functions and residuals of MRI and PET images to obtain the fused image. The fusion process involves a patch-deep fusion technique instead of a pixel-deep fusion technique to reduce spatial artifacts introduced by pixel-wise maps. The results of our method are evaluated on various datasets and compared with the existing methods.

8.1 Introduction

Positron emission tomography (PET) produces an image with functional data that depicts the metabolism of various tissues. However, PET images cannot contain structural information about tissues and have limited spatial resolution. On the other hand, magnetic resonance imaging (MRI), a different non-invasive imaging technique, offers strong spatial resolution information about the soft tissue structure. However, gray color information that indicates the metabolic function of certain tissues is absent in MRI images [1]. The fusion of MRI and PET can deliver complementary data useful for better clinical diagnosis [2].

Image fusion is the technique of combining two or more images together to create a composite image that incorporates the data included in each original

[1]Electronics and Communications Engineering, Velagapudi Ramakrishna Siddhartha Engineering College, India

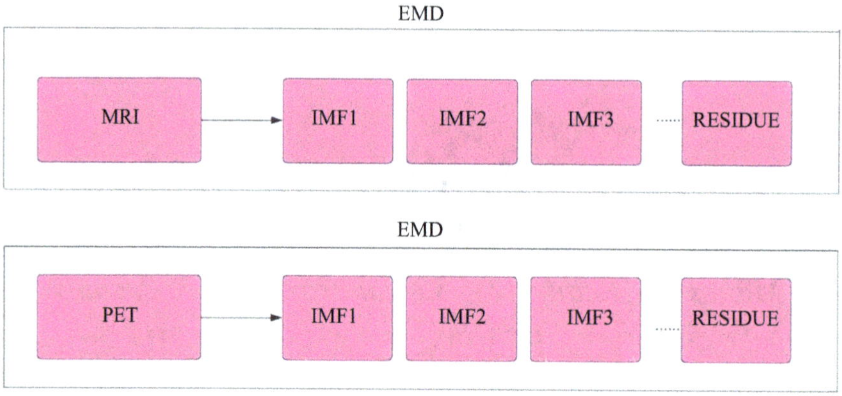

Figure 8.1 Empirical mode decomposition of MRI-PET images

image [3–7]. There are three types of techniques in image fusion, namely, spatial domain fusion, transform domain fusion, and deep learning techniques [8]. Principal component analysis (PCA) and average fusion are simple spatial fusion techniques. In these techniques, the output image is directly obtained by fusing the input images. Due to this, spatial domain fusion techniques produce degradation and distortion in the fused image. Hence, the fused images produced by spatial domain fusion techniques are less efficient compared to transform domain fusion techniques [8].

In transform domain techniques, the input images are first transformed from the spatial domain technique to the frequency domain prior to fusion. Discrete and stationary wavelet transforms are primarily employed in transformed domain techniques. These techniques convert the input image sources into low–low, low–high, high–low, and high–high frequency bands which are referred to as wavelet coefficients. However, these methods suffer from translational invariance problems leading to distorted edges in the fused image [9].

Deep learning techniques for image fusion have been popularized in recent times due to their dominance over the existing spatial and transformed domain techniques. Zhang *et al.* [10] proposed a convolution neural network for estimating the features of input source images. In the obtained image, the input source images are fused region by region. The hierarchical multi-scale feature fusion network is initiated by Lang *et al.* [11]. They used this technique for extracting multi features from input images [11]. In this chapter, we develop an MRI-PET fusion model in a deep learning framework. The degradation in PET low-resolution images is reduced by employing PET-RENN. The input image sources are extracted as IMFs and residual components by applying EMD as described in Figure 8.1. Morphological operations are applied to the IM functions and residues. PETRNN is used to recover higher-resolution images from lower-resolution of PET images [12].

8.2 Preliminaries

8.2.1 Positron emission tomography resolution enhancement neural network (PET-RENN)

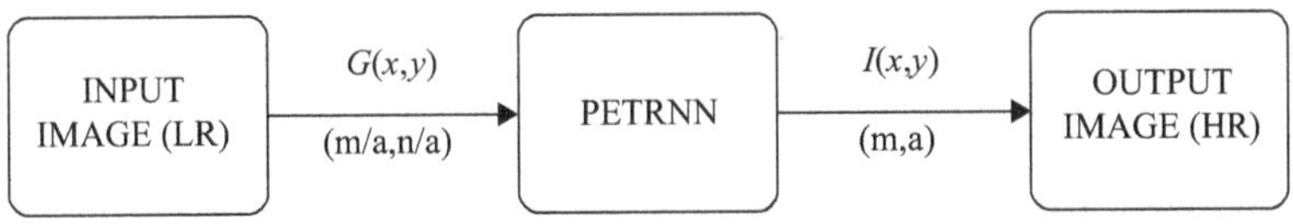

Figure 8.2 Positron emission topography resolution enhancement neural network process on low-resolution images

Due to the rapid progress in image processing technology, there has been an increase in the requirement for higher-resolution scenes and videos. As shown in Figure 8.2, PETRNN technique produces a higher-resolution (HR) image from a lower resolution (LR) image. In our work, PET-RENN technique is used to recover a higher-resolution images from a lower-resolution PET image sources. Let $G(x, y)$ be the input image with a size (m/a, n/a). When PETRNN is applied to the input image, then it is converted to $I(x, y)$ with a size of (m, a).

Jhang *et al.* [13] proposed the PET-RENN technique and explained multiple approaches to the PET-RENN technique. The approaches are construction-based methods, learning-based methods, and interpolation-based methods. Basically learning-based method yields better accurate results. Deep learning-based PET-RENN techniques have been popularized in recent times. Insight these techniques, multiple convolution neural networks are developed to accept the lower resolution of input image sources. After that, these convolution layers convert lower-resolution images to higher-resolution images.

$$(x, y) \rightarrow I(x, y) \tag{8.1}$$

8.3 Multichannel bidimensional EMD through a morphological filter

As far as we know among the BEMD methods currently in use, the EMD approach [8] has the fastest time approach for a greyscale image's decomposition. It employs the envelopes for estimating approach depending on statistics filters. Instead of computing the distance between neighbor maxima/minima, EMD uses the average maxima distance as the filter size [14]. However, it is only intended to interpret. Fringe patterns for single-channel images. In this study, we provide an improved fast empirical mode decomposition method (EFF-EMD) modification-based multi-channel bidimensional EMD method (MF-MBEMD). Here the MF-MBEMD produces the envelope surfaces of a multi-channel (MC) image. This allows for the decomposition of each channel image to extract information with a similar spatial scale. The upper (lower) envelope $V(V_1,...,V_n)$, $(E\ (E_1,...,E_n))$ of a multi-channel picture $J(J_1,..., J_n)$ with the size S*H can be created.

$$V(a,b)|_{k=1,\ldots,n} = (Ik \oplus s)(a,b) = \max\ \mathrm{J_k}(c,d), \tag{8.2}$$

$$V(a,\ b)|_{k\ =\ 1,\ldots,n} = (Ik \oplus s)(a,b) = \max\ \mathrm{J_k}(c,d), \tag{8.3}$$

Here, the morphological expansion filter is represented by and the morphological corrosion filter is represented by g. The pixels in the window are represented by t*t centered on the pixel $(a,\ b)$ that signifies Z_{ab}, $(c,\ d)$ signifies Z_{cd} and s indicated by the binary indicator function on Zab. By using the average filter. envelopes can be made smoother. The architecture can be roughly sketched.

$$v'k(a,b)|k = 1,\ldots.n = \frac{1}{t*t} * vk(c,d) \tag{8.4}$$

For input images, we apply s as the size of the window in (8.2) and equipment (8.3) to look at the data channels' feature extraction. Therefore the minimal extreme distance of images are

$$s = \min\{s1,\ldots.,sn\}, \tag{8.5}$$

Here $S_k(k{=}1,\ \ldots,\ n)$ indicates the kth channels average extremum distance J_k, and it is calculated by

$$s = \surd\left(S * \frac{H}{Nk}\right) \tag{8.6}$$

Here, N_k stands for the average of all of J_k is local minima and maxima. In order to compute all local maxima's and minima's of J_k, we utilize 3×3 window which contains values of pixels. It differs from the enhanced rapid EMD approach [8], where the number of extracted maxima is increased and the extremum window's dimensions are the same as the standard deviation of the extremum from the preceding iteration. As a result, our method can extract considerably fined feature scales from each channel image and obtains more extremes with each iteration.

8.4 Proposed method

In this section, we discuss the proposed EMD fusion of MRI-PET images using deep networks with the help of a block diagram, as outlined in Figure 8.3.

8.4.1 EMD

Let $M(x, y)$ be the input image of MRI and $G(x, y)$ be the input image of PET. But PET images are suffered from low resolution. Hence, we apply the PETRNN technique to recover the high resolution of PET images from a lower resolution. We obtained the image $I(x, y)$. When we apply the EMD technique to

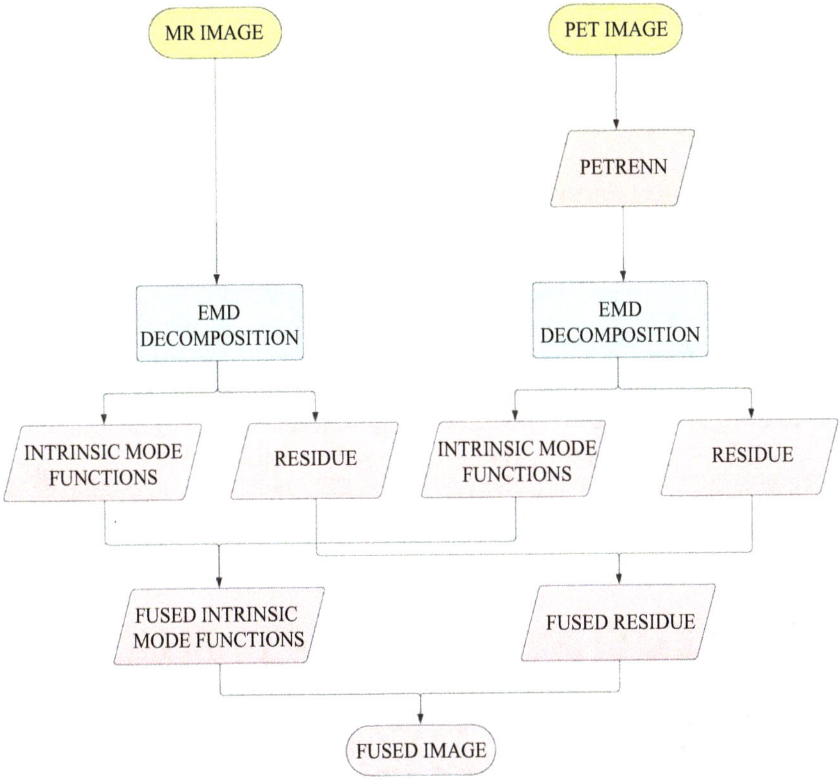

Figure 8.3 Block diagram of the proposed method

the input images $M(x, y)$, $I(x, y)$ then the input images splits into IMFs and a residual component.

$$EM(I) \rightarrow [I_{IMF1} I_{IMF2} \quad \cdots \quad \cdots \quad \cdots \quad \ldots I_{residue}] \tag{8.7}$$

$$EM(M) \rightarrow [M_{IMF1} M_{IMF2} \quad \cdots \quad \cdots \quad \cdots \quad M_{residue}] \tag{8.8}$$

8.4.2 Fusion rule

By applying the fusion technique to the *IM* functions and residue of *I* and *IM* functions and residues of *M*, the fusion technique gives the fused intrinsic mode functions and fused residue.

By applying image fusion to the fused *IM* functions and fused residue, we obtain the fused image.

$$I_{IMF1} + M_{IMF1} \rightarrow F_{IMF1} \tag{8.9}$$

$$I_{IMF2} + M_{IMF2} \rightarrow F_{IMF2} \tag{8.10}$$

$$M_{residue} + I_{residue} \rightarrow F_{residue} \tag{8.11}$$

By applying image fusion to the fused *IM* functions and fused residue, we obtain fused image.

$$F_{IMF1} + F_{IMF2} + \ldots\ldots + F_{residue} \rightarrow FUSED\ \ IMAGE \tag{8.12}$$

8.5 Experiments and results

In this chapter, we performed fusion on various data sets and calculated metrics like peak signal-to-noise ratio (PSNR), structural similarity index (SSIM), and mean square error (MSE) for estimating the performance of an algorithm, and then we discussed how framework are selected to our technique. After getting results, we compared the efficiency of an employed algorithm with the other techniques like coupled deep learning method (CDL) and coupled featured learning method (CFL). All this process and output images which were shown are performed by using MATLAB® 2021 on a laptop with Intel Core i4 CPU and 4.0 GB RAM. Figure 8.4 shows the testing data sets of MRI-PET. The fused results of our method are shown in Figures 8.5–8.8.

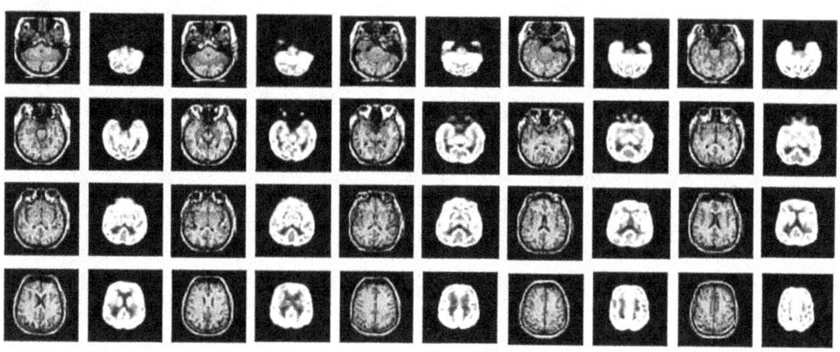

Figure 8.4 Testing datasets of MRI-PET images

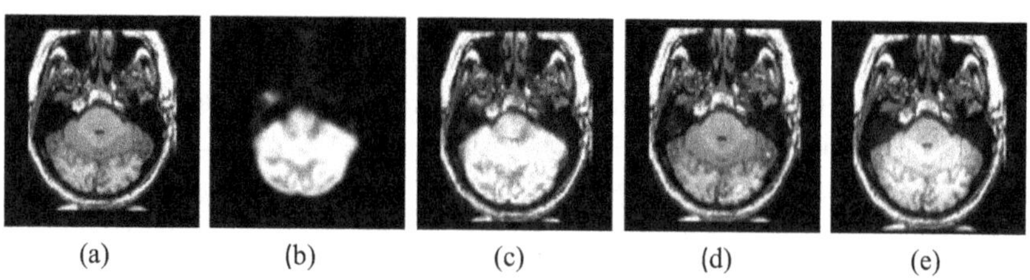

Figure 8.5 Fused results of various techniques on test set-1: (c), (d), and (e) images are fused images of (a) MRI and (b) PET. (c) is a couple dictionary learning fusion method image (CDL), (d) is coupled featured learning fusion method image (CFL), and (e) is our method.

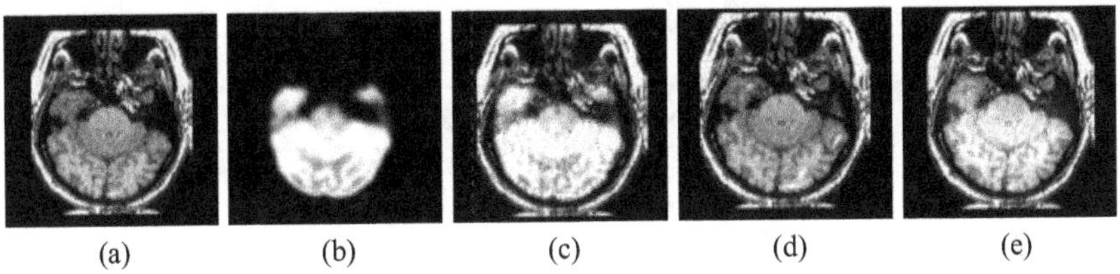

Figure 8.6 *Fused results of various techniques on test set-2: (c), (d), and (e) images are fused images of (a) MRI and (b) PET. (c) is a couple dictionary learning fusion method image (CDL), (d) is coupled featured learning fusion method image (CFL), and (e) is our method.*

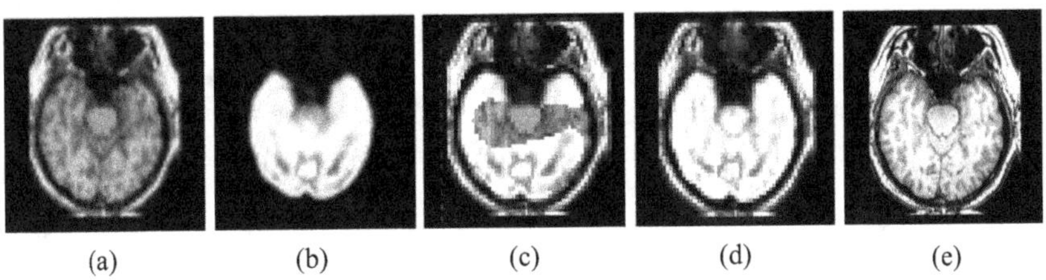

Figure 8.7 *Fused results of various techniques on test set-3: (c), (d), and (e) images are fused images of (a) MRI and (b) PET. (c) is a couple dictionary learning fusion method image (CDL), (d) is coupled featured learning fusion method image (CFL), and (e) is our method.*

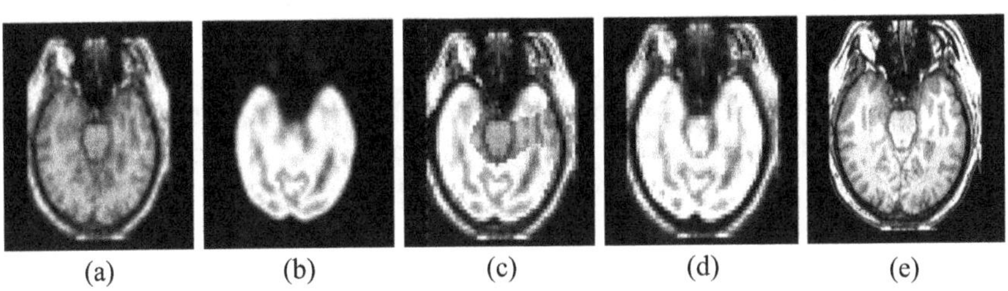

Figure 8.8 *Fused results of various techniques on test set-4: (c), (d), and (e) images are fused images of (a) MRI and (b) PET. (c) is couple dictionary learning fusion method image (CDL), (d) is coupled featured learning fusion method image (CFL), and (e) is our method.*

8.5.1 Objective metrics

Table 8.1 shows the model metrics of various techniques.

8.5.2 Selected specifications

To remove the useful information from the input images, we fuse all of the IMFs using the energy-based maximum selection criterion. For multi-focus images, the

Table 8.1 Calculation of objective metrics

Method	PSNR	SSIM	MSE
CDL	18.58	0.516	0.0419
CFL	15.70	0.514	0.0526
Proposed	19.83	0.519	0.0545

*K*th decomposition level MFMBEMD is set to 1 so that it can capture them effectively. For multi-modal images the *K* (decomposition level) is fixed into 2 by considering the useful information is focused at the top two IMFs of MFMBEMD.

Overlapping number of *N* rows/columns: the overlapping *N* number of rows and columns is higher than it provides fewer spatial artifacts in our method and it costs high to compute. In our trials, we used multi-modal data sets and multi-focus data sets, and we set the overlapping rows or columns number n to $M/6$ and $M-2$, respectively.

More number of trials signifies such selections can generate the best outcomes.

Division of *M* block size: To obtain good fused images we vary the *M* value from 1 to 50 from that we select the *M* value, which shows better performance using average values of three fusion metrics (Table 8.1). For multi-focus and multi-model data sets, we set the overlapping rows or columns number n to $M/6$ and $M-2$, respectively. In our trials, we choose $M = 31$ for grey scale multi-focus in order to get good outcomes.

8.6 Conclusion

We introduced a unique EMD-based image fusion method approach based on deep networks for generating superior fusion images. With multi-bidimensional EMD we generate multiple IM functions and a residual component from the input source images. This enables us to extract salient information from PET and MR images.

References

[1] Zhu, P., Liu, L., and Zhou, X. (2021). Infrared polarization and intensity image fusion based on bivariate BEMD and sparse representation. *Multimedia Tools and Applications*, 80(3), 4455–4471.

[2] Bevilacqua, M., Roumy, A., Guillemot, C., and Alberi-Morel, M. L. (2012). Low-complexity single-image super-resolution based on nonnegative neighbor embedding. In: *Proceedings British Machine Vision Conference*, pp. 135.1–135.10.

[3] Pan, J. and Tang, Y.Y. (2016). A mean approximation based bidimensional empirical mode decomposition with application to image fusion. *Digital Signal Processing*, 50, 61–71.

[4] Li, H., He, X., Tao, D., Tang, Y., and Wang, R. (2018). Joint medical image fusion, denoising and enhancement via discriminative low-rank sparse dictionaries learning. *Pattern Recognition*, 79, 130–146.

[5] Ronneberger, O., Fischer, P., and Brox, T. (2015, October). U-net: convolutional networks for biomedical image segmentation. In *International Conference on Medical Image Computing and Computer-assisted Intervention* (pp. 234–241). Springer, Cham.

[6] Daneshvar, S. and Ghassemian, H. (2010). MRI and PET image fusion by combining IHS and retina-inspired models. *Information Fusion*, 11(2), 114–123.

[7] Ma, J., Liang, P., Yu, W., *et al.* (2020). Infrared and visible image fusion via detail preserving adversarial learning. *Information Fusion*, 54, 85–98.

[8] Ardeshir Goshtasby, A. and Nikolov, S. (2007). Guest editorial: Image fusion: advances in the state of the art. *Information Fusion*, 8(2), 114–118.

[9] Ma, J., Ma, Y., and Li, C. (2019). Infrared and visible image fusion methods and applications: a survey. *Information Fusion*, 45, 153–178.

[10] Liu, Y., Liu, S., and Wang, Z. (2015). A general framework for image fusion based on multi-scale transform and sparse representation. *Information Fusion*, 24, 147–164.

[11] Li, H., Qi, X., and Xie, W. (2020). Fast infrared and visible image fusion with structural decomposition. *Knowledge-Based Systems*, 204, 106182.

[12] Yeh, M.H. (2012). The complex bidimensional empirical mode decomposition. *Signal Process* 92(2), 523–541.

[13] Zhang, J., Chen, D., Liang, J., *et al.* (2014). Incorporating MRI structural information into bioluminescence tomography: system, heterogeneous reconstruction and in vivo quantification. *Biomedical Optics Express*, 5(6), 1861–1876.

[14] Zhang, Y., Brady, M., and Smith, S. (2001). Segmentation of brain MR images through a hidden Markov random field model and the expectation-maximization algorithm. *IEEE Transactions on Medical Imaging*, 20(1), 45–57.

Chapter 9

A convolutional neural network for scoring of sleep stages from raw single-channel EEG signals

A. Ravi Raja[1], Sri Tellakula Ramya[1], M. Rajalakshmi[2] and Duddukuru Sai Lokesh[1]

Sleep disorders have increased rapidly and expeditious development in computer technologies in modern society. Inadequate quality of sleep will lead to many neurological diseases. One of the symptoms of many neurological diseases is a sleep disorder. Obtaining polysomnogram (PSG) signals using traditional methods and manual scoring is time-consuming. Automated sleep pattern monitoring can facilitate the reliable detection of sleep-related disorders. This research paper develops a deep-learning model for automated scoring of sleep stages from single-channel EEG using a one-dimensional convolution neural network (CNN).

The CNN is utilized on an electroencephalogram signal in its raw form to introduce a supervised model that predicts five classes of sleep stages. The input of the network layer is a 30-s epoch, which is further classified into two epochs. Our model is trained and evaluated based on data received from the Sleep Heart Health Study (SHHS). SHHS dataset includes polysomnography records of both healthy and unhealthy persons. The proposed model obtained performance metrics with accuracy and a kappa coefficient of 0.87 and 0.81. Class-wise sleep patterns are also visualized by using the patterns extracted from the network.

9.1 Introduction

For good health, sleep is a crucial factor in everyone's life. A complex biological process in which the condition of both body and mind are inactive or state of irresponsiveness is termed as sleep. Healthy sleep improves human health in a physical way and makes the person more stable in their respective mental states or mental processes. However, nowadays, many large portions of the population are unable to sleep regularly. Improper quality of sleep will weaken your body and give

[1]ECE Department, V R Siddhartha Engineering College, India
[2]Department of Mechatronics Engineering, Thiagarajar College of Engineering, India

rise to various sleep disorders like sleep apnea, hypersomnia, Restless-leg-syndrome, and other breathing-related disorders.

The mental illness from depression, anxiety, work stress, overthinking, some health-related problems, and nerve disorders. Such conditions are the origin of sleep disorders. The mechanism used for diagnosing a person's sleep pattern and preventing sleep disorders is polysomnography. A polysomnogram is a system that consists of physiological signals like an electromyogram (EMG), electrooculogram (EOG), electroencephalogram (EEG), electrocardiogram (ECG), and other environmental signals. These signals are used for monitoring the sleep patterns of an individual in person.

In this modern-day, in many countries' sleep difficulties affect many people. From research by Sahmitha Panda [1], roughly 42.6% of individuals in South India experience sleep disorders. In Canada, around 4–7 million people suffer from sleep disorders. American Sleep Association research [2] tells us that adults between 50 and 70 million in the US country have a sleep disorder. Therefore, many methods came into existence to identify and analyze sleep-related disorders by diagnosing their sleep patterns. Monitoring the sleep and then evaluating sleep patterns is essential to identify sleep disorders. To diagnose a sleep problem, we must analyze an individual's normal sleep quality using the polysomnogram approach. The recorded polysomnogram signals of various subjects will help us in identifying whether the subject is healthy or not. One of the key steps in ruling out sleep disorders is to classify the sleep stages of subject signals which are recorded.

The extraction of sleep stages is carried out traditionally in the presence of field experts in respective polysomnogram eras. The manual scoring of signal recordings will be subject to human errors, and it also consumes more time than the automated scoring of sleep. The main advantage of automated sleep scoring is they can automatically record sleep scoring without the need for any field expert. So automated identification and classification methods are introduced in order to drastically reduce time and produce dependable outcomes. Field experts study the various time series records of different subjects, and each time segment should be assigned to a sleep stage. The assignment of time is given according to standardized rules of classification [3] "Rechtschaffen–Kales" (R&K) rules and also guidelines given by "American-Academy-of-Sleep-Medicine (AASM)."

The polysomnography record is segmented into two epochs that follow each other, 20–30 s for each epoch. The dividing of recorded signals into epochs is defined as sleep staging. This sleep staging process can be performed on a particular channel subset or the entire polysomnogram along with a suitable classifying algorithm. Hypnogram is a graph that represents the successive sleep stages of an individual over a particular period or during night sleep. The hypnogram is simple in representation, and it is advantageous in identifying and diagnosing sleep disorders. For sleep staging, only EEG—single channel is used for this study.

9.2 Background study

By using EEG signals, many researchers created an automatic sleep-stage scoring using two different step methodologies. The first step is to extract different features like features of the time-domain, non-linear domain features, and features of frequency-domain from waveforms [4]. The second step is to classify the trained data from extracted features. For the detection ad extraction of sleep stages, some classifier methods such as Decision Trees [5], support vector machine (SVM) [6], random forest classifier, and CNN [7] are used for better results. Shen-Fu Liang *et al.* [8] used multiscale entropy (MSE) in a combination of autoregressive features models for sleep stage scoring. The single-channel "C3-A2" of EEG signal is used for evaluating sleep score, and around 8,480 each 30-s epoch is considered for evaluation of performance. This method obtained a sensitivity of 76.9% and a kappa coefficient of 0.65.

Guohun Zhu *et al.* [9] used different visibility-graph approaches for feature extraction and also for the classification of sleep stages is done using a SVM. Mean degrees on the visibility-graph (VG) along with the horizontal-visibility graph (HVG) are analyzed for sleep stage classification. Luay Fraiwan *et al.* [10] used a time-frequency approach and employed entropy measurement method (EMM) for feature extraction. The features were extracted from EEG signals which are frequency domains represented using Renyi's entropy method. This method showed performance accuracy and kappa coefficient as 0.83 and 0.76, respectively.

Hassan *et al.* [11] used the ensemble empirical mode decomposition method for feature extraction. Decision Trees, along with bootstrap aggregating, are employed to classify the sleep stage. This study detected the highest accuracy only for two stages (Sleep stage S1 and rapid eye movement (REM)). Hassan *et al.* [12] proposed a method to extract spectral features by decomposing EEG signals into segments and employee tunable Q-factor wavelet transform (TQFWT). Using the random forest classifier method, it reported a performance accuracy of around 90.3%. Sharma *et al.* [13] applied discrete energy separation methodology for instantaneous frequency response and iterative filtering on single-channel EEG signals. This author used many classifiers for comparison and reported the highest accuracy of existing classification methods. Hsu *et al.* [14] used the classifier method, recurrent neural methodology for classification based on energy features extracted.

Most of the studies reported the use of neural networks classification methods. These methods obtain trained data, and this data can be used for feature extraction as well as classification. Tsinalis [15] proposed a CNN method for classification. Supratak [16] implemented a convolutional-neural-network along with a bi-directional long-short-term memory network (Bi-LTSM). In this study, we introduced a supervised deep-learning method for categorizing sleep stages that used only one EEG channel as input. Convolutional neural networks are also used in other domains to produce reliable results. The other domains include image recognition [17], natural language processing [18], and other pattern recognition.

Recently, various applications adopt a convolutional neural method for Brain-computer interface [19], Seizure detection [20], evaluating cognitive performance [21], motor imaging [22], and evaluating sleep stages. This chapter aims to report that CNN methods are applicable and suitable to produce relative sleep-scoring performance using a dataset. The proposed method trains the data for feature extraction. Later the trained data is evaluated using classification, and performance parameters are applied to a dataset.

9.3 Methodology

9.3.1 Sleep dataset

A multicenter cohort research dataset called SHHS [23] is used in this proposed work. "American-National-Heart-Lung and Blood-Institute" initiated this dataset study to determine cardiovascular diseases which are associated with breathing. In this chapter, the dataset study consists of two different records of polysomnography signals. The first polysomnographic record, SHHS-1 is only used in this proposed work because it consists of signals which are sampled at 125–128 Hz frequency. SHHS-1 dataset includes around 5,800 polysomnographic records of all patients. The polysomnographic record includes various channels such as C4-A1, and C3-A2 EEG channels (two EEG channels), 1-ECG channel, 1-EMG channel, 2-EEG channels, and other plethysmography channels.

These polysomnographic records are manually scored by field specialists relying on Rechtschaffen-Kales (R&K) rules.

Each record in this dataset was scored manually per 30 s epoch for sleep stages. They are several sleep stages according to R&K rules, such as Wake-stage, N1-stage, N2-stage, N3-stage, and N4-stage, which is also referred to as non-REM sleep stage and REM sleep stage. Detailed information about sleep scoring manually is provided in this chapter [24].

9.3.2 Preprocessing

A significant "wake" phase, first before the patient falls asleep and the other after he or she awakes, is recorded in most polysomnographic data. These waking periods are shortened in length, such as the number of epochs before and after sleep does not exceed that of most commonly represented other sleep stage classes, Because the accessible EEG signal is symmetrical, such that those signals produce equivalent results. The EEG channel named C4-A1 is used in the following proposed work. Stages N4 and N3 are consolidated into a single sleep-stage N3, as indicated in the AASM guidelines [25]. Even though they may be anomalies, some patients who have no epoch associated with a particular sleep stage are omitted. Table 9.1 shows the summary of the number of epochs (and their proportional

Table 9.1 Summary of the dataset SHHS-1 as per class

Sleep stage	**Total epoch**	**Total number of equivalent days**
Wake	1,514,280	525
N1	201,4312,169,452	70,753
N2	2,169,452	753
N3	719,690	250
REM	779,548	271
Total	5,384,401	1,871

importance) of every stage and total epochs. Classes are extremely uneven, as per the PSG study. Stage N1 has a deficient representation. The EEG readings are not preprocessed in any manner.

9.3.3 CNN classifier architecture

A complete CNN comprises several layers of convolution, more than one fully connected convolutional layer, softmax, flattens, and dropout layers that produce output probability for each class. Figure 9.1 depicts the convolutional layer architecture for one-dimensional (1-D) input signals. Every layer L adds biases B^L to the subset $Y^{(L-1)}$ feature map inputs by convolving them with a collection of trainable kernels (also termed as filters) W^L.W^L has shaped-{K^L, $N^{(L-1)}$, N^L}. N^{L-1} is the number of feature map inputs, and N^L is the number of feature map outputs. Kernel width is given by K^L. Because input had only 1 channel and N (0) equals 1. Consider $\mathbf{w}^{\mathbf{L}}{}_{\mathbf{ij}}$ represents the slices of W^L which extend from given input-feature maps "*i*" to output-feature maps "*j*." $Y^L{}_j$ signifies the *j*th feature map within Y^L. As a function of this expression, it is given in (9.1)

$$Y_j^L = \sigma * g_{P(L)}\left(\sum\nolimits_{i=1}^{N^{L-1}} Y_i^{L-1} * w_{ij}^L + B_j^L\right) \quad (9.1)$$

Here g is the $P^{(L)}$ strode convolutional sub-sampling operation, and σ is a non-linear activation function that is given element-by-element. *Is the one-dimensional convolutional operator. Now, the CNN structure, which is illustrated in Figure 9.2, is discussed. They are around 4,000 samples in 30 s at 125 Hz frequency. The unfiltered EEG signals of such epochs to be categorized are combined with the samples of two subsequent and consecutive epochs as input to the network. These subsequent and succeeding epochs were incorporated to improve scoring procedures, which sometimes refer to previous and subsequent epochs when the

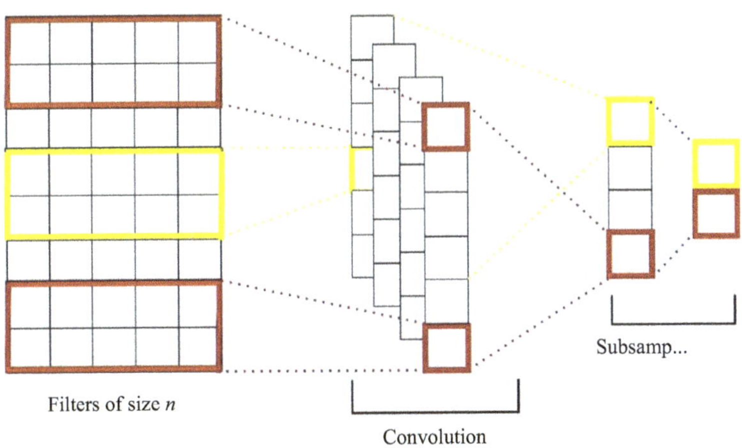

Figure 9.1 Architecture of one-dimensional convolutional-layer

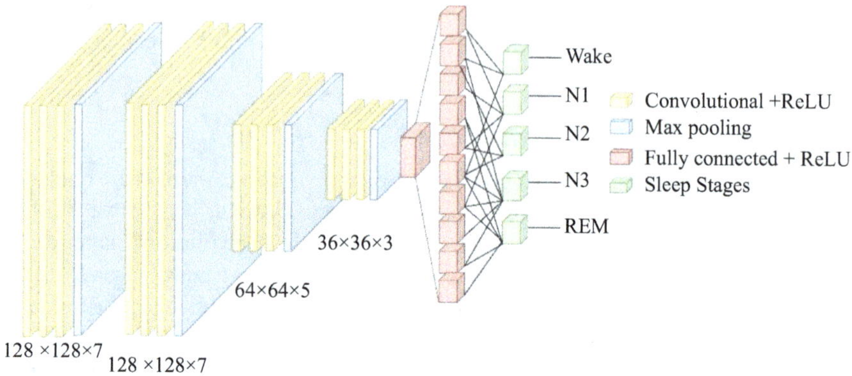

Figure 9.2 Detailed architecture of proposed 1D CNN

current epoch creates a margin for uncertainty. As an instance, we used a collection of four epochs. Because we use all feasible cases, some of them overlap.

There is no feature extraction. We deploy 12 convolutional layers, fully connected layers with a filter size of 256, and the fully-connected layer with a filter size of 5 with non-linear soft-max activation commonly referred to as multinomial logistic regression. Except for the last layer, the activation linear function is a leaky rectified-linear activation unit [25] with a negative slope equal to 0.1. Figure 9.2 shows an overview of system architecture. The size of the convolutional part output is precisely correlated to the size of inputs, number of convolutional layers, and respective strides when implementing a CNN model on a defined time series. Most weights will be in fully connected layers if the outcome across the last convolutional layer becomes too large. We deployed 6–15 layers and 2–5 strides throughout this study. We further tested filters of sizes 3, 7, and 5 and decided on size 7, even though there was minimal variation in performance between sizes 7 and 5. We experimented with various features and decided to continue with a feature map of size 128 for the first six layers and 256 for the last six layers. Moreover, we tested with various number of preceding epochs ranging from 1 to 5 and discovered two prior epochs are a workable approach for the proposed architecture.

9.3.4 Optimization

As a cost function, multiclass cross-entropy was performed, and minimum batch size training was used for optimizing the parameters of the weights and biases. Consider w be all the trainable parameters where **s** signifies the size of the minimum batch of the training sample. Take $ß = \{\mathbf{y_k}^{(0)}, k \in [[1, s]]\}$ is a minimum batch of training samples, with $\{m_k, k \in [[1, B]]\}$ representing one-hot-encoded target class and the range $\{ x_k, k \in [[1,s]] \}$ representing the output of networks connected with the $y_k^{(0)}$ in ß. The minibatch cost C expression is written in (9.2).

$$C(w, ß) = -\sum_{k=1}^{s} m_k^T \log x_k(w) \tag{9.2}$$

Minimizing cross-entropy by using the softmax linear function relates to optimizing the log-likelihood of such a predicted model class as equal to an actual class. A gradient is traditionally calculated via error-back propagation. Adam [26], a technique for optimizing based on a first-order gradient that leverages estimations of low-order features, is used for optimization. Moreover, with a decent dataset like SHHS-1, the entire training data does not really exist in a standard workstation's memory. Therefore, data should be uploaded from inference (or disc) throughout training.

To ensure gradient continuity, randomization should be included in the training data streaming process. However, keeping all training samples in separate files in order to rearrange them is just time-consuming. We took the middle-ground approach and used 50 monitoring channels to input the data from different patients in a random sequence. Following that, a batching queue shuffles and organizes training samples to form a minibatch of a particular size $s = 128$.

9.4 Criteria for evaluation

The database taken is divided into three main parts: training, testing, and validation, with proportions varying from 0.5, 0.3, and 0.2, respectively. The cost of validation is recorded throughout training, and for every twenty thousand (20,000) training sample batches, a single run on the validation set is performed. For testing, the model with the least validation cost is used. Confusion matrix, Cohen's Kappa, classification accuracy, and F1-score are among the evaluation criteria inherited to assess the model's performance. Cohen's Kappa evaluates the agreement between both the field expert and classifier methodology. Cohen's Kappa also corrects the chance agreement that occurred.

$$\kappa = \frac{p_o - p_e}{1 - p_e} \tag{9.3}$$

The observed agreement ratio is given by p_o, while the chance-agreement probability is p_e. A multiclass F1 score is a weighted average of individual classes F1 scores, for (9.3). The weighting of the macro F1 score is uniform. Metrics are determined generally for micro F1 score by determining the total number of true-positive (TP), false-positives (FP), and true-negatives (TN). A positive predicted value (PPV), called precision, and the true-positive rate (TPR), termed recall, is represented by an individual class's F1 score as in (9.4) and (9.5)

$$\text{F1 score} = 2 \cdot \frac{\text{PVV} * \text{TPR}}{\text{PPV} + \text{TPR}} \tag{9.4}$$

$$\text{where PVV} = \frac{\text{TP}}{\text{TP} + \text{FP}} \text{ and TRP} = \frac{\text{TP}}{\text{TP} + \text{FN}} \tag{9.5}$$

Sensitivity and specificity are commonly reported by many medical researchers. We also included precision because specificity is not particularly useful in a multiclass environment. The total is a weighted macro average over classes, and these metrics are presented per class altogether. Additionally, we studied how to visualize our trained neural network and learned about sleep phases during the classification process. There are many different ways to visualize neural networks that have been learned [27,28].

9.5 Training algorithm

The training procedure is a strategy we developed to efficiently train the proposed model end-to-end using back-propagation by avoiding the problem of class imbalance (in other words, training to categorize the majority of sleeping phases) that can occur while working with a significant sleep database. The approaches pre-train this model's representation by learning components before fine-tuning the entire model using two distinct rates of learning. Our model is trained to output probability for two distinct classes using the softmax layer function.

9.5.1 Pre-training

The initial step in training is to monitor the model's representation learning-based section's pre-training with the training set to ensure that the proposed model doesn't over-adapt the many sleeping phases. The two CNNs are retrieved from the proposed model and layered with a softmax layer. This stacked softmax layer is deployed to pre-train the two CNNs in this stage, and its parameters are deleted after the pre-training is accomplished. The softmax layer is eliminated after the conclusion of the pre-training. The training set of class balance is produced by replicating minority stages of sleep in the actual training dataset until every stage of sleep has the same amount of data (in other words oversampling).

9.5.2 Supervised fine-tuning

The second stage is implemented to use a sequential training data set to do supervised fine-tuning on the entire model. This phase includes both implementing the rules of state transition into the proposed model and making appropriate modifications to the pre-trained model. The pre-trained network model parameter was overly tuned to the time series sample, which is unbalanced in terms of class when we utilized the exact training set to fine-tune the entire network model. As a result, at the conclusion of the fine-tuning, the model began to over-fit most of the sleep stage phases. The training set, which is sequential, is generated by organizing the actual training set chronologically among all individual subjects.

9.5.3 Regularization

Further, to avoid overfitting issues, we used two regularization strategies. The dropout layer [29,30] is a method that periodically sets the values of input to zero (i.e., units of the dropout layer along with their connections) with a predefined probability overtraining period. As illustrated in Figure 9.2, the dropout layers with 0.5 probabilities were applied across the model. This dropout layer was only needed for the purpose of training, and it was deleted from the network during the testing period so that consistent output could be produced.

TensorFlow, a deep learning toolkit based on Google TensorFlow libraries [31], was used to build the proposed model. This library enables us to distribute computational techniques over several CPUs, such as validation and training activities. It takes around two days to train the entire model. At a rate of about 30 epochs per second, interference is performed.

9.6 Results

Table 9.2 shows the confusion-matrix derived from the test dataset. Table 9.3 displays the recall, F1-score, and precision for multiclass and graphical representation is represented in Figure 9.3. The sleep stage N1 is the most misclassified, with only 25% of valid classifications. With 93% of valid classifications, sleep stage Wake was the most accurately classified sleep stage.

N2, REM, and N3 are the following sleep stages, with 89%, 87%, and 80%, respectively. The accuracy of overall multiclass classification is 87%, with a kappa-coefficient of 0.81.

Sleepstage N1 is never nearly confused with N3 and is frequently confused with sleep stage N1 (25%), sleep stage N2 (37%), and sleep stage REM (24%). Sleep stage REM is sometimes (4%) mistaken with sleep stage-N3 and rarely with the other sleep stages. Sleep stage-N2, on the other hand, is frequently confused with sleep stage-N3 (21%) and nearly never with other sleep stages.

Table 9.2 Confusion matrix analysis on the test data

Sleep stage	Wake stage	N1 stage	N2 stage	N3 stage	REM stage
Wake stage	91%	51%	13%	46%	24%
N1 stage	15%	25%	37%	24%	24%
N2 stage	47%	25%	89%	21%	61%
N3 stage	39%	0%	22%	78%	78%
REM stage	27%	7%	11%	4%	8%

Table 9.3 The performance of metrics evaluated on the dataset

Sleep stage	Precision	Recall	F1 score	Support
Wake stage	0.93	0.91	0.92	6,483
N1 stage	0.44	0.25	0.31	2,117
N2 stage	0.81	0.89	0.85	6,846
N3 stage	0.87	0.78	0.82	1,287
REM stage	0.85	0.80	0.72	2,530
Total	0.86	0.87	0.80	19,263

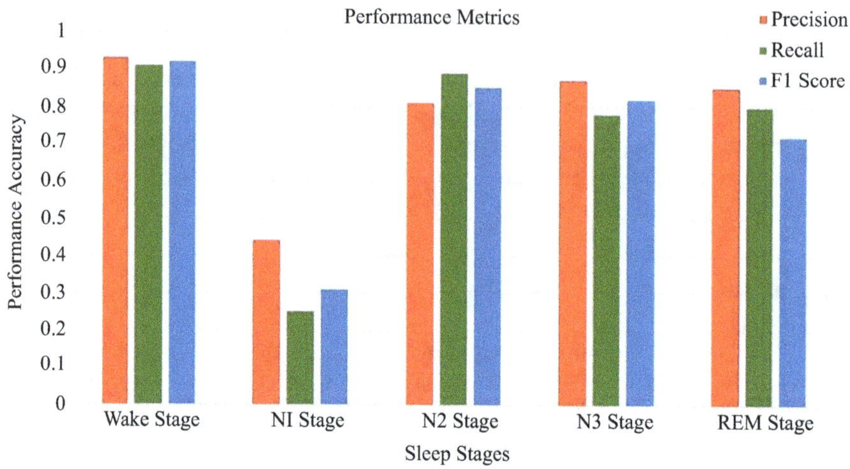

Figure 9.3 Graphical representation of performance metrics

9.7 Discussion

9.7.1 Major findings

This research shows that utilizing a single EEG channel and CNN trained on raw samples makes it feasible to categorize the sleep phases with their performance metrics comparable to other approaches. The training of data is completed from beginning to end, without the need for any specialist expertise in the selection of features or preprocessing of the signal. This is beneficial because the model may learn the features which are most appropriate for task classification. We considered applying a bandpass FIR filter to preprocess the signals, but it probably does not help because the convolution layer could be capable of learning appropriate filters. One more benefit is that this methodology is easy to adapt to different applications or mediums.

Although training a giant CNN is significantly more challenging, the inference is relatively inexpensive and may be performed on a portable device or a home computer once the model has been trained. When it comes to the kind of errors that the model produces, we have seen that they generally correspond to sleep phases that are close together. N3 is frequently confused with sleep stage N2 but nearly never with sleep stage N1. Likewise, while sleep stage N1 is characterized as a stage with the least inter-human agreement, it might be mistaken as REM, N2, or as Wake, all of which contain patterns comparable to sleep stage N1 but nearly never with sleep stage N3.

Lastly, REM is more likely to be confused with sleep stage N2 than sleep stage Wake. One probable explanation is that eye movement is a significant commonality between Wake and REM, yet the EEG channel C4-A1 generation leaks relatively little frontal activity of eye movement.

9.7.2 The problem of class imbalance

Our dataset, like any other sleep-scoring dataset, has a severally unbalanced distribution of class. We tried using oversampling to adjust this. Despite the fact that sleep stage N1 and N3 measures are significantly improved, overall performance as measured by Cohen's Kappa was not really. The above-mentioned results are based on a standard price and sampling techniques. More study is required to solve the imbalance of classes in classification. It is possible that ensemble deep learning [32] or specific CNN approaches [33] will be useful.

9.7.3 Comparison

Table 9.4 summarizes the performance metrics and characteristics found in recent signal channel EEG sleep scoring studies. It is difficult to compare studies in the sleep scoring research since they do not all apply the same database, scoring methods, or number of patients, and they do not all balance the classes in the same manner. The number of hours after and before the night of wake epochs is retained in the PhysioNet Sleep-edfx database [34]. The wake-sleep stage has a substantially bigger number of epochs than the other phase of sleep. Some researchers [35] reduce the number of wake epochs, whereas others include all wake epochs in the evaluation of their performance metrics, which disproportionately benefits the conclusion. To compare various studies objectively, we start with reported confusion matrices, and if the Wake sleep stage is the most popular class, then we adjust it to make as the second most popular class in sleep stages. Sleep stages N4 and N3 are combined into a single sleep stage N3, in one study [36] when only a confusion matrix of 6-class is provided.

Table 9.4 also lists some of the additional study features, including the channel of the EEG signal, the database used, sleep scoring rules, and the methodology used in their study. Although the expanded Sleep-edfx has been long accessible, several recent types of research employ the sleep-edfx database. We got improved results on the sleep-edfx database, which was unexpected. This is because human raters are not flawless, and fewer technicians scored Sleep-EDF than expanded Sleep-edfx, and methodologies evaluating Sleep-EDF can quickly learn the rater's classification technique. Our algorithm, on the other hand, is examined on 1,700 records at test time scored by various approaches. This ensures that the system does not over-dependent on a small group of professionals' rating styles. The study by Arnaud [37] provided support for our proposed study.

This approach demonstrated that this method is comparative in the evaluation of performance and that the network has been trained to detect significant observed patterns. A method for sleep-scoring is desirable as it enables the system to be light. Implementing multichannel CNN models perform greater than one channel, which also offers new possibilities. Our finding revealed that our proposed model was capable of learning features of the model for scoring sleep stages from various raw single-channel EEGs without modifying the model's algorithm for training and

Table 9.4 Summary of performance criteria of various methods using single-channel EEG signal

Reference	Database	Signal used	Rules used	Model	Performance accuracy	Kappa coefficient	F1 score
Tsinalis	Sleep-EDF	Fpz-Cz	R&K	Convolutional neural network	0.75	0.65	0.75
Fraiwan	Custom	C3-A1	AASM	RandomForest classifier	0.83	0.77	0.83
Hassan	Sleep-EDF	Pz-Oz	R&K	Empirical mode decomposition	0.83	0.76	0.83
Zhu	Sleep-EDF	Pz-Oz	R&K	Support vector machine	0.85	0.79	0.85
Suprtak	MASS	Fpz-Cz	AASM	CNN-LTSM	0.86	0.80	0.86
Hassan	Sleep-EDF	Pz-Oz	R&K	EMD-bootstrap	0.86	0.82	0.87
Proposed work	SHHS-1	C4-A1	AASM	Convolutional neural network	0.89	0.81	0.87

model architecture. In the future, to improve classification accuracy, convolutional architectures such as residual connection [38] and separable convolutions depth-wise [39] with multichannel datasets are proposed to develop.

Author contributions

All authors contributed equally.

References

[1] S. Panda, A.B. Taly, S. Sinha, G. Gururaj, N. Girish, and D. Nagaraja, "Sleep-related disorders among a healthy population in South India", *Neurol. India*, 60(1), 68–74, 2012.

[2] "American Sleep Association Research, Sleep and Sleep disorder statistics", https://www.sleepassociation.org/about-sleep/sleep-statistics

[3] K-Rechtschaffen, *A Manual of Standardized Terminology Techniques and Scoring System for Sleep Stages of Human Subjects*, Washington, DC: Public Health Service, US Government Printing Office, 1971.

[4] M. Radha, G. Garcia-Molina, M. Poel, and G. Tononi, "Comparison of feature and classifier algorithms for online automatic sleep staging based on a single EEG signal", in: *Annual International Conference of the IEEE Engineering in Medicine and Biology Society*, 2014.

[5] L. Fraiwan, K. Lweesy, N. Khasawneh, H. Wenz, and H. Dickhaus, "Automated sleep stage identification system based on time-frequency analysis of a single EEG channel and random forest classifier", *Comput. Methods Progr. Biomed.*, 108, 10–19, 2012.

[6] D.D. Koley, "An ensemble system for automatic sleep stage classification using single channel EEG signal", *Comput. Biol. Med*, 42, 1186–1195, 2012.

[7] O. Tsinalis, P.M. Matthews, and Y. Guo, "Automatic sleep stage scoring using time-frequency analysis and stacked sparse autoencoders", *Ann. Biomed. Eng.*, 44, 1587–1597, 2016.

[8] S.-F. Liang, C.-E. Kuo, Y.-H. Hu, Y.-H. Pan, and Y.-H. Wang, "Automatic stage scoring of single-channel sleep EEG by using multiscale entropy and autoregressive models", *IEEE Trans. Instrum. Meas.*, 61(6), 1649–1657, 2012.

[9] G. Zhu, Y. Li, and P.P. Wen, "Analysis and classification of sleep stages based on difference visibility graphs from a single-channel EEG signal", *IEEE J. Biomed. Health Inf.* 18(6), 1813–1821, 2014.

[10] L. Fraiwan, K. Lweesy, N. Khasawneh, H. Wenz, and H. Dickhaus, "Automated sleep stage identification system based on time-frequency transform and spectral features", *J. Neurosci. Methods*, 271, 107–118, 2016.

[11] A.R. Hassan and M.I.H. Bhuiyan, "Computer-aided sleep staging using complete ensemble empirical mode decomposition with adaptive noise and bootstrap aggregating", *Biomed. Signal Process. Control*, 24, 1–10, 2016.

[12] A.R. Hassan and M.I.H. Bhuiyan, "A decision support system for automatic sleep staging from EEG signals using tunable Q-factor wavelet transform and spectral features", *J. Neurosci. Methods*, 271, 107–118, 2016.
[13] R. Sharma, R.B. Pachori, and A. Upadhyay, "Automatic sleep stages classification based on iterative filtering of electroencephalogram signals", *Neural Comput. Appl.*, 28, 1–20, 2017.
[14] Y.-L. Hsu, Y.-T. Yang, J.-S. Wang, and C.-Y. Hsu, "Automatic sleep stage recurrent neural classifier using energy features of EEG signals", *Neurocomputing*, 104, 05–114, 2013.
[15] O. Tsinalis, P.M. Matthews, Y. Guo, and S. Zafeiriou, "Automatic sleep stage scoring with single-channel EEG using convolutional neural networks", 2016, arXivpree-prints.
[16] A. Supratak, H. Dong, C. Wu, and Y. Guo, "DeepSleepNet: a model for automatic sleep stage scoring based on raw single-channel EEG", 2017, arXiv preprint arXiv:1703.04046.
[17] A. Krizhevsky, I. Sutskever, and G.E. Hinton, "ImageNet classification with deep convolutional neural networks", *Adv. Neural Inf. Process. Syst.*, 1, 1097–1105, 2012.
[18] R. Collobert and J. Weston, "A unified architecture for natural language processing: deep neural networks with multitask learning", in: *Proceedings of the 25th International Conference on Machine Learning*, ICML, ACM, New York, NY, USA, 2008.
[19] H. Cecotti and A. Graser, "Convolutional neural networks for p300 detection with application to brain–computer interfaces", *IEEE Trans. Pattern Anal. Mach. Intell.*, 33(3), 433–445, 2011.
[20] M. Hajinoroozi, Z. Mao, and Y. Huang, "Prediction of driver's drowsy and alert states from EEG signals with deep learning", in: *IEEE 6th International Workshop on Computational Advances in Multi-Sensor Adaptive Processing (CAMSAP)*, IEEE, pp. 493–496, 2015.
[21] A. Page, C. Shea, and T. Mohsenin, "Wearable seizure detection using convolutional neural networks with transfer learning", in: *IEEE International Symposium on Circuits and Systems (ISCAS)*, IEEE, pp. 1086–1089, 2016.
[22] Z. Tang, C. Li, and S. Sun, "Single-trial EEG classification of motor imagery using deep convolutional neural networks", *Optik* 130, 11–18, 2017.
[23] S.F. Quan, B.V. Howard, C. Iber, *et al.*, "The sleep heart health study: design, rationale, and methods", *Sleep,* 20 (12) 1077–1085, 1997.
[24] Sleep Data – National Sleep Research Resource – *NSRR*, https://sleepdata.org/.
[25] R.B. Berry, R. Brooks, C.E. Gamaldo, S.M. Harding, C. Marcus, and B. Vaughn, "AASM manual for the scoring of sleep and associated events", *J. Clin. Sleep Med.* 13(5), 665–666, 2012.
[26] D. Kingma and J. Ba, "Adam: a method for stochastic optimization", 2014, arXiv:1412.6980.

[27] D. Erhan, Y. Bengio, A. Courville, and P. Vincent, "Visualizing higher-layer features of a deep network", Technical Report 1341, University of Montreal, p. 3, 2009.

[28] M.D. Zeiler, D. Krishnan, G.W. Taylor, and R. Fergus, "Deconvolutional networks", in: *IEEE Conference on Computer Vision and Pattern Recognition (CVPR)*, IEEE, 2010, pp. 2528–2535, 2010.

[29] N. Srivastava, G. Hinton, A. Krizhevsky, I. Sutskever, and R. Salakhutdinov, "Dropout: a simple way to prevent neural networks from overfitting", *J Mach Learn Res.*, 15, 1929–1958, 2014.

[30] W. Zaremba, I. Sutskever, and O. Vinyals, "Recurrent neural network regularization", 2014, *arXivpree-prints*.

[31] M. Abadi, A. Agarwal, P. Barham, *et al.*, "TensorFlow: large-scale machine learning on heterogeneous distributed systems", 2016, *arXivpree-prints*.

[32] T.G. Dietterich, "Ensemble methods in machine learning", *Mult. Classif. Syst*. 1857, 1–15, 2000.

[33] C. Huang, Y. Li, C. Change Loy, and X. Tang, "Learning deep representation for imbalanced classification", in: *Proceedings of the IEEE Conference on Computer Vision and Pattern Recognition*, 2016, pp. 5375–5384.

[34] A.L. Goldberger, L.A.N. Amaral, L. Glass, *et al.*, "PhysioBank, PhysioToolkit, and PhysioNet components of a new research resource for complex physiologic signals", *Circulation* 101(23), 215–220, 2000.

[35] A.R. Hassan and M.I.H. Bhuiyan, "Automatic sleep scoring using statistical features in the EMD domain and ensemble methods", *Biocybern. Biomed. Eng*., 36(1), 248–255, 2016.

[36] A.R. Hassan and M.I.H. Bhuiyan, "Automated identification of sleep states from EEG signals by means of ensemble empirical mode decomposition and random under sampling boosting", *Comput. Methods Progr. Biomed*., 140, 201–210, 2017.

[37] A. Sors, S. Bonnet, S. Mirek, L. Vercueil, and J.-F. Payen. "A convolutional neural network for sleep stage scoring from raw single-channel EEG". *Biomed. Signal Process. Control*, 42, 107–114, 2018.

[38] K. He, X. Zhang, S. Ren, and J. Sun, "Deep residual learning for image recognition", in: *Proceedings of the IEEE Conference on Computer Vision and Pattern Recognition*, 2016, pp. 770–778.

[39] F. Chollet, "Xception: Deep Learning with Depthwise Separable Convolutions", 2016, arXiv preprint arXiv:1610.02357.

Chapter 10

Fundamentals, limitations, and the prospects of deep learning for biomedical image analysis

T. Chandrakumar[1], *Deepthi Tabitha Bennet*[1] *and Preethi Samantha Bennet*[1]

The use of artificial intelligence (AI) in healthcare has made great strides in the past decade. Several promising new applications are proving to be useful in medicine. The most important and significant of them are image analysis and classification using deep learning. This has led to several intelligent disease detection systems assisting doctors. They are a boon not only for the doctors as it reduces their workload effectively and efficiently, but also for the patients with accurate and fast results. Hence it becomes necessary to understand the concepts along with their limitations and future prospects. This will help in designing and applying image analysis across a wide range of medical specialties. We discuss the basic concepts of deep learning neural networks and focus on the applications in the specialties of radiology, ophthalmology, and dermatology. In addition to a thorough literature survey, we have also built a representational system for each of these specialties and presented the results. We also discuss the benefits for the patients along with the limitations of such intelligent systems. We have built a neural network intelligent system in each of these specialties and presented the results with details of the dataset and models used. We have got high-performance metrics of AUC up to 90% in our radiological system, accuracies of 92% in ophthalmology, and 98% in dermatology. With enough amount of data, highly efficient and effective disease detection systems can be built, to perform as aides to healthcare professionals for screening and monitoring for various diseases and disorders. More image datasets should be available in the public domain for further research, improved models, and better performance metrics. Also applications should use parallel processing of data to reduce time taken. Healthcare professionals should be fully trained to adapt to the use of intelligent decision-making systems to assist them in patient care.

[1]Thiagarajar College of Engineering Madurai, India

10.1 Introduction

The life expectancy of humans has more than doubled in the past 200 years. If we consider recent time periods, global data shows an increase of about 6.6 years in life expectancy from 2000 to 2019. While this is a very great achievement, a result of advances in medicine and healthcare, the healthy life expectancy (HALE) is rising at a lower rate (5.4 years) [1]. It becomes necessary to ably support the already stretched healthcare community, to effectively serve a growing population, and to ensure adequate healthcare for all. Intelligent systems are already proving their efficacy, accuracy, and speed in several aspects of healthcare, from diagnostics to complex surgeries.

Artificial intelligence (AI), especially deep learning (DL) models are making great strides in diagnostics by screening images. Figure 10.1 shows the usage of AI in healthcare is much higher than all other technologies. Several studies have proven AI systems perform at least as well as qualified professionals after suitable training. In some studies, AI/DL systems outperform the experts too! [2].

Automation of disease detection started making steady progress in healthcare with the introduction of machine learning (ML) models. As computing power and data storage capabilities increased over the years, newer models and various deep learning models have come into greater significance in healthcare.

Image capture devices have also become much better with higher-resolution images aiding disease detection. Databases which can store large volumes of data, which is required for storing large numbers of high-quality images have supported the development of deep learning models.

We are now entering what can be described as a golden era of intelligent systems aiding humans, especially in healthcare. Deep learning for medical image analysis is a vast domain in itself.

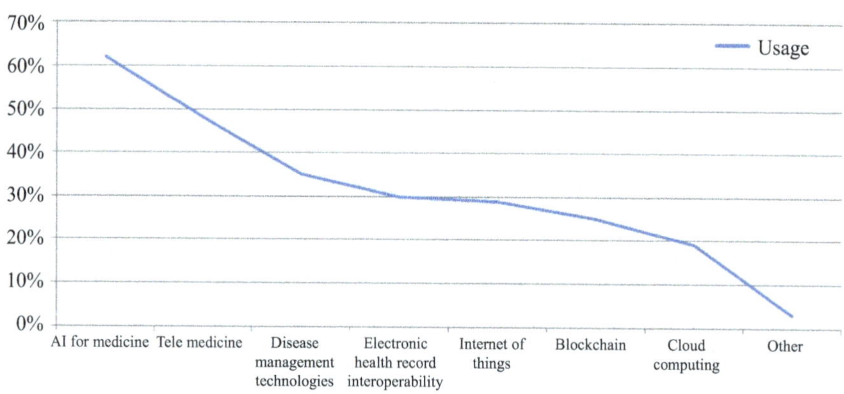

Figure 10.1 Comparison of the usage of AI with other technologies in healthcare. Source: [3].

In this chapter, the fundamentals of deep learning and the current specialties in healthcare where DL applications are already performing successfully, are presented. We will also present exciting future trends in addition to the challenges and limitations in biomedical image analysis using deep learning. The structure of this paper is shown in Figure 10.2.

This chapter has three major sections:

1. Demystifying deep learning—a simple introduction to DL
2. Current trends and what we can expect in the future
3. Challenges and limitations in building biomedical DL systems

The structure of this chapter is shown in Figure 10.2.

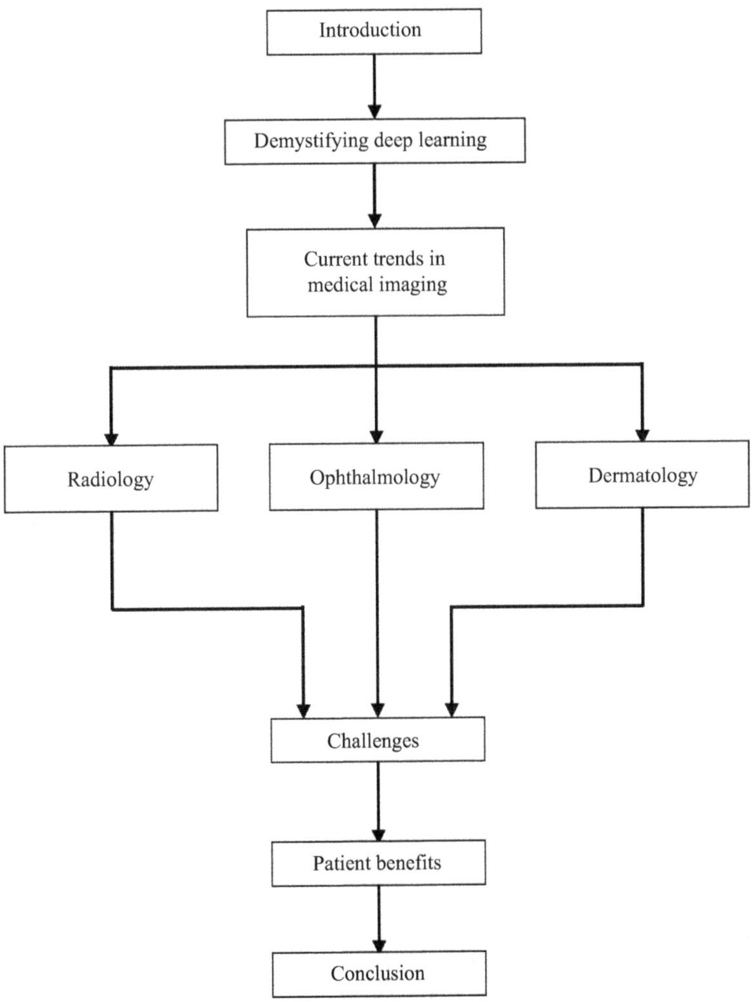

Figure 10.2 Structure of this chapter

The goal of the first section is to explain the concepts of deep learning with reference to AI and ML. It also highlights the differences between the three techniques (AI, ML, and DL).

The current trends section is further subdivided into *overview, radiology, ophthalmology, and dermatology.* These three specialties are chosen for this section, based on their current success in DL applications for image analysis.

In radiology, AI plays a major role in disease diagnosis from X-rays, mammograms, and CT/MRI images. X-rays are easy to obtain and involve minimal radiation exposure. X-rays are mainly used for imaging bones and the lungs. Recently, X-rays for assessing the severity of COVID-19 with lung involvement were widely used globally. Several studies conducted for these applications have found AI diagnostics and decision support systems to be at least as good as that of doctors and trained specialists.

In ophthalmology, the AI analysis of images mainly refers to using retinal fundus images (RFI) and optical coherence tomography (OCT) to detect various diseases, not just ophthalmological diseases including diabetic retinopathy, glaucoma, etc., but even neurological diseases. Recent studies show good results in the early detection of Alzheimer's disease just by the AI image analysis of the retinal fundus. Papilledema and hence any swelling of the brain can also be detected. In addition to this, the direct visualization of the microvasculature in the retinal fundus is now proving to be useful in predicting and detecting systemic diseases like chronic kidney failure and cardiovascular diseases.

In dermatology, AI has achieved great success in analyzing skin photographs and diagnosing diseases including detecting skin cancers, dermatitis, psoriasis, and onychomycosis. Research is still ongoing to enable patients to just upload an image and get an instant and reliable diagnosis.

The third section presents the challenges and future work needed before the widespread use of AI in medical diagnostics. Challenges including normalization of images from various sources, a large database of images for training, and the need to consider and ensure patient safety, legal and ethical issues are presented.

10.2 Demystifying DL

As this chapter will deal with AI, ML, and DL, it is necessary to first define these three terms. AI is the superset which encompasses both ML and DL (Figure 10.3). Although DL can be considered as the subset of ML, it differs from conventional ML algorithms or techniques, in that, DL uses a large volume of data to learn insights from the data by itself. These patterns are then used to make predictions on any similar data or unseen data.

AI and ML usually describe systems that follow a fixed set of rules to make predictions. These rules were predefined by an expert in that field. These were not considered as a data-driven approach, but just automation based on a few predefined sets of instructions.

Intelligent systems designed for deep learning networks are inspired by the human brain. The architecture of deep learning systems, closely resembles human brain structure. The basic computational unit of a neural network (NN) is called a perceptron. This closely resembles a human neuron. Similar to the electrical pulses traveling through the neurons, the perceptron uses signals to provide suitable outputs

Similar to the neurons combining together to form the human neural network, the perceptron combines to form an intelligent system. The NNs used for DL are comprised of an input layer, an output layer, and several hidden layers as shown in Figure 10.4.

Figure 10.5 shows a schematic diagram of a deep learning system which can be customized according to the application. Further study and a detailed explanation of the concepts of deep learning can be found in [5].

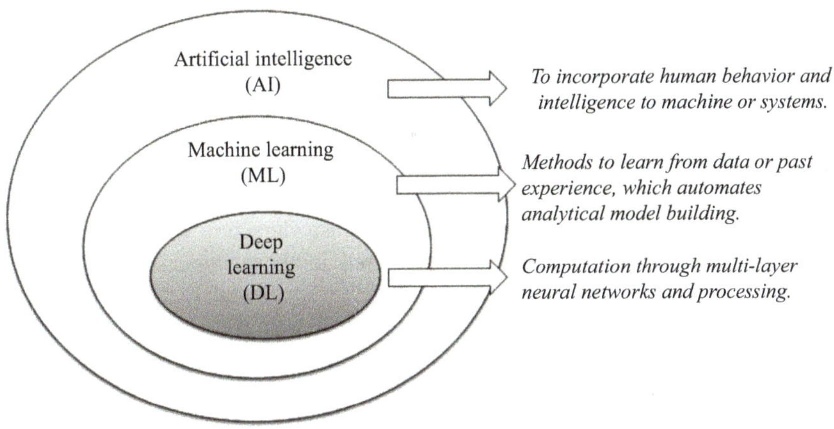

Figure 10.3 Concepts of AI, ML, and DL. Source: [5].

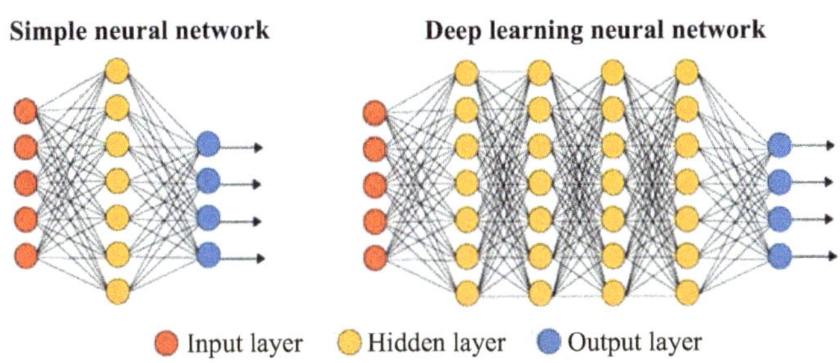

Figure 10.4 Architecture of neural networks. Source: [4].

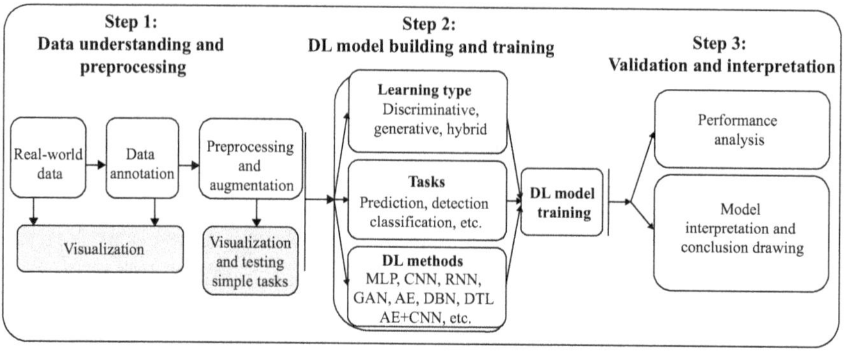

Figure 10.5 Schematic diagram of a deep learning system. Source: [5].

10.3 Current trends in intelligent disease detection systems

10.3.1 Overview

Automation of disease detection based on image classification started out with machine learning models, especially support vector machine (SVM)-based classifiers. The advent of neural networks has significantly reduced training times and decision times with high-performance metrics. This has led to a diverse and wide range of applications of classification and disease detection in healthcare. The successful applications are mainly concentrated in these three specialties: radiology, ophthalmology, and dermatology. These three are chosen as the specialties in focus for our discussion, mainly because of the existing literature documenting good performance metrics. Several other specialties have also started using AI for aiding healthcare professionals in various tasks including disease detection and triaging of patients. The current trends are discussed here with existing literature along with the results we obtained with our models.

10.3.2 Radiology

Radiology is the foremost of all specialties in healthcare to primarily use image analysis. From simple X-rays to mammograms, CT/MRI, and PET scans, the images in diagnostic radiology are used to non-invasively visualize the inner organs and bones in our body. Radiological visualization is used by almost all other specialties in healthcare. These image-based diagnoses play a pivotal role not only in disease detection but also guide subsequent treatment plans.

Radiology was one of the first few specialties in medicine to use digitized images and adapt AI/ML methods, and more recently computer vision (CV) techniques using advanced neural networks. A recent study in radiology shows that AI applications are used for the following tasks in diagnostic radiology. Perception (70%) and reasoning (17%) tasks are the primary functionalities for AI tools (Figure 10.6).

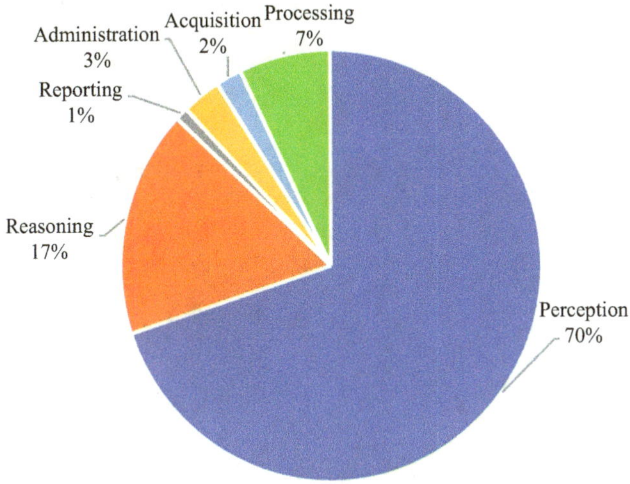

Figure 10.6 AI in diagnostic radiology. Source: [6].

Most of the existing AI applications in radiology are for CT, MRI, and X-ray modalities (29%, 28%, and 17%, respectively) [6].

Most of the current applications are for any one of these modalities and focus on any one anatomical part. Very few applications work for multiple modalities and multiple anatomical regions. The AI applications to analyze images of the brain have the highest share of about 27%, followed by the chest and lungs at 12% each. Mammograms to detect cancer have also achieved good results in screening programs.

Several monotonous and repetitive tasks like segmentation, performed by radiologists are successfully performed by intelligent systems in a much shorter time. This has saved up several man-hours for the doctors and has enabled quicker results for the patients. In some applications, smaller lesions and finer features are detected by the DL system better than by human diagnosticians, leading to high accuracy in disease detection.

10.3.2.1 Literature review

A review of existing literature which studies the applications of AI in radiology is presented with significant results and suggested future work in Table 10.1.

10.3.2.2 Radiology—the proposed deep learning system

The intelligent system we built to demonstrates the application of AI in radiology to detect diseases using chest X-rays. The details of the system with the results obtained are given below. Figure 10.7 shows the schematic of the proposed system, and Figure 10.8 shows the actual model plot.

Dataset used: NIH chest X-ray dataset [19]

Table 10.1 Literature survey—AI in radiology

Reference	Dataset	Dataset size	Disease detected	Algorithms used	Significant results	Limitations/future work
[7]	Own data	5,232 images	Bacterial Pneumonia, Viral Pneumonia	InceptionNet V3	Best Model: **InceptionNet V3** Pneumonia/Normal Accuracy: 92.8% Sensitivity: 93.2% Specificity: 90.1% AUC: 96.8% Bacterial/Viral Accuracy: 90.7% Sensitivity: 88.6% Specificity: 90.9% AUC: 94.0%	Images from different devices (different manufacturers) for training and testing to make the system Universally useful
[7,8]		5,232 images	Pneumonia	Xception, VGG16	Best Model: **VGG16** Accuracy: 87% Sensitivity: 82% Specificity: 91%	N/A
[7,9]		5,856 images	Pneumonia	VGG16, VGG19, DenseNet201, Inception_ResNet_V2, Inception_V3, Resnet50, MobileNet_V2, Xception	Best Model: **ResNet50** Accuracy: 96.61% Sensitivity: 94.92% Specificity: 98.43% Precision: 98.49% F1 score: 96.67%	More datasets and advanced feature extraction techniques maybe used – You-Only-Look- Once (YOLO), and U-Net.
[10,11]		273 images	COVID-19	Inception V3 Combined with MLP	Best model: **InceptionNet V3** combined with MLP Sensitivity: 93.61% Specificity: 94.56% Precision: 94.85% Accuracy: 94.08% F1 score: 93.2% Kappa value: 93.5%	The FM-HCF-DLF model – other classifiers can be tried (instead of MLP).

(Continues)

[12,13]	LIDC-IDRI database	3,500 images	Pneumonia, lung cancer	AlexNet, VGG16, VGG19, ResNet50, MAN- SoftMax, MAN-SVM	Best Model: **MAN-SVM** Accuracy: 97.27% Sensitivity: 98.09% Specificity: 95.63% Precision: 97.80% F1 score: 97.95%	EFT implementation for Local Binary Pattern (LBP) based feature extraction
[13,14]		112,120 images	Atelectasis, cardiomegaly, consolidation, edema, effusion, emphysema, fibrosis, hernia, infiltration, mass, nodule, pleural thickening, pneumonia, pneumothorax	CheXNeXt (121-layer DenseNet)	Mass detection: Sensitivity: 75.4% Specificity: 91.1% Nodule detection: Sensitivity: 69.0% Specificity: 90.0%	Both CheXNeXt and the radiologists did not consider patient history or review previous visits. If considered, it is known to improve the diagnostic performance of radiologists.
[14]	Own data (ChestX-ray8)	108,948 images	Atelectasis, cardiomegaly, Effusion, infiltration, mass, nodule, pneumonia, pneumothorax	AlexNet GoogLeNet, VGGNet-16, ResNet50	Mean accuracy: 82.8% Best model: **ResNet50** Accuracy: Atelectasis: 70.69% Cardiomegaly: 81.41% Effusion: 73.62% Infiltration: 61.28% Mass: 56.09% Nodule: 71.64% Pneumonia: 63.33% Pneumothorax: 78.91%	Dataset can be extended to cover, more disease classes and also to integrate other clinical information

(Continues)

Table 10.1 (Continued)

Reference	Dataset	Dataset size	Disease detected	Algorithms used	Significant results	Limitations/future work
[11,15]	Kaggle	1,215 images	COVID-19, bacterial pneumonia, viral pneumonia	ResNet50. ResNet101	Best Model: **ResNet101** Accuracy: 98.93% Sensitivity: 98.93% Specificity: 98.66% Precision: 96.39% F1-score: 98.15%	The system could be extended to detect other viruses (MERS, SARS, AIDS, and H1N1)
[16]	Radiology assistant, Kaggle	380 images	COVID-19	SVM classifier (with Linear, Quadratic, Cubic, and Gaussian kernel)	Best model: **SVM (Linear kernel)** Accuracy: 94.74% Sensitivity: 91.0% Specificity: 98.89% F1 score: 94.79% AUC: 0.999	Other lung diseases can be considered
[17,18]		5,606 images	Atelectasis, pneumonia, hernia, edema, emphysema, cardiomegaly, fibrosis, pneumothorax, consolidation, pleural thickening, mass, effusion, infiltration, nodule	VDSNet, vanilla gray, vanilla RGB, hybrid CNN and VGG, modified capsule network	Best model: **VDSNet** Recall: 0.63 Precision: 0.69 $F\beta$ (0.5) score: 0.68 Validation accuracy: 73%	Image augmentation for increasing the accuracy

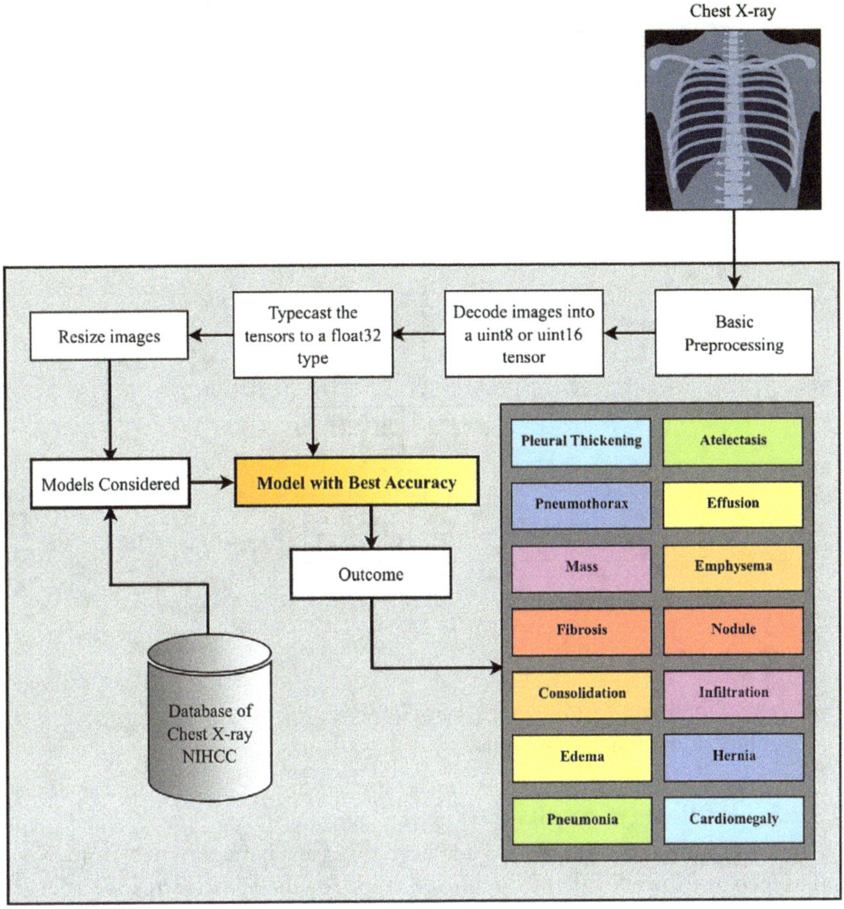

Figure 10.7 Radiology—schematic diagram of proposed intelligent system

Number of images—112,120 preprocessing techniques applied:

- Drop the column "img_ind"
- Decode images into a uint8 or uint16 tensor
- Typecast the tensors to a float32 type
- Resize images to target size
- Basic data augmentation

Diseases detected: 14 diseases including cardiomegaly, hernia, infiltration, nodule, and emphysema

Model: SEResNet: a variation of ResNet with additional squeeze and excitation blocks

Batch size: 96

No. of epochs: 50 optimal epoch: 13 optimizer: Adam

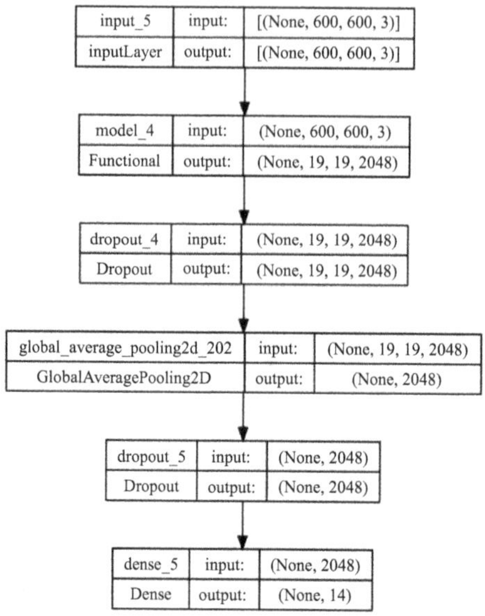

Figure 10.8 Radiology—the plot of the NN model

Cross-validation: k-fold learning rate: 0.001
Loss function: "binary_crossentropy"

10.3.2.3 Radiology: results obtained

The multi-classification model to identify disease characteristics from X-ray images performs well and has achieved high levels of performance metrics. AUC values range from 0.71 to 0.9. Figure 10.9 shows the accuracy/epoch plot, and Figure 10.10 shows the AUC values of various diseases detected by the model.

10.3.3 Ophthalmology

Ophthalmology was a natural forerunner in adapting AI screening tools for image analysis, mainly because it relies on several images for disease detection and monitoring. Retinal imaging, which includes retinal fundus imaging (RFI) and optical coherence tomography (OCT) is used for diagnosing several diseases of the eye, brain, and even systemic diseases like diabetes and chronic kidney disease.

Diabetic retinopathy (DR) is caused by damage to the retina which in turn is caused by diabetes mellitus. It can be diagnosed and assessed using retinal fundus images. Early diagnosis and intervention can save vision. Similarly, age-related macular degeneration (AMD) is also avoidable if diagnosed early. Again the diagnosis is based on retinal images.

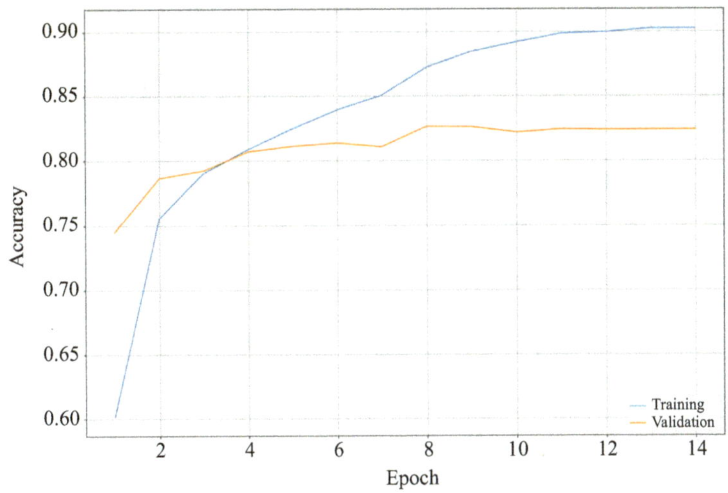

Figure 10.9 Radiology—training and validation accuracy plot

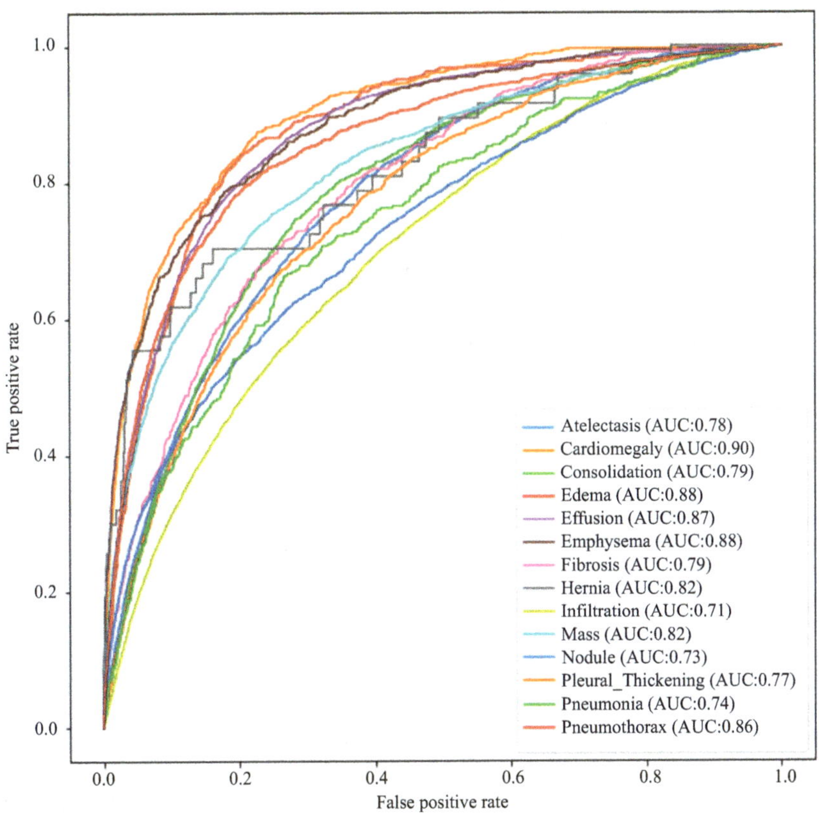

Figure 10.10 Radiology—AUC plot for diseases detected

10.3.3.1 Literature review

An exhaustive literature review was carried out about the use of AI in ophthalmology and the main features are listed in Table 10.2. This shows the significant progress of AI in ophthalmological image classification systems.

10.3.3.2 Ophthalmology: the proposed deep learning system

In ophthalmology, we built an intelligent image classification system which uses OCT images to detect diseases of the eye. The schematic diagram is given in Figure 10.11, and the model plot of the neural network showing various layers is given in Figure 10.12.

Dataset: Labelled Optical Coherence dataset [30] Number of images: 84,495
Preprocessing:
Encode labels to hot vectors
Resize images to target size
Basic data augmentation

Diseases detected: choroidal neovascularization (CNV), diabetic macular edema (DME), and age-related macular degeneration (AMD)
Model: InceptionNet V3 (transfer learning)
No. of epochs: 50
Optimal epochs: 12
Optimizer: Adam
Cross-validation: k-fold
Learning rate: 0.001
Loss function: "categorical_crossentropy"

10.3.3.3 Ophthalmology—results obtained

The proposed system was optimized and tested with labelled OCT images. The accuracy values of training and validation are plotted in Figure 10.13. Figure 10.14 lists the performance metrics of the model with an average accuracy of 0.92. This is quite high for multi-disease detection systems.

10.3.4 Dermatology

Dermatology has hugely successful applications of artificial intelligence for a wide range of diagnoses, from common skin conditions to screening for skin cancer. Almost all of these applications are based on image recognition models and are also used to assess and manage skin/hair/nail conditions. Google is now introducing an AI tool which can analyze images captured with a smartphone camera. Patients themselves or non-specialist doctors can use this tool to identify and diagnose skin conditions. This is very useful for telehealth applications too. AI systems can prove invaluable in the early detection of skin cancer, thereby saving lives [31]. Detection, grading, and monitoring are the main uses of AI systems mainly in melanoma, psoriasis, dermatitis, and onychomycosis. They are also now used for acne grading and also monitoring ulcers by automatic border detection and area calculations.

Table 10.2 Literature survey—AI in ophthalmology

Reference dataset	Dataset size	Disease detected	Algorithms used	Significant results	Limitations/future work
[20] Cirrus HD-OCT, Cirrus SD-OCT images	1,208 images	Glaucoma, myopia	gNet3D	Best model: **gNet3D** AUC: 0.88	SD-OCT scans with low SS are included. Risk factors like pseudo-exfoliation/ pigment dispersion/ secondary mechanisms are not considered.
[21] 3D OCT-2000, Topcon images	357 images	Glaucoma	CNN, random forest	Best model: **RF**, AUC: 0.963	N/A
[22] 3D OCT-1000, 3D OCT-2000 images	71 images	Age-related macular degeneration	AMDnet, CNN, VGG16, SVM	Best model: **AMDnet,** AUC: 0.89	Models generalization to patients with early or intermediate AMD is not known
[23] SD-OCT images	1,621 images	Age-related macular degeneration	CNN, transfer learning	AMD detection: Best model: **CNN** Sensitivity: 100% Specificity: 91.8% Accuracy: 99% Exudative changes detection: Best model: **transfer learning Model** Sensitivity: 98.4% Specificity: 88.3% Accuracy: 93.9%	Patients who had other associated diseases were excluded. Unclear if the results can be used in general .
[24] SS-OCT,	260 images	Multiple sclerosis	SVM (linear, polynomial, radial basis, sigmoid), decision tree, random forest	Best model: **Decision tree**	MS cohort should be modified to consider patients with at least only one year of disease duration as opposed to the average duration of 7.12 years

(Continues)

Table 10.2 (Continued)

Reference dataset	Dataset size	Disease detected	Algorithms used	Significant results	Limitations/future work
DRI OCT Triton images				Wide protocol: Accuracy: 95.73% AUC: 0.998 Macular protocol: Accuracy: 97.24% AUC: 0.995	
[25] SD-OCT	6,921 images	Glaucomatous	ResNet 3D deep-learning system, ResNet 2D deep-learning system	Best model: **ResNet 3D system**	Performance in external validations was reduced compared to primary validation. Only gradable images and cases of glaucomatous optic neuropathy with corresponding visual field defects were included.
images		Optic neuropathy		AUC: 0·969 Sensitivity: 89% Specificity: 96% Accuracy: 91%	
[26] Cirrus SD-	20,000 images	Age-related macular degeneration	ReLayNet (for segmentation), Inception-v3, InceptionresNet50	Best Model: **Inceptionres-Net50**	No explicit definitions of features were given, so the algorithm may use features previously not recognized or ignored by humans. The images were from a single clinical site.
OCT images				Accuracy: 86–89%	
[27] Zeiss	463 volumes	Diabetic retinopathy	VGG19,	Best model: **VGG19**	Only images with a signal strength of 7 or above were considered, which maybe sometimes infeasible in patients with pathology.

(Continues)

PlexEite 9,000 images			ResNet50, DenseNet	Sensitivity: 93.32% Specificity: 87.74% Accuracy: 90.71%	
[28] Zeiss Cirrus	35,900 images	Age-related macular degeneration (dry, inactive wet, active wet)	VGG16,	Best model: **InceptionV3**	N/A
HD-OCT			InceptionV3, ResNet50	Accuracy: 92.67%	
4000, Optovue				Sensitivity (dry): 85.64% Sensitivity (inactive wet): 97.11%	
RTVue-XR				Sensitivity (active wet): 88.53%	
Avanti images				Specificity (dry): 99.57% Specificity (inactive wet): 91.82% Specificity (active wet): 99.05%	
[29] Cirrus OCT,	8,529 volumes	Age-related macular degeneration	Logistic regression	Best model: **Logistic regression**	OCT angiography to detect subclinical MNV not included, which could be significant in assessing progression risk with drusen
Zeiss images				0.5–1.5 mm area – AUC: 0.66 0–0.5 mm area – AUC: 0.65	

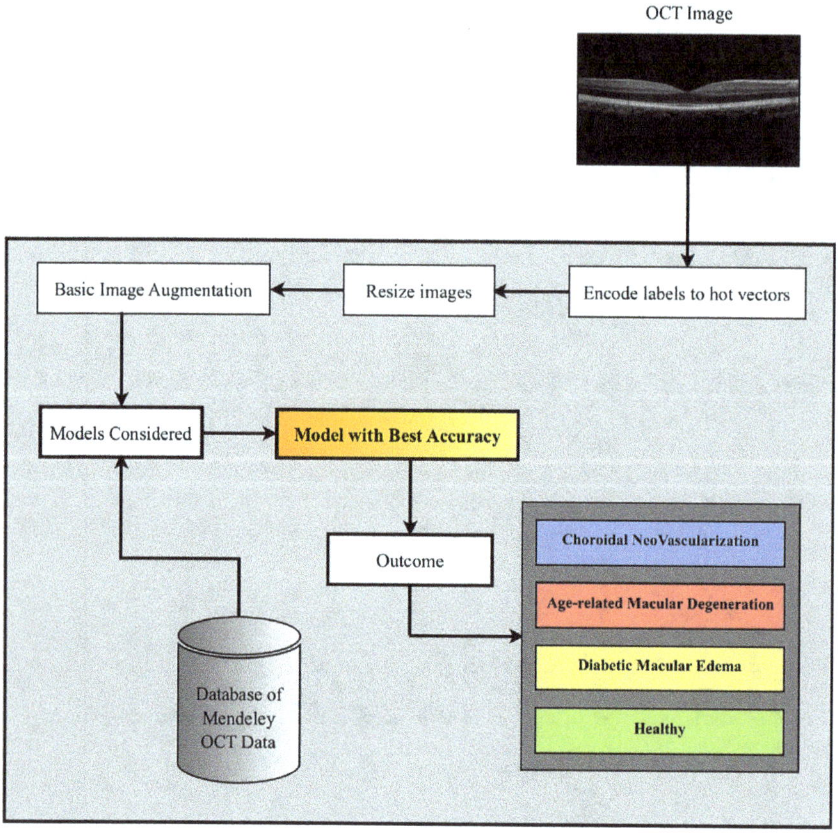

Figure 10.11 Ophthalmology—schematic diagram of proposed intelligent system

10.3.4.1 Literature review

We will present existing literature in Table 10.3, highlighting the applications of AI and DL in the field of dermatology.

10.3.4.2 Dermatology—the proposed deep learning system

The proposed neural network we built is a multi-disease detection system using dermatoscopic images to detect several skin conditions including skin cancers. The schematic diagram of the system is given in Figure 10.15, and the model plot of the convolutional neural network is given in Figure 10.16.

Dataset used: HAM 10,000 dataset [42]

Number of images: 10,015 **Preprocessing**:

* Replace null "age" values with mean
* Convert the data type of "age" to "int32"
* Convert the images to pixel format, and adding the pixel values to the dataframe
* Basic data augmentation

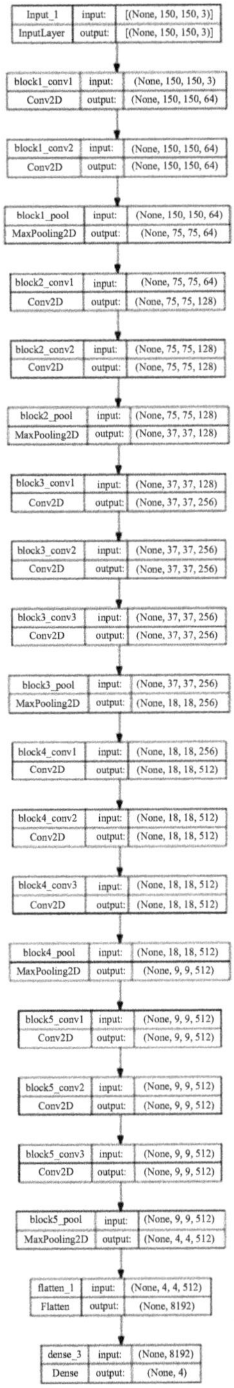

Figure 10.12 Ophthalmology—plot of the NN model

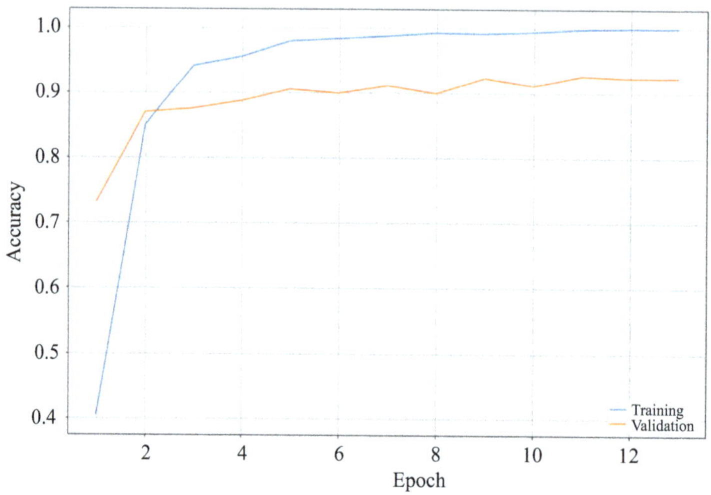

Figure 10.13 Ophthalmology—training and validation accuracy plot

	Precision	Recall	F1 - score	Support
Normal	0.90	0.90	0.90	42
CNV	0.93	0.98	0.95	42
DME	0.92	0.86	0.89	42
DRUSEN	0.93	0.95	0.94	42
Accuracy		0.92		168
Macro average	0.92	0.92	0.92	168
Weighted average	0.92	0.92	0.92	168

Figure 10.14 Ophthalmology—performance metrics

Diseases detected: Seven skin diseases including melanoma and carcinoma

Model: Convolutional neural network (CNN)

Batch Size: 64 No. of epochs: 50

Optimal Epoch: 26 Optimizer: Adam

Cross-Validation: k-fold Learning rate: 0.001

Loss Function: "sparse_categorical_crossentropy"

Table 10.3 Literature survey—AI in dermatology

Reference	Dataset	Dataset size	Disease detected	Algorithms used	Significant results	Limitations/future work
[32]	DermNet	2,475 images	Melanoma	DT, RF, GBT, CNN	Best model: **CNN** Accuracy: 88.83% Precision: 91.07% Recall: 87.68% F1-Score: 89.32%	Tested only on one dataset, of limited size.
[33]	HAM10000	10,015 images	Actinic keratoses, basal cell carcinoma, benign keratosis-like lesions, dermatofibroma, melanoma, melanocytic nevi, vascular lesions	CNN, RF, DT, LR, LDA, SVM, KNN, NB, Inception V3	Best model: **CNN** Accuracy: 94% Precision: 88% Recall: 85% F1-Score: 86%	Model can be improved by hyper- parameter fine-tuning
[34]	ISIC	N/A	Skin cancer	CNN, GAN, KNN, SVM	Best model: **CNN** Accuracy: 92% Precision: 92% Recall: 92% F1-Score: 92%	N/A
[4,35]		120 images	Melanoma	KNN	Best model: **KNN** Accuracy: 98%	Ensemble learning methods or evolutionary algorithms can be considered for faster and more accurate results
[36]	Available on request	120 images	Herpes, dermatitis, psoriasis	SVM, GLCM	Best model: **SVM** Accuracy (Herpes): 85% Accuracy (Dermatitis): 90% Accuracy (Psoriasis): 95%	Very limited dataset, with only 20 images for each class

(Continues)

Table 10.3 (Continued)

Reference	Dataset	Dataset size	Disease detected	Algorithms used	Significant results	Limitations/future work
[37]	Own data	80 images	Melanoma, eczema, psoriasis	SVM, AlexNet	Best model: **SVM** Accuracy (melanoma): 100% Accuracy (eczema): 100% Accuracy (psoriasis): 100%	Very limited dataset, with only 20 images for each class. Overfitting is likely the reason for such high accuracies.
[38]	ISIC	640 images	Melanoma	KNN, SVM, CNN Majority voting	Best model: **CNN** Accuracy: 85.5%	Semi-supervised learning could be used to overcome lack of enough labeled training data
[39]	ISBI 2016 Challenge dataset for Skin Lesion Analysis	1,279 images	Melanoma	VGG16 ConvNet (i) trained from scratch; (ii) pre-trained on a larger dataset (iii) fine-tuning the ConvNets	Best model: **VGG16 ConvNet with fine-tuning** Accuracy: 81.33% Sensitivity: 0.7866 Precision: 0.7974 Loss: 0.4337(on test data)	Larger dataset can be used to avoid overfitting. Additional regularization and fine-tuning of hyper-parameters can be done.
[40]	HAM10000	10,015 images	Skin cancer	AlexNet, ResNet, VGG-16, DenseNet, MobileNet, DCNN	Best model: **DCNN** Accuracy (Train): 93.16% Accuracy (Test): 91.43% Precision: 96.57% Recall: 93.66% F1-Score: 95.09%	A user-friendly CAD system can be built.
[41]	Subset of DermNet	N/A	N/A	Inception V2, Inception V3, MobileNet, ResNet, Xception	Best model: **Inception V3** Precision: 78% Recall: 79% F1-Score: 78%	N/A

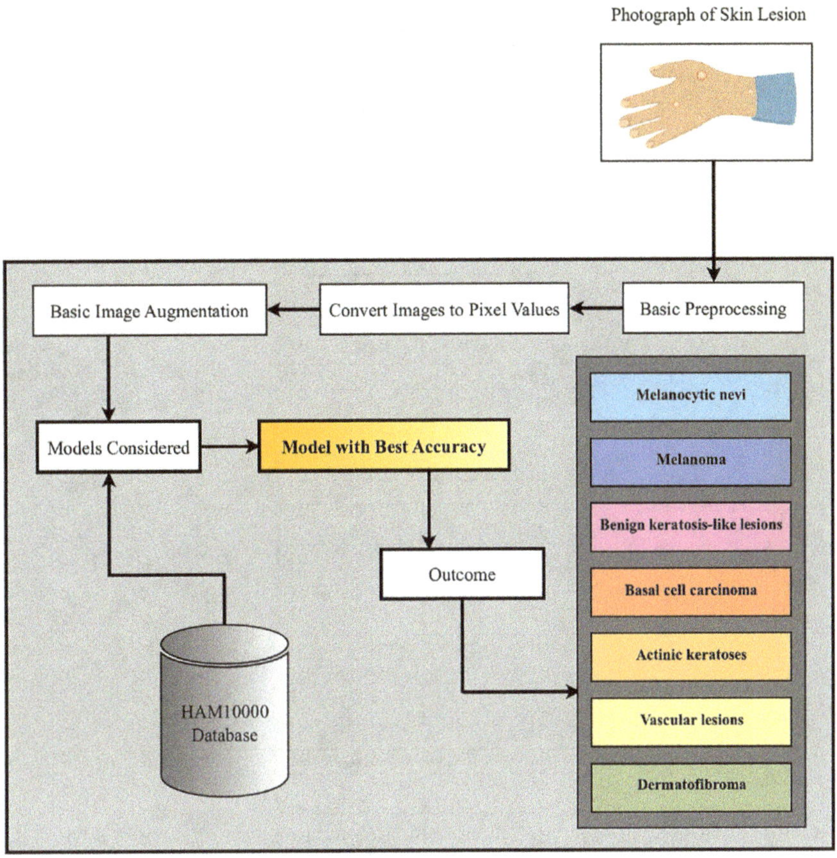

Figure 10.15 Dermatology—schematic diagram of proposed intelligent system

10.3.4.3 Dermatology—results obtained

The CNN model for detecting dermatological diseases from images performs very well with an accuracy of 0.99. The accuracy versus epoch plot for both training and validation is shown in Figure 10.17. The performance metrics we obtained for our system are tabulated in Figure 10.18.

10.4 Challenges and limitations in building biomedical image processing systems

Historically, the main challenges in building biomedical decision systems or analysis systems using machine learning were mainly focused on the computing speed and resources required. Also the databases required for the storage of high-resolution images were the main limitations. But now with the high-power computing we have available, and with the huge and distributed databases, those two

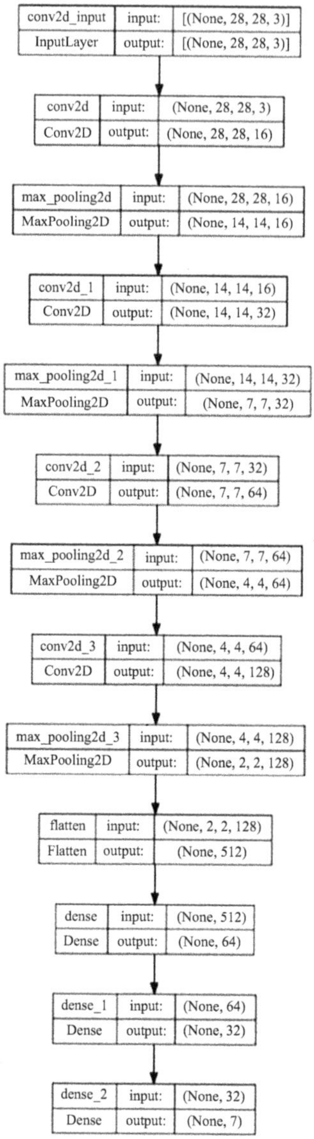

Figure 10.16 Dermatology—plot of the NN model

limitations have become obsolete. But several other limitations do exist for biomedical imaging and analysis systems. We will see a few in this section.

The first challenge in medical image processing with artificial intelligence is the availability of data. While certain fields and subdomains like ophthalmology and diabetic retinopathy have large volumes of data available in the public domain other rarer diseases and other fields have very limited datasets. So, most of the literature is based

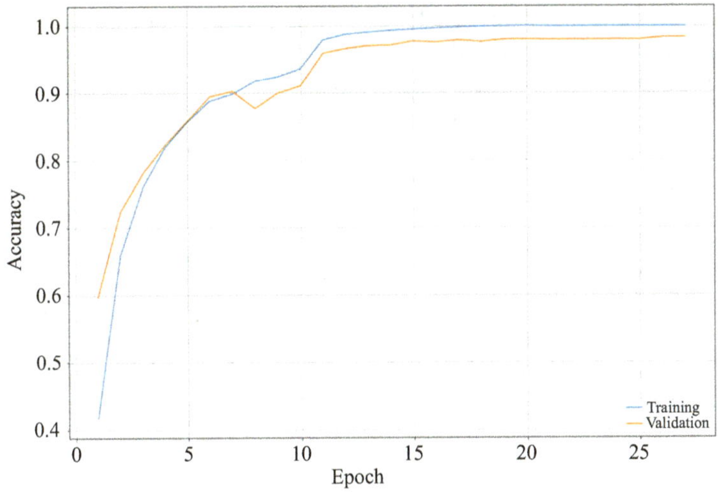

Figure 10.17 Dermatology—training and validation accuracy plot

	Precision	Recall	F1 - score	Support
Melanocytic nevi	0.99	0.87	0.93	1385
Melanoma	0.93	1.00	0.96	1328
Benign keratosis-like lesions	0.95	1.00	0.98	1294
Basal cell carcinoma	0.99	1.00	0.99	1325
Actinic keratoses	1.00	1.00	1.00	1270
Vascular lesions	1.00	1.00	1.00	1293
Dermatofibroma	1.00	1.00	1.00	1257
Accuracy		0.98		9152
Macro average	0.98	0.98	0.98	9152
Weighted average	0.98	0.98	0.98	9152

Figure 10.18 Dermatology—performance metrics

on a few sets of images. More availability of diverse data would ensure more versatile and robust models which can work with different inputs [43]. Freely available datasets in the public domain are needed for further progress in this field.

Ethics and legalities in collecting and using data have to be strictly followed by international standards. All images must be de-identified and obtained with consent. Patient privacy has to be preserved properly.

Another concern in using artificial intelligence for disease detection is the lack of explainability of the models, i.e., we do not know on what features the models base their decisions on. The upcoming explainable artificial intelligence (XAI) or explainable machine learning (XML) may solve this problem to a certain extent as it helps us to understand how the intelligent systems process the data and base their decisions on and what to base their decisions on [44]. This eliminates the black box approach which is currently prevalent, where we input the data and have the decision as the output. Also, the decision systems have to take into account, the other data about the patient in addition to the image being analyzed. Intelligent decision systems must take into account age, previous medical history, and other co-morbidities in addition to the images in the decision-making process.

Very often the collected data has an inbuilt bias. This can also affect the training of the model and hence performance. This can be avoided by carefully planning and monitoring the data collection process.

Each country or region has regulatory bodies for approving medical devices. The intelligent systems for users in disease detection/decision-making systems also have to undergo stringent checks and tests and approval has to be sort from suitable regulatory bodies before using for patient benefits.

Training medical experts in using AI systems efficiently will lead to them adapting intelligent systems quickly and easily in their regular practice. A recent survey shows less than 50% of medical experts in radiology, ophthalmology, and dermatology have at least average knowledge of AI applications in their specialty [45] (Figure 10.19).

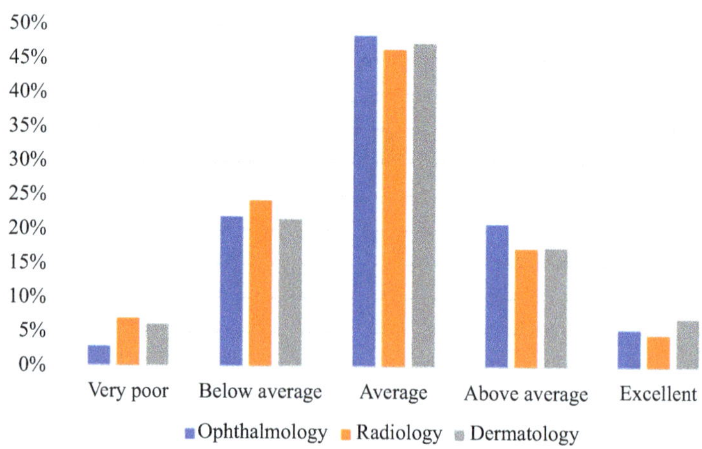

Figure 10.19 Self-assessment of knowledge of AI in their respective fields. Source: [45].

10.5 Patient benefits

The use of AI for medical image analysis will be a definite boon to healthcare professionals, both by saving time and confirming diagnoses for documentation purposes. They will be a bigger boon to the patients in the following aspects:

- Efficient screening programs
- Fast and reliable diagnoses
- Eliminates inter and intra-observer variations
- Detection of finer patterns which may not be obvious to the human eye

The goal is to maximize benefits for both the patients and the healthcare professionals while striving to minimize risks and challenging.

10.6 Conclusions

It is evident that intelligent disease detection systems, using image analysis in healthcare have had a huge boost with the widespread use of big data and deep learning systems. High-power computing systems have also reduced the time taken to a fraction of the original time required. The success is evident in radiology, ophthalmology, and dermatology as described in this chapter holds huge possibilities for all other specialties in healthcare. With enough training and in the right hands, AI will be a great tool beneficial to both medical experts and patients.

Further work suggestions include experimenting with newer neural networks (like vision transformers (ViT), EfficientNets, etc.), other than CNN-based networks to see if computational times and resources can be reduced and high-performance metrics achieved. Also, diverse datasets from different cameras/devices and from different ethnic groups can be curated to train better and more robust models.

References

[1] https://www.who.int/data/gho/data/themes/mortality-and-global-health-estimates/ghe-life-expectancy-and-healthy-life-expectancy retrieved on 18.12.2022.

[2] Pham, TC., Luong, CM., Hoang, VD. *et al.* AI outperformed every dermatologist in dermoscopic melanoma diagnosis, using an optimized deep-CNN architecture with custom mini-batch logic and loss function. *Sci Rep*, 11, 17485 (2021).

[3] Kumar, Y., Koul, A., Singla, R., and Ijaz, M. F. (2022). Artificial intelligence in disease diagnosis: a systematic literature review, synthesizing framework and future research agenda. *J Ambient Intell Humanized Comput*, 14, 1–28.

[4] Savalia, S. and Emamian, V. (2018). Cardiac arrhythmia classification by multi-layer perceptron and convolution neural networks. *Bioengineering*, 5 (2), 35. https://doi.org/10.3390/ bioengineering5020035

[5] Sarker, I.H. (2021). Deep learning: a comprehensive overview on techniques, taxonomy, applications and research directions. *SN Comput Sci*, 2, 420. https://doi.org/10.1007/s42979-021-00815-1

[6] Rezazade Mehrizi, M. H., van Ooijen, P., and Homan, M. (2021). Applications of artificial intelligence (AI) in diagnostic radiology: a technography study. *Eur Radiol*, 31(4), 1805–1811.

[7] Kermany, D. S., Goldbaum, M., Cai, W., *et al.* (2018). Identifying medical diagnoses and treatable diseases by image-based deep learning. *Cell*, 172(5), 1122–1131.

[8] Ayan, E. and Ünver, H. M. (2019, April). Diagnosis of pneumonia from chest X-ray images using deep learning. In *2019 Scientific Meeting on Electrical-Electronics & Biomedical Engineering and Computer Science (EBBT)* (pp. 1–5). IEEE.

[9] El Asnaoui, K., Chawki, Y., and Idri, A. (2021). Automated methods for detection and classification pneumonia based on x-ray images using deep learning. In *Artificial Intelligence and Blockchain for Future Cybersecurity Applications* (pp. 257–284). Springer, Cham.

[10] Shankar, K. and Perumal, E. (2021). A novel hand-crafted with deep learning features based fusion model for COVID-19 diagnosis and classification using chest X-ray images. *Complex Intell Syst*, 7(3), 1277–1293.

[11] https://github.com/ieee8023/covid-chestxray-dataset

[12] Bhandary, A., Prabhu, G. A., Rajinikanth, V., *et al.* (2020). Deep-learning framework to detect lung abnormality – a study with chest X-Ray and lung CT scan images. *Pattern Recogn Lett*, 129, 271–278.

[13] Rajpurkar, P., Irvin, J., Ball, R. L., *et al.* (2018). Deep learning for chest radiograph diagnosis: a retrospective comparison of the CheXNeXt algorithm to practicing radiologists. *PLoS Med*, 15(11), e1002686.

[14] Wang, X., Peng, Y., Lu, L., Lu, Z., Bagheri, M., and Summers, R. M. (2017). Chestx-ray8: hospital-scale chest x-ray database and benchmarks on weakly-supervised classification and localization of common thorax diseases. In *Proceedings of the IEEE Conference on Computer Vision and Pattern Recognition* (pp. 2097–2106).

[15] Jain, G., Mittal, D., Thakur, D., and Mittal, M. K. (2020). A deep learning approach to detect Covid-19 coronavirus with X-ray images. *Biocybernet Biomed Eng*, 40(4), 1391–1405.

[16] Ismael, A. M. and Şengür, A. (2021). Deep learning approaches for COVID-19 detection based on chest X-ray images. *Expert Syst Appl*, 164, 114054.

[17] Bharati, S., Podder, P., and Mondal, M. R. H. (2020). Hybrid deep learning for detecting lung diseases from X-ray images. *Informat Med Unlock*, 20, 100391

[18] https://www.kaggle.com/nih-chest-xrays/data

[19] https://nihcc.app.box.com/v/ChestXray-NIHCC/folder/36938765345

[20] Russakoff, D. B., Mannil, S. S., Oakley, J. D., *et al.* (2020). A 3D deep learning system for detecting referable glaucoma using full OCT macular cube scans. *Transl Vis Sci Technol*, 9(2), 12–12.

[21] An, G., Omodaka, K., Hashimoto, K., *et al.* (2019). Glaucoma diagnosis with machine learning based on optical coherence tomography and color fundus images. *J Healthcare Eng*, 2019.

[22] Russakoff, D. B., Lamin, A., Oakley, J. D., Dubis, A. M., and Sivaprasad, S. (2019). Deep learning for prediction of AMD progression: a pilot study. *Invest Ophthalmol Visual Sci*, 60(2), 712–722.

[23] Motozawa, N., An, G., Takagi, S., *et al.* (2019). Optical coherence tomography-based deep-learning models for classifying normal and age-related macular degeneration and exudative and non-exudative age-related macular degeneration changes. *Ophthalmol Therapy*, 8(4), 527–539.

[24] Perez del Palomar, A., Cegonino, J., Montolio, A., *et al.* (2019). Swept source optical coherence tomography to early detect multiple sclerosis disease. The use of machine learning techniques. *PLoS One*, 14(5), e0216410.

[25] Ran, A. R., Cheung, C. Y., Wang, X., *et al.* (2019). Detection of glaucomatous optic neuropathy with spectral-domain optical coherence tomography: a retrospective training and validation deep-learning analysis. *Lancet Digital Health*, 1(4), e172–e182.

[26] Saha, S., Nassisi, M., Wang, M., Lindenberg, S., Sadda, S., and Hu, Z. J. (2019). Automated detection and classification of early AMD biomarkers using deep learning. *Sci Rep*, 9(1), 1–9.

[27] Heisler, M., Karst, S., Lo, J., *et al.* (2020). Ensemble deep learning for diabetic retinopathy detection using optical coherence tomography angiography. *Transl Vis Sci Technol*, 9(2), 20–20.

[28] Hwang, D. K., Hsu, C. C., Chang, K. J., *et al.* (2019). Artificial intelligence-based decision-making for age-related macular degeneration. *Theranostics*, 9(1), 232.

[29] Waldstein, S. M., Vogl, W. D., Bogunovic, H., Sadeghipour, A., Riedl, S., and Schmidt-Erfurth, U. (2020). Characterization of drusen and hyperreflective foci as biomarkers for disease progression in age-related macular degeneration using artificial intelligence in optical coherence tomography. *JAMA Ophthalmol*, 138(7), 740–747.

[30] Kermany, D., Zhang, K., and Goldbaum, M. (2018), Labeled Optical Coherence Tomography (OCT) and Chest X-Ray Images for Classification, Mendeley Data, V2, doi: 10.17632/ rscbjbr9sj.2

[31] Liopyris, K., Gregoriou, S., Dias, J. *et al.* (2022). Artificial intelligence in dermatology: challenges and perspectives. *Dermatol Ther (Heidelb)* **12**, 2637–2651. https://doi.org/10.1007/s13555-022-00833-8

[32] Allugunti, V. R. (2022). A machine learning model for skin disease classification using convolution neural network. *Int J Comput Program Database Manag*, 3(1), 141–147.

[33] Shetty, B., Fernandes, R., Rodrigues, A. P., Chengoden, R., Bhattacharya, S., and Lakshmanna, K. (2022). Skin lesion classification of dermoscopic images using machine learning and convolutional neural network. *Sci Rep*, 12 (1), 1–11.

[34] Wang, X. (2022, December). Deep learning-based and machine learning-based application in skin cancer image classification. *J Phys: Conf Ser,* 2405 (1), 012024. IOP Publishing.

[35] Hatem, M. Q. (2022). Skin lesion classification system using a K-nearest neighbor algorithm. *Vis Comput Ind Biomed Art*, 5(1), 1–10.

[36] Wei, L. S., Gan, Q., and Ji, T. (2018). Skin disease recognition method based on image color and texture features. *Comput Math Methods Med*, 10, 1–10.

[37] ALEnezi, N. S. A. (2019). A method of skin disease detection using image processing and machine learning. *Proc Comput Sci*, 163, 85–92.

[38] Daghrir, J., Tlig, L., Bouchouicha, M., and Sayadi, M. (2020, September). Melanoma skin cancer detection using deep learning and classical machine learning techniques: a hybrid approach. In *2020 5th International Conference on Advanced Technologies for Signal and Image Processing (ATSIP)* (pp. 1–5). IEEE.

[39] Lopez, A. R., Giro-i-Nieto, X., Burdick, J., and Marques, O. (2017, February). Skin lesion classification from dermoscopic images using deep learning techniques. In *2017 13th IASTED International Conference on Biomedical Engineering (BioMed)* (pp. 49–54). IEEE.

[40] Ali, M. S., Miah, M. S., Haque, J., Rahman, M. M., and Islam, M. K. (2021). An enhanced technique of skin cancer classification using deep convolutional neural network with transfer learning models. *Mach Learn Appl*, 5, 100036.

[41] Patnaik, S. K., Sidhu, M. S., Gehlot, Y., Sharma, B., and Muthu, P. (2018). Automated skin disease identification using deep learning algorithm. *Biomed Pharmacol J*, 11(3), 1429.

[42] Tschandl, P. (2018). "The HAM10000 dataset, a large collection of multi-source dermatoscopic images of common pigmented skin lesions", https://doi.org/10.7910/DVN/DBW86T, Harvard Dataverse, V3.

[43] Daneshjou, R., Vodrahalli, K., Novoa, R. A., *et al.* (2022). Disparities in dermatology AI performance on a diverse, curated clinical image set. *Sci Adv*, 8(31), eabq6147

[44] Singh, A., Sengupta, S., and Lakshminarayanan, V. (2020). Explainable deep learning models in medical image analysis. *Journal of Imaging*, 6(6), 52

[45] Scheetz, J., Rothschild, P., McGuinness, M. *et al.* (2021). A survey of clinicians on the use of artificial intelligence in ophthalmology, dermatology, radiology and radiation oncology. *Sci Rep* 11, 5193. https://doi.org/10.1038/s41598-021-84698-5

Chapter 11

Impact of machine learning and deep learning in medical image analysis

Kirti Rawal[1]*, Gaurav Sethi*[1] *and Gurleen Kaur Walia*[1]

In the whole world, people are suffering from a variety of diseases. In order to detect these diseases, several medical imaging procedures are used in which images from different parts of the body are captured through advanced sensors and well-designed machines. These medical imaging procedures increase the expectations of patients in achieving better healthcare services from medical experts. Till now, various image processing algorithms such as neural networks (NN), convolutional neural networks (CNN), and deep learning are used for image analysis, image representation, and image segmentation. Yet, these approaches are not giving promising results in some applications of the healthcare sector. So, this chapter gives an overview of state-of-the-art image processing algorithms as well as highlights its limitations. Most deep learning algorithm implementations focus on the images of digital histopathology, computerized tomography, mammography, and X-rays. This work offers a thorough analysis of the literature on the classification, detection, and segmentation of medical image data. This review aids the researchers in considering necessary adjustments to deep learning algorithm-based medical image analysis. Further, the applications of medical image processing using Artificial Intelligence (AI), machine learning (ML), and deep learning in the healthcare sector are discussed in this chapter.

11.1 Introduction

Medical image processing plays an important role in identifying a variety of diseases. Earlier the datasets which were available for analyzing the medical images were very small. Nowadays, large datasets are available for interpreting medical images. To analyze these large image datasets, various highly experienced medical experts or radiologists are required. The number of patients outnumbered the number of available medical experts. Further, there is a high probability that the analysis done by medical experts is more prone to human errors. In order to avoid this problem, various machine learning algorithms are used to automate the process of medical image analysis [1]. Various image feature extraction and feature

[1]School of Electronics and Electrical Engineering, Lovely Professional University, India

selection methods are used for analyzing the medical images where the system is developed to train the data. Nowadays, neural network (NN), convolutional neural networks (CNN), and deep learning methods give a remarkable effect in the field of science. These methods not only give improvements in analyzing medical images but also use artificial intelligence to automate the detection of various diseases [2]. With the advent of machine learning algorithms, the medical images are possible to be analyzed more accurately as compared to the existing algorithms.

Zhu *et al.* [3] used a Memristive pulse coupled neural network (M-PCNN) for analyzing the medical images. The results in this chapter proved that the network can be further used for denoising medical images as well as for extracting the features of the images. Tassadaq Hussain [4] proposed an architecture for analyzing medical images or videos. Rajalakshmi *et al.* [5] proposed a model for the retina which is used for detecting the light signal through the optic nerve. Li *et al.* [6] exploited deep neural networks and hybrid deep learning models for predicting the age of humans by using 3D MRI brain images. Maier *et al.* [7] give an overview of analyzing medical images using deep learning algorithms. Selvikvag *et al.* [8] used machine learning algorithms such as artificial neural networks and deep neural networks on MRI images. Fourcade *et al.* [9] analyzed medical images using deep learning algorithms for improving visual diagnosis in the health sector. The authors also claimed that these novel techniques are not only going to replace the expertise of medical experts but they may automate the process of diagnosing various diseases. Litjens *et al.* [10] used machine learning algorithms for analyzing cardiovascular images. Zhang *et al.* [11] proposed a synergic deep learning model using deep convolutional neural networks (DCNN) for classifying the medical images on four datasets. Further, Kelvin *et al.* [12] discussed several challenges that are associated with a diagnosis of cardiovascular diseases by using deep learning algorithms. Various authors [13–17] used deep learning algorithms for image segmentation, image classification, and pattern recognition, as well as detecting several diseases by finding meaningful interpretations of medical images.

Thus, it is concluded that machine learning algorithms, deep learning algorithms, and artificial intelligence plays a significant role in medical image processing and its analysis. Machine learning algorithms not only extract the hidden information from medical images but also facilitate doctors for predicting accurate information about diseases. The genetic variations in the subjects are also analyzed with the help of machine learning algorithms. It is also observed that machine learning algorithms process medical images in raw form and it takes more time to tune the features. Although it shows significantly good accuracies in detecting the diseases as compared to the conventional algorithms. Deep learning algorithms show promising results and superior performance in the automated detection of diseases in comparison to machine learning algorithms.

11.2 Overview of machine learning methods

A brief introduction to the various machine learning algorithms used for analyzing medical images is given in this section. Learning algorithms are mainly classified

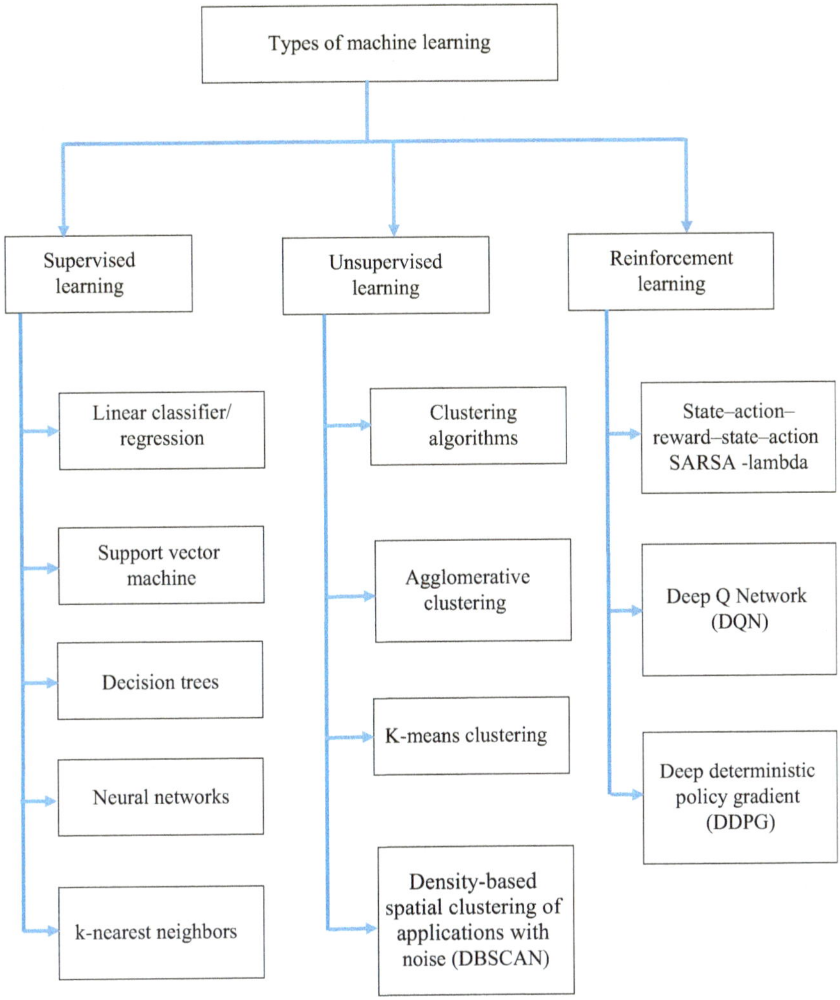

Figure 11.1 Classification of machine learning models

into three parts such as supervised learning, unsupervised learning, and reinforcement learning as shown in Figure 11.1.

11.2.1 Supervised learning

In supervised learning, a set of independent variables is used to predict the dependent variables. These variables are further used for generating the function by mapping the input to get the required output [16]. The machine is trained by using the labeled data until the desired accuracy is achieved. Several examples of supervised learning are the following.

11.2.1.1 Linear regression

In linear regression, the real values are calculated based on continuous variables. The relationship among two variables is estimated by fitting the best-fit line. This best-fit line is known as a regression line. The equation of a line is defined using the following equation:

$$Y = a\ X + b \tag{11.1}$$

where Y is the dependent variable, X is the independent variable, a is the slope, and b is the intercept. The coefficients a and b are calculated by minimizing the sum of the squared difference of distances between data points and the regression line.

Thus, in linear regression, data is trained for predicting the single output value.

11.2.1.2 Logistic regression

In logistic regression, a set of independent variables is exploited for finding the discrete values, i.e., 0 and 1. Thus, data is fitted to a logit function for finding the probability of occurrence. Basically, supervised learning problems are further classified into regression and classification.

Classification is used to define the output as a category like orange or green color, healthy or diseased subject. When the input is labeled into two categories, it is known as binary classification. When more than one class is identified, it is a multiclass classification.

It suffers from some limitations. Only linear relationships can use linear regression. It is prone to underfitting and sensitive to outliers. Only independent data is handled by linear regression.

11.2.1.3 Support vector machine

Support vector machine (SVM) is a classification algorithm that plots the data in n-dimensional space where each feature is having the value of a particular coordinate. If there are two features, then data is plotted in two-dimensional space where each point has two coordinates or support vectors. Afterward, the line which acts as a classifier is used to classify the data into two groups.

However, it has the following drawbacks. Large data sets are not a good fit for the SVM algorithm. When the target classes are overlapping and the data set includes more noise, SVM does not perform very well. The SVM will perform poorly when there are more training data samples than features for each data point. There is no probabilistic justification for the classification because the support vector classifier places data points above and below the classifying hyperplane.

11.2.2 Unsupervised learning

In unsupervised learning, the variables are not predicted but it is used for clustering populations in different groups [16]. There is only an input variable and no output variable. Thus, unsupervised learning is not associated with the teacher who is having correct answers.

11.2.2.1 Hierarchical clustering

A method called hierarchical clustering creates a hierarchy of clusters. It starts with the data, which is given its own cluster. In this case, two closely related clusters will be in the same cluster. When there is just one cluster left, the algorithm terminates. Its limitations are as follows. It has many arbitrary judgments and rarely offers the best solution. It does not work well with missing data and performs poorly with mixed data types. It performs poorly on very big data sets, and its primary output, the dendrogram, is frequently read incorrectly.

11.2.2.2 K-means clustering

In k-means clustering, data is classified by using several numbers of clusters, i.e., k clusters. For each cluster, k number of points are calculated known as centroids. The clusters are formed with close centroids between each data point. In this way, the new centroids are calculated based on the closest distance between each data point. The process is repeated until the formation of new centroids. The k-means clustering method is further classified into two sub-groups: agglomerative clustering and dendrogram.

11.2.2.2.1 Agglomerative clustering

It starts with a predetermined number of clusters. It distributes all the data among the precise number of clusters. The number K of clusters is not required as an input for this clustering technique. Each piece of data is first formed into a single cluster before the agglomeration process starts. The number of clusters (one in each iteration) is decreased by this method's usage of a distance metric and merging. Finally, all the objects are collected into a single large cluster. Even when groups have a high degree of overall dissimilarity, groups with close pairs may merge earlier than is ideal, which is its main limitation.

11.2.2.2.2 Dendrogram

Each level in the dendrogram clustering algorithm indicates a potential cluster. The height of the dendrogram indicates how similar two joined clusters are to one another.

11.2.2.3 K-nearest neighbors

The simplest machine learning classifier is the K-nearest neighbor. In contrast to other machine learning methods, it does not create a model. It is a straightforward method that categorizes new examples based on a similarity metric and stores all the existing cases. When the training set is huge and the distance calculation is complex, the learning rate is slow. It has the following limitations. Large datasets are problematic since it would be exceedingly expensive to calculate the distances between each data instance. High dimensionality makes it difficult to calculate distance for each dimension, hence it does not perform well in this case. It is vulnerable to noisy and incomplete data.

11.2.3 Reinforcement learning

In reinforcement learning, specific decisions are made by training the machine. The machine is trained itself by using past experiences and based on that training and knowledge, the best decision is made by the machine.

11.3 Neural networks

A NN is an algorithm which is used to recognize the patterns just like the human brain. It is a widely used machine learning technique which is used by scientists for solving problems more efficiently and accurately. It is not only used for interpreting the data as well as recognizing the patterns in the form of images, signals, sound waves, etc. The human brain is made up of a connection of several brain cells or neurons. These brain cells or neurons send signals to other neurons just like the message. In the same way, the neuron is the key component of a NN which consists of two parameters, i.e., bias and weights. The inputs to these neurons are multiplied by weights and added together. These added inputs are then given to the activation function which converts these inputs to the outputs. The four layers in the NN are the input layer, the first hidden layer, the second hidden layer, or more hidden layers and output layer. The more the number of layers, the more will be the accuracy of the NN [18,19]. The basic elements of the NN are the following:

1. Neurons
2. Weights
3. Connection between neurons
4. Learning algorithm

The data are required to be trained for predicting the accurate output of the NN. For training the data, there is a need to assign a label for each type of data. After the initialization of weights, all the nodes are activated in the hidden layers which further activates the output layer, i.e., the final output. The initialization of weights done in the above process is random which leads to inaccurate output. The accuracy of the algorithm can be increased by optimizing the weights. The algorithm which is used to optimize the weights is known as the backpropagation algorithm.

In the backpropagation algorithm, the primary initialization of weights is done randomly and afterward, and it is compared with the ideal output i.e., equal to the label. The cost function is used here for calculating the error. The cost function is minimized for optimizing the weights by using the technique known as gradient descent.

NNs have some drawbacks. A lot of computer power is needed for artificial NNs. Understanding NN models is challenging. Careful consideration must be given to data preparation for NN models. It can be difficult to optimize NN models for manufacturing. A lot of computer power is needed for artificial NNs. Understanding NN models is challenging. Careful consideration must be given to data preparation for NN models. It can be difficult to optimize NN models for manufacturing.

11.3.1 Convolutional neural network

The CNN uses convolutional filters for transforming the two-dimensional image into a three-dimensional image [20,21]. It gives superior performance in analyzing two-dimensional images. In CNN, the convolution operation is performed over each image. When more hidden layers are added to the NN, it becomes a deep neural network. Any complex data can be analyzed by adding more layers to the deep neural network [22]. It shows superior performance in various applications of analyzing medical images such as identifying cancer in the blood, etc. [23–25]. It has the following limitations. The position and orientation of an object are not encoded by CNN. It is not capable of spatial invariance with respect to the input data.

11.4 Why deep learning over machine learning

Acquiring the images from the machines and then interpreting them is the first step for analyzing the medical images. The high-resolution medical images (CT, MRI, X-rays, etc.) are extracted from machines with high accuracy. For interpreting these images accurately, various machine learning algorithms are used. But, the main drawback of machine learning algorithms is that they require expert-crafted features for interpreting the images. Also, these algorithms are not reliable because of the huge variation in data of each subject. Here the role of deep learning comes which uses a deep neural network model. In this model, multiple layers of neurons are used, weights are updated, and then, finally, the output is generated. The steps of making a machine learning model are shown in Figure 11.2.

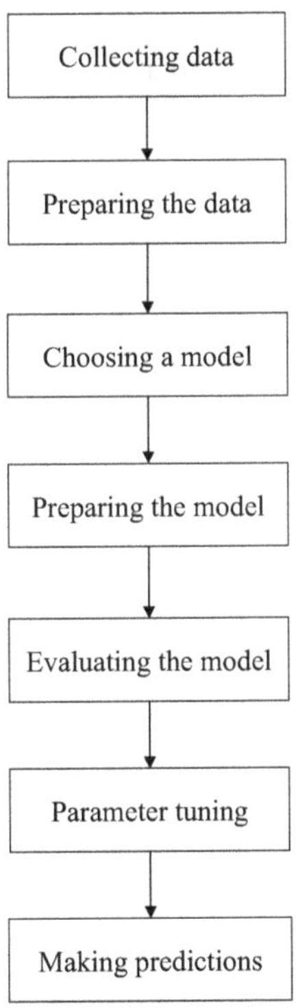

Figure 11.2 Steps of making a machine learning model

11.5 Deep learning applications in medical imaging

Deep learning is quickly establishing itself as the cutting-edge foundation, generating improved results in a variety of medical applications as shown in Figure 11.3. It has been concluded by various authors than deep learning showed superior performance to machine learning methods. These accomplishments created interest to explore more in the area of medical imaging, along with deep learning in medical applications such as digital histopathology, computerized tomography, mammography, and X-rays.

11.5.1 Histopathology

Histopathology is the study of human tissues under a microscope using a sliding glass to determine various diseases including kidney cancer, lung cancer, breast cancer, and others. In histopathology, staining is utilized to visualize a particular area of the tissue [26].

Deep learning is rapidly emerging and improving histopathology images. The challenges in analyzing multi-gigabyte whole slide imaging (WSI) images for developing deep learning models were discussed by Dimitriou *et al.* [27]. In their discussion of many public "Grand Challenges," Serag *et al.* [28] highlight deep learning algorithm innovations in computational pathology.

11.5.2 Computerized tomography

Images of various parts of the body are generated by CT using computers and rotating X-ray equipment. Different areas of the body's soft tissues, blood arteries, and bones can be seen on a CT scan. CT has a high detection efficiency and can spot even small

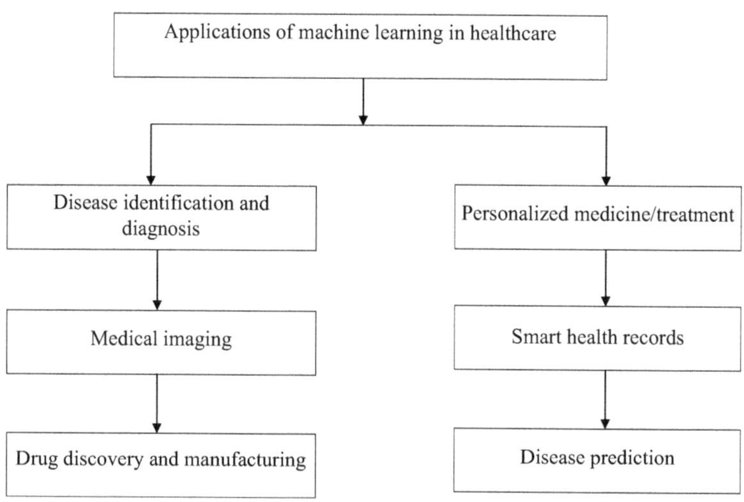

Figure 11.3 Application of machine learning in healthcare

lesions. CT scans also identify pulmonary nodules [29]. To make an early diagnosis of lung cancer, malignant pulmonary nodules must be identified [30,31].

Li *et al.* [32] proposed deep CNN for identifying semisolid, solid, and ground-glass opacity nodules. Balagourouchetty *et al.* [33] suggested a GoogLeNet-based ensemble FCNet classifier for classifying liver lesions.

Three modifications are made to the fundamental Googlenet architecture for feature extraction. To detect and classify the lung nodules, Masood *et al.* [34] presented a multidimensional region-based fully convolutional network (mRFCN), which exhibits 97.91% classification accuracy. Using supervised MSS U-Net and 3DU-Net, Zhao and Zeng (2019) [35] suggested a deep-learning approach to autonomously segment kidneys and kidney cancers from CT images. Further, Fan *et al.* [36] and Li *et al.* [37] used deep learning-based methods for COVID-19 detection from CT images.

11.5.3 Mammograph

Mammograph (MG) is the most popular and reliable method in order to find breast cancer. MG is used to see the structure of the breasts in order to find breast illnesses [38]. A small fraction of the actual breast image is made up of cancers, making it challenging to identify breast cancer on mammography screenings. There are three processes in the analysis of breast lesions from MG: detection, segmentation, and classification [39]. Active research areas in MG include the early detection and automatic classification of masses. The diagnosis and classification of breast cancer have been significantly improved during the past ten years using deep learning algorithms.

Fonseca *et al.* [40] proposed a breast composition categorization model by using the CNN method. Wang *et al.* [41] introduced a novel CNN model to identify Breast Arterial Calcifications (BACs) in mammography images. Without involving humans, a CAD system was developed by Ribli *et al.* [42] for identifying lesions. Wu *et al.* [43] also developed a deep CNN model for the classification of breast cancer. A deep CNN-based AI system was created by Conant *et al.* [44] for detecting calcified lesions.

11.5.4 X-rays

The diagnosis of lung and heart conditions such as hypercardiac inflation, atelectasis, pleural effusion, and pneumothorax, as well as tuberculosis frequently involves the use of chest radiography. Compared to other imaging techniques, X-ray pictures are more accessible, less expensive, and dose-effective, making them an effective tool for mass screening [45].

It was suggested to develop the first deep CNN-based TB screening system by Hwang *et al.* [46] in 2016. Rajaraman *et al.* [47] proposed modality-specific ensemble learning for the detection of abnormalities in chest X-rays (CXRs).

The abnormal regions in the CXR images are visualized using class selective mapping of interest (CRM). For the purpose of detecting COVID-19 in CXR pictures, Loey *et al.* [48] suggested a GAN with deep transfer training. More CXR images were created using the GAN network as the COVID-19 dataset was not available. To create artificial CXR pictures for COVID-19 identification, a CovidGAN model based on the Auxiliary Classifier Generative Adversarial Network (ACGAN) was created by Waheed *et al.* [49].

11.6 Conclusion

Deep learning and machine learning algorithms showed promising results in analyzing medical images as compared to conventional machine learning algorithms. This chapter discusses several supervised, unsupervised, and reinforcement learning algorithms. It gives a broad overview of deep learning algorithm-based medical image analysis. After 10–20 years, it's anticipated that most daily tasks would be automated with the use of deep learning algorithms. The replacement of humans in the upcoming years will be the next step, especially in diagnosing medical images. For radiologists of the future, deep learning algorithm can support clinical choices. Deep learning algorithm enables untrained radiologists to make decisions more easily by automating their workflow. By automatically recognizing and categorizing lesions, a deep learning algorithm is designed to help doctors diagnose patients more accurately. By processing medical image analysis more quickly and efficiently, deep learning algorithm can assist doctors in reducing medical errors and improving patient care. As the healthcare data is quite complex and nonstationary, it is important to select the appropriate deep-learning algorithm to deal with the challenges of medical image processing. Thus, it is concluded that there are numerous opportunities to exploit the various machine learning and deep learning algorithms for enhancing the use of medical images in the healthcare industry.

Conflict of interest

None.

References

[1] Zhou Z., Rahman Siddiquee M.M., Tajbakhsh N., and Liang J. 'UNet++: a nested U-Net architecture for medical image segmentation'. *In Proceedings of the Deep Learning in Medical Image Analysis and Multimodal Learning for Clinical Decision Support—DLMIA 2018*, Granada, Spain, 2018. Springer International Publishing: New York, NY, 2018; 11045, pp. 3–11.

[2] Litjens G., Kooi T, Bejnordi B. E., *et al.* 'A survey on deep learning in medical image analysis'. *Medical Image Analysis*. 2017; **42**:60–88.

[3] Song Z., Lidan W., and Shukai D. 'Memristive pulse coupled neural network with applications in medical image processing'. *Neurocomputing*. 2017; **27**:149–157.

[4] Hussain T. 'ViPS: a novel visual processing system architecture for medical imaging'. *Biomedical Signal Processing and Control*. 2017; **38**:293–301.

[5] Rajalakshmi T. and Prince S. 'Retinal model-based visual perception: applied for medical image processing'. *Biologically Inspired Cognitive Architectures*. 2016; **18**:95–104.

[6] Li Y., Zhang H., Bermudez C., Chen Y., Landman B.A., and Vorobeychik Y. 'Anatomical context protects deep learning from adversarial perturbations in medical imaging'. *Neurocomputing.* 2020; **379**:370–378.

[7] Maier A., Syben C., Lasser T., and Riess C. 'A gentle introduction to deep learning in medical image processing'. *Zeitschrift für Medizinische Physik.* 2019; **29**(2):86–101.

[8] Lundervold A.S. and Lundervold A. 'An overview of deep learning in medical imaging focusing on MRI'. *Zeitschrift fur Medizinische Physik.* 2019; **29**(2):102–127.

[9] Fourcade A. and Khonsari R.H. 'Deep learning in medical image analysis: a third eye for doctors'. *Journal of Stomatology, Oral and Maxillofacial Surgery.* 2019; **120**(4):279–288.

[10] Litjens G., Ciompi F., Wolterink J.M., *et al.* State-of-the-art deep learning in cardiovascular image analysis. *JACC: Cardiovascular Imaging.* 2019; **12** (8):1549–1565.

[11] Zhang J., Xie Y., Wu Q., and Xia Y. 'Medical image classification using synergic deep learning'. *Medical Image Analysis.* 2019; **54**:10–19.

[12] Wong K.K.L., Fortino G., and Abbott D. 'Deep learning-based cardiovascular image diagnosis: a promising challenge'. *Future Generation Computer Systems.* 2020; **110**:802–811.

[13] Sudheer KE. and Shoba Bindu C. 'Medical image analysis using deep learning: a systematic literature review. In *Emerging Technologies in Computer Engineering: Microservices in Big Data Analytics. ICETCE 2019.* Communications in Computer and Information Science Springer: Singapore, 2019, p. 985.

[14] Ker J., Wang L., Rao J., and Lim T. 'Deep learning applications in medical image analysis'. *IEEE Access.* 2018; **6**:9375–9389.

[15] Zheng Y., Liu D., Georgescu B., Nguyen H., and Comaniciu D. '3D deep learning for efficient and robust landmark detection in volumetric data'. In. *LNCS*, Springer, Cham, 2015; **9349**:565–572.

[16] Suzuki K. 'Overview of deep learning in medical imaging'. *Radiological Physics and Technology.* 2017; **10**:257–273.

[17] Suk H.I. and Shen D. 'Deep learning-based feature representation for AD/MCI classification'. *LNCS Springer, Heidelberg. Medical Image Computing and Computer Assisted Intervention.* 2013; **16**:583–590.

[18] Esteva A., Kuprel B., Novoa R.A., *et al.* 'Dermatologist-level classification of skin cancer with deep neural networks'. *Nature.* 2017; **542**:115–118.

[19] Cicero M., Bilbily A., Colak E., *et al.* Training and validating a deep convolutional neural network for computer-aided detection and classification of abnormalities on frontal chest radiographs. *Investigative Radiology.* 2017; **52**:281–287.

[20] Zeiler M.D. and Fergus R. 'Visualizing and understanding convolutional networks'. In: Fleet, D., Pajdla, T., Schiele, B., Tuytelaars, T. (eds.), Computer Vision – ECCV 2014. ECCV 2014. *Lecture Notes in Computer Science*, 8689; 2014 Springer, Cham.

[21] Pearl J. *Probabilistic Reasoning in Intelligent Systems: Networks of Plausible Inference*. San Mateo, CA: Morgan Kaufmann; 1988.
[22] LeCun Y., Bengio Y., and Hinton G. 'Deep learning'. *Nature*. 2015; **521**:436–444.
[23] Cireşan D.C., Giusti A., Gambardella L.M., and Schmidhuber J. 'Mitosis detection in breast cancer histology images with deep neural networks'. *Proceedings of the International Conference on Medical Image Computing and Computer-assisted Intervention*. Springer, 2013; pp. 411–418.
[24] Hinton G., Deng L., Yu D., *et al.* 'Deep neural networks for acoustic modeling in speech recognition: the shared views of four research groups'. *IEEE Signal Processing Magazine*. 2012; **29**(6):82–97.
[25] Russakovsky O., Deng J., Su H., *et al.* 'Imagenet large scale visual recognition challenge'. *International Journal of Computer Vision*. 2015; **115**: 211–252.
[26] Gurcan M.N., Boucheron L.E., Can A., Madabhushi A., Rajpoot N.M., and Yener B. 'Histopathological image analysis: a review'. *IEEE Reviews in Biomedical Engineering*. 2009; **2**:147–171.
[27] Dimitriou N., Arandjelović O., and Caie P.D. 'Deep learning for whole slide image analysis: an overview'. *Frontiers in Medicine*. 2019; **6**:1–7.
[28] Serag A., Qureshi H., McMillan R., Saint Martin M., Diamond J., and Hamilton P. 'Translational AI and deep learning in diagnostic pathology'. *Frontiers in Medicine*. 2019; **6**:1–15.
[29] Ma J., Song Y., Tian X., Hua Y., Zhang R., and Wu J. 'Survey on deep learning for pulmonary medical imaging'. *Frontiers in Medicine*. 2020; **14** (4):450–469.
[30] Murphy A., Skalski M., and Gaillard F. 'The utilisation of convolutional neural networks in detecting pulmonary nodules: a review'. *The British Journal of Radiology*. 2018; **91**(1090):1–6.
[31] Siegel R.L., Miller K.D., and Jemal A. 'Cancer statistics'. *CA: a Cancer Journal for Clinicians*. 2019; **69**(1):7–34.
[32] Li W., Cao P., Zhao D., and Wang J. 'Pulmonary nodule classification with deep convolutional neural networks on computed tomography images'. *Computational and Mathematical Methods in Medicine*. 2016; **2016**:6215085.
[33] Balagourouchetty L., Pragatheeswaran J. K., Pottakkat, B., and Rajkumar G. 'GoogLeNet based ensemble FCNet classifier for focal liver lesion diagnosis'. *IEEE Journal of Biomedical and Health Informatics*. 2020; **24** (6):1686–1694.
[34] Masood A., Sheng B., Yang P., *et al.* 'Automated decision support system for lung cancer detection and classification via enhanced RFCN with multilayer fusion RPN'. *IEEE Transactions on Industrial Informatics*.2020; **16**:7791–7801.
[35] Zhao W. and Zeng Z. *Multi Scale Supervised 3D U-Net for Kidney and Tumor Segmentation*. 2019; 1–7.

[36] Fan D.-P., Zhou T., Ji G.P., *et al.* 'Inf-Net: automatic COVID-19 lung infection segmentation from CT scans'. *IEEE Transactions on Medical Imaging*. 2020; **39**(8):2626–2637.

[37] Li L., Qin L., Xu Z., *et al.* 'Artificial intelligence distinguishes COVID-19 from community acquired pneumonia on chest CT'. *Radiology*. 2020; **296** (2):E65–E71.

[38] Gardezi S.J.S., Elazab A., Lei B., and Wang T. 'Breast cancer detection and diagnosis using mammographic data: systematic review'. *Journal of Medical Internet Research*. 2019; **21(7)**:1–22.

[39] Shen L., Margolies L.R., Rothstein J.H., Fluder E., McBride R., and Sieh W. 'Deep learning to improve breast cancer detection on screening mammography'. *Scientific Reports*. 2019; **9**(1):1–13.

[40] Fonseca P., Mendoza J., Wainer J., *et al.* 'Automatic breast density classification using a convolutional neural network architecture search procedure'. In *Proceedings of Medical Imaging 2015: Computer Aided Diagnosis*, 2015; p. 941428.

[41] Wang J., Ding H., Bidgoli F.A., *et al.* 'Detecting cardiovascular disease from mammograms with deep learning'. *IEEE Transactions on Medical Imaging*. 2017; **36**(5):1172–1181.

[42] Ribli D., Horvath A., Unger Z., Pollner P., and Csabai I. 'Detecting and classifying lesions in mammograms with deep learning'. *Scientific Reports*. 2018; **8**(1):4165.

[43] Wu N., Phang J., Park J., *et al.* 'Deep neural networks improve radiologists' performance in breast cancer screening'. *IEEE Transactions on Medical Imaging*. 2020; **39**:1184–1194.

[44] Conant E.F., Toledano A.Y., Periaswamy S., *et al.* 'Improving accuracy and efficiency with concurrent use of artificial intelligence for digital breast tomosynthesis'. *Radiology: Artificial Intelligence*. 2019; **1**(4):e180096.

[45] Candemir S., Rajaraman S., Thoma G., and Antani S. 'Deep learning for grading cardiomegaly severity in chest x-rays: an investigation'. In *Proceedings of IEEE Life Sciences Conference (LSC)*. 2018, pp. 109–113.

[46] Hwang S., Kim H.-E., Jeong J., and Kim H.-J. 'A novel approach for tuberculosis screening based on deep convolutional neural networks'. In *Proceedings of Medical Imaging 2016: Computer Diagnosis*. 2016; 9785, p. 97852W.

[47] Rajaraman S. and Antani S.K. 'Modality-specific deep learning model ensembles toward improving TB detection in chest radiographs'. *IEEE Access*. 2020; **8**:27318–27326.

[48] Loey M., Smarandache F., and Khalifa N.E.M. 'Within the lack of chest COVID-19 X-ray dataset: a novel detection model based on GAN and deep transfer learning'. *Symmetry*. 2020; **12**(4):651.

[49] Waheed A., Goyal M., Gupta D., Khanna A., Al-Turjman F., and Pinheiro P. R. 'CovidGAN: data augmentation using auxiliary classifier GAN for improved Covid-19 detection'. *IEEE Access*. 2020; **8**:91916–91923.

Chapter 12

Systemic review of deep learning techniques for high-dimensional medical image fusion

Nigama Vykari Vajjula[1], Vinukonda Pavani[2], Kirti Rawal[3] and Deepika Ghai[3]

In recent years, the research on medical image processing techniques plays a major role in providing better healthcare services. Medical image fusion is an efficient approach for detecting various diseases in different types of images by combining them to make a fused image in real-time. The fusion of two or more imaging modalities is more beneficial for interpreting the resulting image than just one image, particularly in medical history. The fusion of two images refers to the process of the combined output generated from multiple sensors to extract more useful information. The application of deep learning techniques has continuously proved to be more efficient than conventional techniques due to the ability of neural networks to learn and improve over time.

Deep learning techniques are not only used due to reduced acquisition time but also to extract more features for the fused image. So, in this chapter, the review of image fusion techniques proposed in recent years for high-dimensional imaging modalities like MRI (magnetic resonance imaging), PET (positron emission tomography), SPECT (single photon emission-computed tomography), and CT (computed tomography) scans are presented. Further, a comparative analysis of deep learning algorithms based on convolutional neural networks (CNNs), generative models, multi-focus and multi-modal fusion techniques along with their experimental results are discussed in this chapter. Afterward, this chapter gives an overview of the recent advancements in the healthcare sector, the possible future scope, and aspects for improvements in image fusion technology.

12.1 Introduction

Raw image data in most cases have limited information. For example, the focus location is different and the object closest or farther from the image appears to be blurred. In cases of medical diagnosis, it is confusing and difficult for the doctor to identify the problem and provide better care. Image fusion technology is increasingly

[1]DRDO New Delhi, India
[2]Department of Biomedical Engineering, Manipal hospital, India
[3]Lovely Professional university, India

being applied in diagnosing diseases as well as analyzing patient history. There are different types of classification schemes that work toward fighting these anomalies and getting as much data from the image as possible [1].

It is evident from recent studies that image fusion techniques like multi-level, multi-focus, multimodal, pixel-level, and others can aid medical practitioners to arrive at a better, unbiased decision based on the quantitative assessment provided by these methods. Image fusion can be studied primarily in four categories such as (i) signal level, (ii) pixel level, (iii) feature level, and (iv) decision level which will be further explored in the upcoming sections as shown in Figure 12.1.

High-dimensional imaging modalities like CT, MRI, PET, and SPECT are prevalent imaging techniques which are used in medical diagnosis where the information is captured from several angles. In clinical settings, there are many problems in the comparison and synthesis of image formats such as CT with PET, MRI with PET, and CT with MRI. So, in order to produce more fruitful information for medical diagnosis, it is necessary to combine images from multiple sources. Although, it is very difficult to show clear views of organs in our body for identifying life-threatening diseases like cancer. Tumors in the brain can be detected by fusing MRI and PET images. Further, abdomen-related problems can be identified by fusing the SPECT and CT scans, and fusion of ultrasound images with MRI gives the vascular blood flow analysis [2]. This procedure is termed as multimodal image fusion which will further be discussed in this chapter.

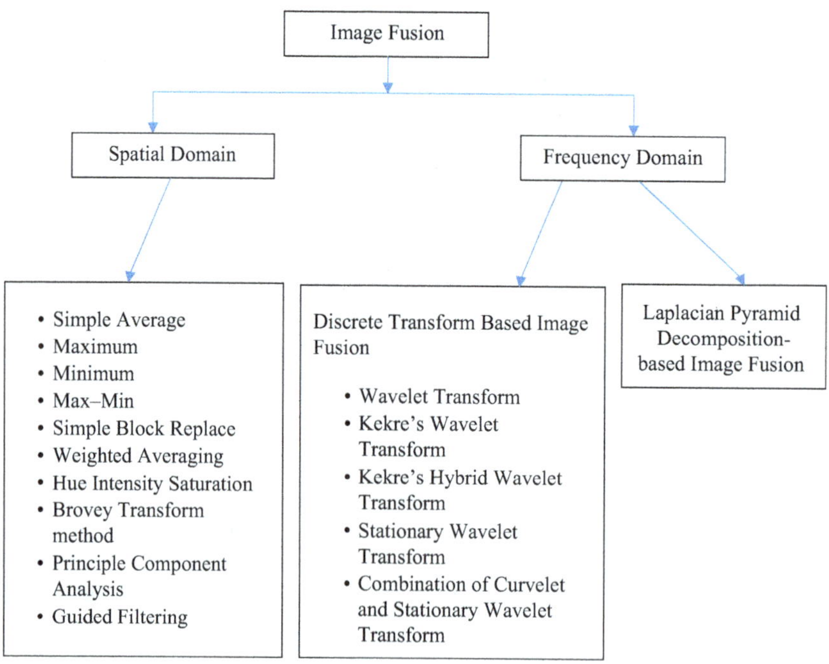

Figure 12.1 Image fusion

Numerous image fusion methods were designed to identify a specific disease. These techniques are mainly directed toward solving the three major challenges in fusing medical images such as image reconstruction, feature extraction, and feature fusion. Most of the authors [1,3,4] concentrated on different applications of image fusion technology, but have missed recent techniques on medical image fusion like multi-spectral imaging, morphological component analysis, and U-Net models of hyperspectral images. So, in this chapter, the most recent and effective solutions for medical image fusion by using deep learning algorithms have been investigated.

12.2 Basics of image fusion

All objects in focus cannot be obtained because of the limited depth of the focal point of the image. In order to obtain images in which all objects are concentrated, it is necessary to merge multiple focused image fusion processes to produce a clear view for humans as well as for the perception of machines. The fusion techniques are classified into two parts: (i) spatial domain (pixel-level) and (ii) transform domain [1]. In the former method, the source images are combined into a single image, and in the latter, the images are converted to the frequency domain where images are fused using the results of Fourier and inverse Fourier transforms.

To obtain significant information from the results, the role of each input image is essential [3]. Spatial domain methods are said to retain more original information compared to other feature or decision-level fusions. Some of the image fusion algorithms using spatial domains are simple averaging, Brovey method, principal component analysis (PCA), and intensity hue saturation (IHS). Since we directly deal with image pixels, it is possible that the values of each pixel can be easily manipulated to obtain the desired results.

Another drawback of these methods is that the fused images are incorporated with spatial distortion which increases the noise and leads to misregistration [3]. The problem of spatial distortion can become critical while dealing with classification steps [4].

Frequency domain methods and multi-resolution analysis methods are further used to overcome the problem of spatial distortion. The methods such as discrete wavelet transform, Laplacian transform, and curvelet transform have shown significantly better results in fusing the images as well as avoiding spectral and spatial distortion [4]. Hence, the applications of transform-level fusion techniques range from medical image analysis and microscope imaging to computer vision and robotics. The fusion techniques are further divided into three levels such as pixel level, feature level, and decision level as shown in Figure 12.2.

12.2.1 Pixel-level medical image fusion

Pixel-level medical image fusion method uses the original information from the source images. It has been used in various applications, such as remote sensing, medical diagnosis, surveillance, and photography applications. The main advantage of these pixel-level methods is that they are fast and easier to implement. However, the

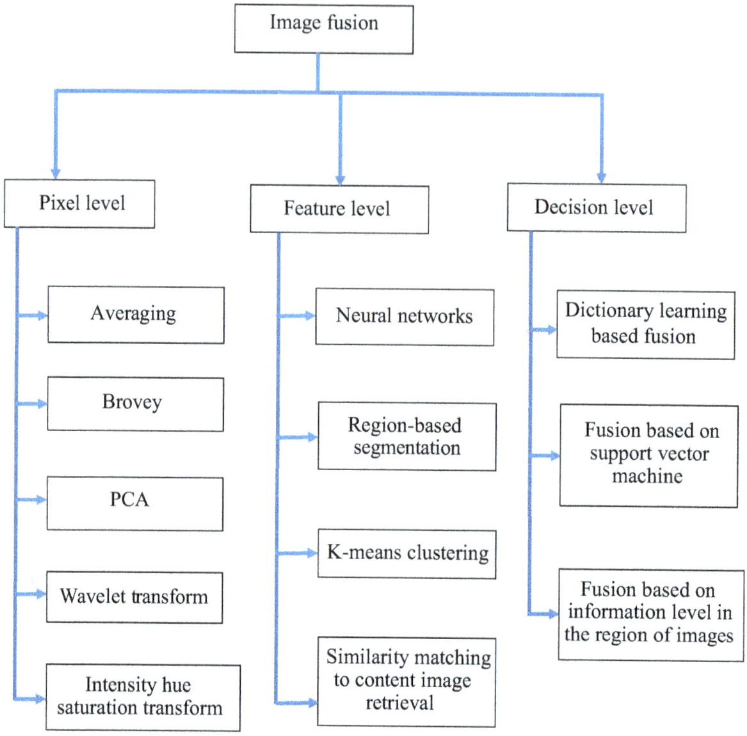

Figure 12.2 Level classification of image fusion methods

limitation is that they rely heavily on the accurate assessment of weights for different pixels. If the estimation is not accurate, then it limits the performance of fusion.

Image fusion can be done by taking the average pixel intensity values for fusing the images. These are also called averaging methods and don't require prerequisite information about the images. There are also techniques based on prior information which can be more beneficial in terms of medical diagnosis.

While dealing with the fusion of medical images, radiologists must be aware of the input images such as PET with CT or MRI with PET. Pixel-based techniques also use fuzzy logic to handle imprecise information from the images received from the radiologists. The models that can be built using fuzzy logic are mentioned in detail in [5]. Fuzzy inference system (FIS) is the multimodal image fusion techniques used in medical diagnosis. By selecting the right parameters to compute these models, good results can be obtained with less computational cost [3]. The main disadvantages of these approaches are the requirement of large data for processing which further decreases the contrast of the fused image.

12.2.2 Transform-level medical image fusion

To improve on the information loss caused by spatial domain techniques, transform-based methods like multi-scale decomposition are used. In any transformation process of medical image fusion, there are three steps involved which are clearly described in [3]. Fourier transform and wavelet transforms are one of the famous techniques used for medical image processing. Wavelet transform covers the time domain information that cannot be obtained from Fourier transform [6].

Another significant transformation method is the contourlet transform which has better efficiency and directionality [7]. The contourlet transform is different from other transforms by accepting input at every scale. It also obtains high levels of efficiency in image representation which further produces redundancy. Despite its drawbacks, contourlet transform is popular due to its fixed nature which is the main feature in improving the efficiency of image fusion.

12.2.3 Multi-modal fusion in medical imaging

As introduced in the previous section, there are many imaging modalities that can be used in the fusion of medical images. Multi-modal approaches are especially used to keep the qualities of the individual modality and combine them with other advanced techniques to get more information. Magnetic resonance imaging is particularly known for its clear images of soft tissues, nervous system, and fat distribution, but it is difficult to identify bones because of their low proton density. However, computer tomography works based on an X-ray principle to produce sectional images of the organ and bones depending on the requirement. Basically, MRI contains more information and hence give superior performance than CT images. CT images provide clearer images of the human body. The multimodal fusion of MRI and CT scans can address the shortcomings of these individual modalities and give the best results from both scans. This also eliminates the need for the development of new devices and aids in cost reduction. Similarly, SPECT is an imaging modality to visualize the blood flow in arteries, the metabolism of human tissues, and identifying malignant tumors. However, the SPECT images have low resolution in comparison to PET images.

Multi-modal image fusion can be used to resolve all the issues in the best possible manner. The most well-known methods among these in medical diagnosis are MRI-PET, MRI-CT, and MRI-SPECT fusions. The work of Bing Huang *et al.* [7] provides a great insight into each of the fusion techniques, their trends, and comparative analysis in various diagnostic applications.

12.3 Deep learning methods

Deep learning takes advantage of artificial neural networks that have the capability to understand and process input information and make decisions. This capability of neural networks to predict, examine, and understand information from a given dataset makes it different from traditional approaches. The training and testing of these neural networks and observing the changes in predictions over time will enable several applications for solving various problems. Given the incredible features of neural networks, it becomes a very tedious task to justify their superiority in comparison to other imaging techniques. As these methods often depend on the quality of the training and testing images which varies with different imaging conditions.

The traditional fusion methods [1,3,4] make use of mathematical transformations for manually analyzing the fusion rules in spatial and transform domains. The drawbacks of these techniques have been very apparent and there was a need to introduce deep learning algorithms for adopting innovative transformations in feature extraction and feature classification. Deep learning is a way to improve medical image fusion, by taking advantage of better-level measurements and well-designed loss functions to obtain more targeted features. There are numerous methods proposed over time which address the problems with the previous ones or introduce entirely new methods. Some methods are better than others because they can be used for batch processing (processing multiple images at once) and it results in images with better detail and clarity. In addition, it advances the proficiency of detecting diseases and reduces the time to recover from the suggested cures.

The initial steps of developing a deep learning model involve pre-processing a large number of images and then dividing them into training and testing data sets. Afterward, the model for fusing the images and the related optimized factors are created. The final step is to test the model by inputting several sets of images and batch-processing multiple group images. The two famous methods in recent years for achieving effective medical image fusion are CNN- and generative adversarial network (GAN)-based techniques. Here, we focus on deep learning methods that are particularly useful in medical diagnosis and imaging modalities.

12.3.1 Image fusion based on CNNs

The existing image fusing techniques have some disadvantages, such as needing artificial design and having a small correlation between the different features. Further, CNN has been applied for fusing the images in 2017 [8–11]. The proposed methods discuss the potential for a convolutional neural network to be successful in the field of image fusion. The convolutional layer is responsible for feature extraction and weighting the average in order to produce the output image [12–14].

The U-Net architecture was one of the most influential papers released in 2015 [15]. Although there are advanced algorithms that may perform better at segmentation, U-Net is still very popular since it can achieve good results on limited training samples. This network was trained end-to-end, which was said to have outperformed the previous best method at the time (a sliding window CNN). With U-Net, the input image first goes through a contraction path, where the input size is downsampled, and then it has an expansive path, where the image is upsampled. In between the contraction and expansive paths, it has to skip connections. This architecture was developed to not only understand what the image was but also to get the location of the object and identify its area. Since it was developed keeping medical images in mind, we can say that it can perform well even with some unlabeled/unstructured data. This model is widely used for segmenting the images but fusing the images is a new area of experimentation for improving the spatial resolution of hyperspectral images [16].

IFCNN is another general image fusion framework that comprises three main components: (i) feature extraction, (ii) feature fusion, and (iii) image reconstruction [17]. For training the model, the image dataset has been generated. The perceptual loss has been introduced for generating fused images that are more similar to the ground-truth fused images.

12.3.2 Image fusion by morphological component analysis

Morphological component analysis (MCA) can be used to create a more complete representation of an image than traditional methods. It can be used to obtain sparse representations of multiple components in the image [18]. This is because MCA uses the morphological diversity of an image, which means that it can take into account different structures within the image. The advantage of this is that it can generate more clear fused images in comparison to the existing methods.

While using MCA for fusing the images, first the input image needs to be disintegrated into cartoon and textured components. Afterward, both these components are combined as per the well-defined fusion rules. This combination will not only be used for representing the sparse coefficients but also for the entire image. The process in [18] is better used, representing the sparse coefficients due to the in-built characteristics and structures present in the input images.

12.3.3 Image fusion by guided filtering

The proposed method is guided filtering is based on decomposing an image into two layers: large-scale intensity variations are present in the base layer, and small-scale details are present in the detail layer. In the proposed method, the two layers are then fused to obtain spatial consistency. The experimental results have shown that this proposed method can produce better results than existing methods for the fusion of multispectral, multifocus, multimodal, and multi-exposure images. In guided filtering, the average filter is used for creating the two-scale representations of the source images. Both the layers such as the base layer and detail layer of each image are fused together using a weighted average method.

12.3.4 Image fusion based on generative adversarial network (GAN)

GANs are the most straightforward generative models that are capable of learning to generate plausible data. Through GANs, it is possible to generate a neural network to produce samples that implicitly define a probability distribution. Since these models have been widely used for feature extraction, feature fusion, and image registration [17,19]. The idea behind GANs is a discriminator network which works to classify an observation from a training set. These are probably the conventional methods in deep learning for generating new and quality information.

Deep convolutional GANs are a more stable version of GAN models [20]. The recent work on deep convolutional GANs [21] integrates two modules in its network architecture alongside dense blocks which results in medical images with rich information. The proposed method claims to address the weakness of active feature fusion of the traditional methods by manual design through which it can process the intermediate layer to avoid information loss.

GFPPC-GAN [22] introduces GAN by fusing generative facial prior (GFP) and PC images, which an employs adversarial learning process between the PC image and the fused image for improving the quality of information present in the image.

Although GANs can perform exceptionally well in medical image fusion, the intensity level in the pixels in the functional image is far greater than the structural information in the image. Most GAN models introduce a new challenge to medical image fusion using GANs as the probability of feature imbalance can be frequent [23–27].

12.3.5 Image fusion based on autoencoders

Autoencoders are feed-forward neural networks, and they take a particular variable as an input and predict the same function. These are usually used to map high-dimensional data in 2D for visualization. Autoencoders are also known for reducing the size of the input data given that they are paired with another supervised or unsupervised task. Deep autoencoders on the other hand are nonlinear and can learn from more power sources for a given dimensionality compared with linear autoencoders.

Autoencoders are mainly used in remote sensing, satellite imagery, and other image fusion categories and are fairly new approaches especially in medical image processing compared to CNNs and GANs. However, there are new techniques that are aiming to develop autoencoders for image fusion technology. Autoencoders based on multi-spectral image fusion [28] is a deep learning technique which can be very effective for medical image fusion. The intervention of this proposed work is a deep learning-based sharpening method for the fusion of panchromatic and multispectral images.

Deep autoencoders on the other hand have achieved superior performance in comparison with the conventional transform coding methodology [29]. The three interventions used are deep autoencoder with multiple backpropagation (DA MBP), deep autoencoder with RBM (DA RBM), and deep convolutional autoencoder with RBM (DCA RBM) [14,30–36]. The process of image fusion is shown in Figure 12.3.

12.4 Optimization methods

The optimization methods for deep learning in image fusion include noise reduction, image registration, and other pre-processing techniques applied to a variety of images. These large numbers of images in the datasets will be further divided into training datasets and testing datasets as per the application. Afterward, optimization techniques will be used for classifying the images. In model learning, the metrics of the model are learned by assigning labels to various images (training). In the later step, the testing is done for predicting the output for unknown input. The final iteration of the test subjects gives the fused image.

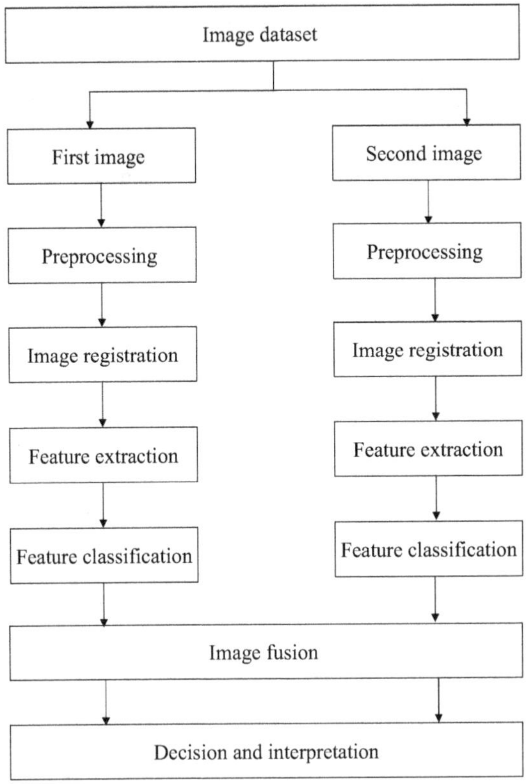

Figure 12.3 Process of image fusion

12.4.1 Evaluation

Operational efficiency is the most significant factor for measuring the performance of fusion using deep learning methods. Experiment results on the public clinical diagnostic medical image dataset show that the GAN-based algorithms have tremendous detail preservation features and it can remove the artifacts which leads to superior performance in comparison to other methods. The GAN- and CNN-based methods are reported to have results that are high in efficiency due to their common characteristics such as simple network architecture and low model parameters. A simple network structure, more appropriate tasks-specific constraints, and optimization methods can be designed, achieving good accuracy and efficiency. The advancement of these algorithms allows researchers to analyze the properties of image fusion tasks before increasing the size of the neural network [37–43].

12.5 Conclusion

Medical image fusion plays an essential role in providing better healthcare services. Due to the advancements in multi-focus image fusion methods, the existing classification methods failed to accurately position all images. It is concluded from the literature that deep learning techniques give superior performance in fusing medical images and provide insights into each of those techniques. Medical image fusion is a technique that can be used for the diagnosis and assessment of medical conditions. In this chapter, we present a summary of the major modalities that are used for medical imaging fusion, their applications in diagnosis, assessment, and treatment, as well as a brief overview of the fusion techniques and evaluations based on the observed data.

References

[1] Deepak Kumar S. and Parsai M.P. 'Different image fusion techniques–a critical review'. *International Journal of Modern Engineering Research (IJMER)*. 2012; 2(5): 4298–4301.

[2] Benjamin Reena J. and Jayasree T. 'An efficient MRI-PET medical image fusion using non-subsampled shearlet transform'. In *Proceedings of the IEEE International Conference on Intelligent Techniques in Control, Optimization and Signal Processing (INCOS)*, 2019. pp. 1–5.

[3] Galande A. and Patil R. 'The art of medical image fusion: a survey'. In *Proceedings of the 2013 International Conference on Advances in Computing, Communications and Informatics (ICACCI)*, 2013. pp. 400–405.

[4] Chetan Solanki K. and Narendra Patel M. 'Pixel based and wavelet based image fusion methods with their comparative study'. In *Proceedings of the National Conference on Recent Trends in Engineering & Technology*, 2011.

[5] Irshad H., Kamran M., Siddiqui A.B., and Hussain A. 'Image fusion using computational intelligence: a survey'. In *Proceedings of the Second International Conference on Environmental and Computer Science*, ICECS '09, 2009. pp. 128–132.

[6] Guihong Q., Dali Z., and Pingfan Y. 'Medical image fusion by wavelet transform modulus maxima'. *Optics Express*. 2001; 9: 184–190.

[7] Bing H., Feng Y., Mengxiao Y., Xiaoying M., and Cheng Z. 'A review of multimodal medical image fusion techniques'. *Computational and Mathematical Methods in Machine Learning*. 2020; 2020: 1–16.

[8] Liu Y., Chen X., Cheng J., and Peng H. 'A medical image fusion method based on convolutional neural networks'. In *Proceedings of the 20th International Conference on Information Fusion (Fusion). IEEE*, 2017, pp. 1–7.

[9] Liu Y., Chen X., Ward R.K., and Wang Z.J. 'Image fusion with convolutional sparse representation'. *IEEE Signal Processing Letters*. 2016; 23(12): 1882–1886.

[10] Liu Y., Chen X., Ward R.K., and Wang Z.J. 'Medical image fusion via convolutional sparsity based morphological component analysis'. *IEEE Signal Processing Letters*. 2019; 26(3): 485–489.

[11] Liu Y., Liu S., and Wang Z. 'A general framework for image fusion based on multi-scale transform and sparse representation'. *Information Fusion*. 2015; 24: 147–164.

[12] Pajares G. and De La Cruz J.M. 'A wavelet-based image fusion tutorial'. *Pattern Recognition*. 2004; 37: 1855–1872.

[13] Li S., Kang X., and Hu J. 'Image fusion with guided filtering'. *IEEE Transactions on Image Processing*. 2013; 22: 2864–2875.

[14] Li S., Yang B., and Hu J. 'Performance comparison of different multi-resolution transforms for image fusion'. *Information Fusion.* 2011; 12(2)12: 74–84.

[15] Olaf R., Philipp F., and Thomas B. 'U-Net: convolutional networks for biomedical image segmentation'. In *IEEE Conference on Computer Vision and Pattern Recognition*. 2015, pp. 1–8.

[16] Xiao J., Li J., Yuan Q., and Zhang L. 'A Dual-UNet with multistage details injection for hyperspectral image fusion'. *IEEE Transactions on Geoscience and Remote Sensing*. 2022; 60: 1–13.

[17] Zhang Y., Liu Y., Sun P., Yan H., Zhao X., and Zhang L. 'IFCNN: a general image fusion framework based on convolutional neural network'. *Information Fusion*. 2020; 54; 99–118.

[18] Jiang Y. and Wang M. 'Image fusion with morphological component analysis'. *Information Fusion*. 2014; 18: 107–118.

[19] Zhao C., Wang T., and Lei B. 'Medical image fusion method based on dense block and deep convolutional generative adversarial network'. *Neural Computing and Applications*. 2020; 33(12): 6595–6610.

[20] Zhiping X. 'Medical image fusion using multi-level local extrema'. *Information Fusion*. 2014; 19: 38–48.

[21] Le, Z., Huang J., Fan F., Tian X., and Ma J. 'A generative adversarial network for medical image fusion'. In *Proceedings of the IEEE International Conference on Image Processing (ICIP)*, 2020. pp. 370–374.

[22] Tang W., Liu Y., Zhang C., Cheng J., Peng H., and Chen X. 'Green fluorescent protein and phase-contrast image fusion via generative adversarial networks'. *Computational and Mathematical Methods in Medicine*. 2019; Article ID 5450373:1–11.

[23] Bavirisetti D.P., Kollu V., Gang X., and Dhuli R. 'Fusion of MRI and CT images using guided image filter and image statistics'. *International Journal of Imaging Systems and Technology*. 2017; 27(3): 227–237.

[24] Burt P.J. and Adelson E.H. 'The Laplacian pyramid as a compact image code'. *IEEE Transactions on Communications*. 1983; 31(4): 532–540.

[25] Ding Z., Zhou D., Nie R., Hou R., and Liu Y. 'Brain medical image fusion based on dual-branch CNNs in NSST domain'. *BioMed Research International*. 2020; 2020: 6265708.

[26] Du J., Li W., Xiao B., and Nawaz Q. 'Union Laplacian pyramid with multiple features for medical image fusion'. *Neurocomputing.* 2016; 194: 326–339.

[27] Eckhorn R., Reitboeck H.J., Arndt M., and Dicke P. 'Feature linking via synchronization among distributed assemblies: simulations of results from cat visual cortex'. *Neural Computation*. 1990; 2(3): 293–307.

[28] Azarang A., Manoochehri H.E., and Kehtarnavaz N. 'Convolutional autoencoder-based multispectral image fusion'. *IEEE Access*. 2019; 7: 35673–35683.

[29] Saravanan S. and Juliet S. 'Deep medical image reconstruction with auto-encoders using Deep Boltzmann Machine Training'. *EAI Endorsed Transactions on Pervasive Health and Technology*. 2020; 6(24): 1–9.
[30] Ganasala P. and Kumar V. 'CT and MR image fusion scheme in non-subsampled contourlet transform domain'. *Journal of Digital Imaging*. 2014; 27(3): 407–418.
[31] Gomathi P.S. and Bhuvanesh K. 'Multimodal medical image fusion in non-subsampled contourlet transform domain'. *Circuits and Systems*. 2016; 7(8): 1598–1610.
[32] Gong J., Wang B., Qiao L., Xu J., and Zhang Z. 'Image fusion method based on improved NSCT transform and PCNN model'. In *Proceedings of the 9th International Symposium on Computational Intelligence and Design (ISCID)*. IEEE, 2016. pp. 28–31.
[33] James A.P. and Dasarathy B.V. 'Medical image fusion: a survey of the state of the art'. *Information Fusion*. 2014; 19: 4–19.
[34] Kaur H., Koundal D., and Kadyan V. 'Image fusion techniques: a survey'. *Archives of Computational Methods in Engineering*. 2021;28 : 1–23.
[35] Keith A. and Johnson J.A.B. Whole brain atlas. http://www.med.harvard.edu/aanlib/. Last accessed on 10 April 2021.
[36] Li B., Peng H., and Wang J. 'A novel fusion method based on dynamic threshold neural p systems and nonsubsampled contourlet transform for multi-modality medical images'. *Signal Processing*. 2021; 178: 107793.
[37] Mankar R. and Daimiwal N. 'Multimodal medical image fusion under non-subsampled contourlet transform domain'. In *Proceedings of the International Conference on Communications and Signal Processing (ICCSP)*. IEEE, 2015. pp. 0592–0596.
[38] Nazrudeen M., Rajalakshmi M.M., and Suresh Kumar M.S. 'Medical image fusion using non-subsampled contourlet transform'. *International Journal of Engineering Research (IJERT)*. 2014; 3(3): 1248–1252.
[39] Polinati S. and Dhuli R. 'A review on multi-model medical image fusion'. In *Proceedings of the International Conference on Communication and Signal Processing (ICCSP)*. IEEE, 2019. Pp. 0554–0558.
[40] Polinati S. and Dhuli R. 'Multimodal medical image fusion using empirical wavelet decomposition and local energy maxima'. *Optik*. 2020; 205: 163947.
[41] Tan W., Thiton W., Xiang P., and Zhou H. 'Multi-modal brain image fusion based on multi-level edge-preserving filtering'. *Biomedical Signal Processing and Control*. 2021; 64: 102280.
[42] Tian Y., Li Y., and Ye F. 'Multimodal medical image fusion based on nonsubsampled contourlet transform using improved PCNN'. In *Proceedings of the 13th International Conference on Signal Processing (ICSP)*. IEEE, 2016. pp. 799–804.
[43] Tirupal T., Mohan B.C., and Kumar S.S. 'Multimodal medical image fusion techniques-a review'. *Current Signal Transduction Therapy*. 2020; 15(1): 1–22.

Chapter 13

Qualitative perception of a deep learning model in connection with malaria disease classification

R. Saranya[1], *U. Neeraja*[1], *R. Saraswathi Meena*[1] *and T. Chandrakumar*[1]

Malaria is a potentially fatal blood illness spread by mosquitoes. Frequent signs of malaria include fever, exhaustion, nausea, and headaches. In extreme circumstances, it may result in coma, jaundice, convulsions, or even death. Ten to fifteen days after being bitten by an infected mosquito, symptoms often start to manifest. People may experience relapses of the illness months later if they are not appropriately treated. Even though malaria is uncommon in areas with a moderate climate, it is nevertheless ubiquitous in countries that are tropical or subtropical.

Plasmodium-group single-celled microorganisms are the primary cause of malaria. It only spreads by mosquito bites from infected *Anopheles* species. Through a mosquito bite, the parasites from the insect's saliva enter the victim's bloodstream. The liver is the destination of the parasites, where they develop and procreate. Humans are capable of transmitting five different *Plasmodium* species. *P. falciparum* is mostly responsible for fatal cases, but *Plasmodium vivax*, *Plasmodium ovale*, and *Plasmodium malaria* typically result in a less severe type of malaria.

Rarely the *Plasmodium knowlesi* species can harm people. Antigen-based fast diagnostic tests or microscopic inspection of blood on blood films are frequently used to detect malaria. Although there are techniques that employ the polymerase chain reaction to find the parasite's DNA, their expense and complexity prevent them from being extensively used in places where malaria is a problem.

As a result, gathering all photographs of a person's parasitized and uninfected cells taken under a microscope will enable classification to determine if the individual is afflicted or not. Convolution neural network (CNN) architecture, one of the methodologies in the field of deep learning, is the technique applied in this case. A part of machine learning is called deep learning. Using data and algorithms, machine learning enables computers to learn autonomously. Reinforcement learning, unsupervised learning, and supervised learning are all components of machine learning.

[1]Department of Applied Mathematics and Computational Science, Thiagarajar College of Engineering, India

13.1 Image classification

The first image classification process involves labeling a picture into a group according to its visual content. For picture classification, there are four fundamental phases. Image pre-processing comes first. The purpose of this method is to enhance some key picture features and suppress undesired distortions in order to improve the image data, which was previously known as an attribute, so that computer vision models may use it to their advantage. Image pre-processing includes the following steps: reading the picture, resizing the image, and data augmentation. The second phase is object detection, which includes the localization of an object, which entails segmenting the picture and locating the object of interest.

Feature extraction and training are the most important phases of the picture classification process. The most crucial stage of image categorization is when the most intersecting patterns from the picture are discovered using deep learning or statistical approaches. Extracting features that could be exclusive to a class will help the model distinguish between them in the future. Model training is the process through which the model learns the characteristics from the dataset. The classification of the image to that relevant class was the last phase of this procedure. Using an appropriate classification approach that compares the picture patterns with the target patterns, this stage places recognized items into predetermined classes.

13.1.1 Deep learning

Deep learning is a branch of machine learning that focuses on a family of algorithms called artificial neural networks that are inspired by the structure and function of the brain. Deep learning has been useful for speech recognition, language translation, and image categorization. It can tackle any pattern recognition problem without any human intervention. Deep learning is powered by artificial neural networks, which include several layers. Such networks include deep neural networks (DNNs), where each layer is capable of carrying out complicated operations like representation and abstraction to make sense of text, voice, and picture data.

Information is fed into deep learning systems as massive data sets since they need a lot of information to get correct findings. Artificial neural networks can categorize data while processing it using the responses to a series of binary yes or false questions involving extremely difficult mathematical computations.

13.2 Layers of convolution layer

13.2.1 Convolution neural network

The reason for CNN instead of artificial neural network is because they can retain spatial information as they take images in the original format [1]. They work with both RGB and grayscale images. An image is represented as an array of pixel values. If it is a grayscale image then it is represented as (height, width, 1) or (height, width) by default. If it is an RGB image then it is represented as (height, width, 3) where 3 is the number of color channels. So grayscale can be also said as

2D array and RGB as 3D array. There are different types of layers in the CNN: convolutional layer, pooling layer, flatten layer, and fully connected layer (dense).

13.2.1.1 Convolution layer

This is the first layer in CNN architecture. There can be many layers of this same type. The first layer takes the images as input and extracts the features from an image while maintaining the pixel values [2]. The important parameters in the convolution layer are the following:

Filters: *The number of filters also known as kernel. This refers to the depth of the feature map considered.*
Kernel size: *This specifies the height and width of the kernel (convolution) window. It takes an integer or a tuple of two integers like (3, 3). The window is typically a square with equal height and breadth. A square window's size can be provided as an integer, such as 3 for a window with the dimensions (3, 3).*
Strides: *The number of pixels that we move the filter over the input image. For the steps along the height and breadth, this requires a tuple. The default setting is (1, 1).*
Padding*: There are two choices Valid or same. valid refers to no padding. Similar outcomes when padding with zeros such that the feature map's size is the same as the input's size when strides are equal to 1.*

13.2.1.2 Activation function

The activation function refers to the non-linear change we do on the input signal. This modified output serves as input to the layer of neurons below. By default, use the "ReLU" activation function in the convolution layer [3–5]. The input shape parameter specifies the height, width, and depth of the input image or simply refers to the size of the image. It is compulsory to add this parameter in a first convolutional layer that is the first layer in the model immediately after the input layer. Then it is not included in other intermediate layers.

13.2.1.3 Convolution operation

The convolution operation is an elementwise multiply-sum operation between an image section and the filter. It outputs the feature map which will be the input for the next pooling layer. The number of distinct picture sections that we were able to produce by sliding the filter(s) on the image is equal to the number of elements in the feature map. If the image is RGB, the filter must have three channels. Because an RGB image has three color channels and three channel filters are needed to do the calculations [6,7].

In this instance, the computation takes place as before on each relevant channel between the picture segment and the filter. The final result is obtained by adding all outputs of each channel's calculations.

13.2.1.4 Pooling layer

The next layer in CNN architecture is the pooling layer. Commonly convolution and pooling layer is used together as a pair. There can be any number of pooling layers in a CNN. Pooling layer extracts the most important features from the output (feature map) by averaging or getting a maximum number. And it reduces the dimension of feature maps and reduces the parameters to be learned and computation complexity in the network.

There are two types of pooling layers: max pooling and average pooling. Max pooling is a pooling operation that selects the most detail from the location of the feature map included via way of means of the filter. Thus, the output after the max pooling layer could be a feature map containing the maximum outstanding capabilities of the preceding feature map. Average pooling is a pooling operation that computes the average of elements of the feature map included via way of means of a filter. Thus the output after average pooling is a feature map consisting of the average features of the preceding feature map. Parameters in the pooling layer are pool_size, strides, and padding.

Padding: *The feature map is given padding to change the size of the combined feature map.*
Pool_size: *This specifies the size of the pooling window and by default, it is (2,2).*

Use a filter with the same number of channels as the feature map if it contains multiple channels. Each channel will get the pooling operations on an autonomous basis.

13.2.1.5 Flatten layers

The output from the pooling layer is flattened into a single column as input for the multilayer perceptron (MLP) that can classify the final pooled feature map into a class label. Between the last pooling layer and the first dense layer, there is a flattened layer.

13.2.1.6 Fully connected layers

These are the final layers in a CNN architecture. The input is a flattened layer and there can be multiple fully connected layers. The final layer does the classification task and an activation function is used in each fully connected layer.

Parameters in the dense layer are units, activation, and input_shape. Units refer to the number of nodes in a layer. The activation function in the last layer will be "softmax" in most classification-related tasks.

13.2.2 Pointwise and depthwise convolution

13.2.2.1 Limitations of spatial separable convolution

Separable convolutions are separating one convolution into two. Commonly, spatial separable convolution deals with spatial dimensions of the image and kernel. In general, it divides the kernel into two smaller kernels. For example, divide a 3×3 kernel into a 3×1 and 1×3 kernel.

Now, to accomplish the same result, we perform two convolutions with three multiplications each (for a total of six), as opposed to one convolution with nine multiplications. The computational complexity decreases with fewer multiplications, allowing the network to operate more quickly.

Not all kernels can be "split" into two smaller kernels, the spatial separable convolution has a major flaw. This is especially problematic during training because the network can only use a tiny portion of the kernels that can be divided into two smaller kernels out of all those that it could have adopted.

13.2.2.2 Depthwise separable convolution

The kernels used by depthwise separable convolutions, in contrast to spatial separable convolutions, cannot be "factored" into two smaller kernels. Thus, it is more frequently employed. Here similar convolutional neural network architecture is followed with a change only in convolution operation where a separable convolution layer is used instead of a normal convolutional layer.

It is so-called depthwise since it deals with the depth dimension that is the number of channels of an image. Separable convolution of this kind can be found in keras.layers.SeparableConv2D. A depthwise separable convolution divides a kernel into two different kernels, each of which performs two convolutions: depthwise and pointwise convolution [8].

Consider an RGB image of $12 \times 12 \times 3$ (height, width, and number of channels) as shown in Figure 13.1. On normal convolution with 5×5 kernel for three channels (i.e., $5 \times 5 \times 3$) as shown in Figure 13.2, the $12 \times 12 \times 3$ image becomes a $8 \times 8 \times 1$ image as shown in Figure 13.3. If we need to increase the number of channels in the output image, we can generate 256 kernels to produce 256 $8 \times 8 \times 1$ images, and then we can stack those images to get an output of $8 \times 8 \times 256$ image.

Unlike normal convolution, a depthwise separable convolution consists of a depthwise convolution and a pointwise convolution that divide this procedure into two sections.

A depthwise convolution applies the convolution along only one spatial dimension (i.e., channel), whereas a normal convolution applies the convolution across all spatial dimensions/channels at each step. This is the main distinction between a normal convolutional layer and a depthwise convolution.

The number of output channels is equal to the number of input channels since we apply one convolutional filter to each output channel. We next apply a pointwise convolutional layer after this depthwise convolutional layer. A normal convolutional layer with a 1×1 kernel is a pointwise convolutional layer.

Since there is a greater decrease in parameters and computations that would offset the additional computational cost of performing two convolutions instead of one, depthwise separable convolutions is more likely to work better on deeper

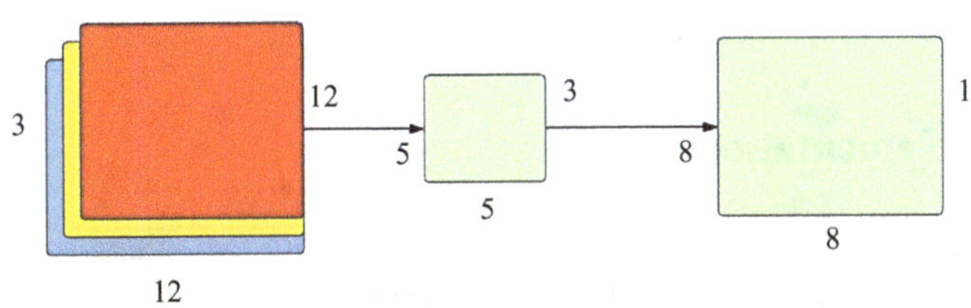

Figure 13.1 A normal convolution layer with $8 \times 8 \times 1$

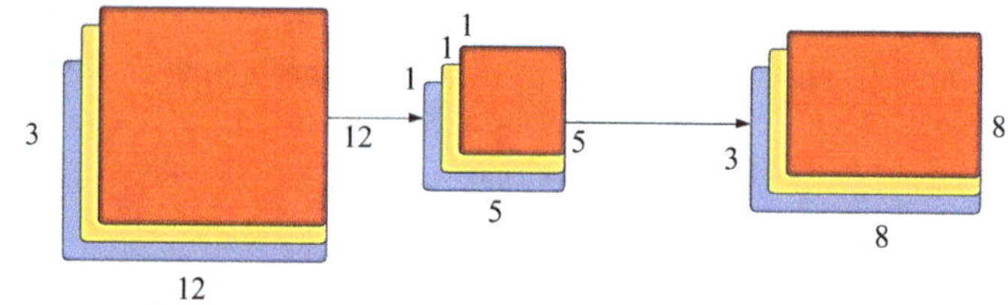

Figure 13.2 Depthwise convolution uses three kernels to transform a 12×12×3 image to a 8×8×3 image

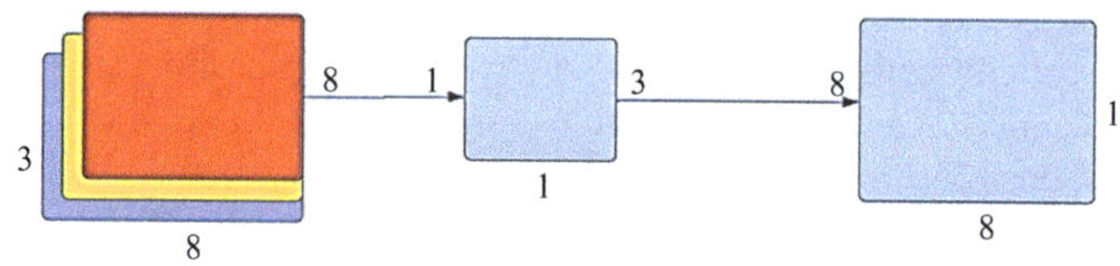

Figure 13.3 Pointwise convolution transforms an image of three channels to an image of one channel

models that may encounter an overfitting problem and on layers with larger kernels.

13.3 Proposed model

The dataset of cells from malaria-affected and unaffected patients is used as input in this system. On the basis of these photographs, predictions about infected and uninfected cells in a person's cells photographed at the microscopic level can be made. Using a CNN, this prediction is put into action. The model is trained using a finite number of iterations once the photos are divided into training and test categories. At each iteration of the training process, the model is trained based on the mean squared error (MSE) value, after which the weights are selected and stored in the model as well as local storage.

The convolution layer, separable layer, max pooling layer, and flatten layer, coupled with a fully connected neural network make up the CNN. The model is put to the test using the test set, and the outcomes of the predictions are used to determine the F1-Score, Precision, and Recall. Based on the picture of the cell, this technique determines whether or not the person seemed to have malaria.

13.4 Implementation

This part implements the whole categorization procedure for malaria cells. The first stage is to gather all the data needed for the procedure. The data in this case was gathered from a variety of sources, including Kaggle and several medical sites

(National Library of Medicine). The training pictures folder and the testing images folder make up the dataset. Create a new folder in this folder called "single prediction" to forecast the class of the image based on the model learned using the data in the training and testing folders. Two subfolders, namely, parasitized and uninfected, may be found in the training and testing folder. Red blood cells at the microscopic level were included in the photos. Images of cells afflicted by the malarial sickness are found in the folder "Parasitized." These pictures demonstrate how the user has been impacted by malaria. Another folder contains human red blood cells that have not been infected with the malarial illness.

Hence, two CNNs are built for the dataset, one of which has a convolution layer, a max pooling layer, a flatten layer, and a fully connected layer, while the other has a pointwise and depthwise layers, a max pooling layer, a flatten layer, and a fully connected layer. This research compares two alternative architectural designs on a crucial picture categorization issue. To begin putting this method into practice, select the appropriate integrated developed environment (IDE). Python 3.7.12 was used in conjunction with Jupyter Notebook version 6.1.4 to develop this approach. Importing sequential from keras.models for a simple stack of layers with each layer having precisely one input tensor and one output tensor, as well as all the necessary packages to implement the architecture from the package keras.layers, such as Dense, Conv2D, MaxPool2D, Flatten, and SeparatableConv2D.

IDE loads the dataset using the "ImageDataGenerator" function. This function's task is to import the dataset in its current state. The dataset import has to be 224×224 pixels in size. The next step is to construct a fully connected layer of a CNN. In this CNN architecture, the convolution layer contains filter sizes of 64, 128, 256, and 512. The input size of the first convolution layer was (224, 224, 3). The input size for each layer is (224, 224, 3), the maximum pooling layer is 2×2, and the CNN concludes with one flattened layer since the photos in the folders are red, green, and blue (RGB) images. The result of the flattened layer is fed to the completely connected layer. Two neurons connect the flattened layer to the output layer, and here, the sigmoid activation function is utilized, whereas the ReLU activation function was used for previous layers.

Similar to the first CNN, the second one was built in a similar manner. With the exception of the convolution layer utilized for the input, the prior design replaced all convolution layers with separable convolution layers. The convolution layer's number of neurons is the same as what was previously employed. In this design, the sigmoid function was utilized as the activation function for the output, while the ReLU function was employed for the remaining layers. The number of parameters in the normal CNN was 1,788,610 but the versa in the separable CNN was 384,194 as shown in Figure 13.4.

After construction, the architecture was put together using the metrics "binary accuracy," the optimizer "adam," and the loss function "MSE." The average of the square of the difference between the actual and anticipated value is used to compute MSE as shown in Figure 13.5.

Adam is an optimization strategy that can be used to iteratively update network weights based on training data as opposed to the standard stochastic gradient

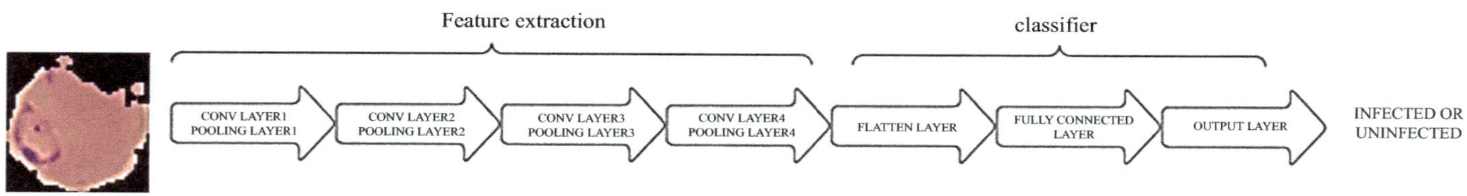

Figure 13.4 Architecture of convolution neural network

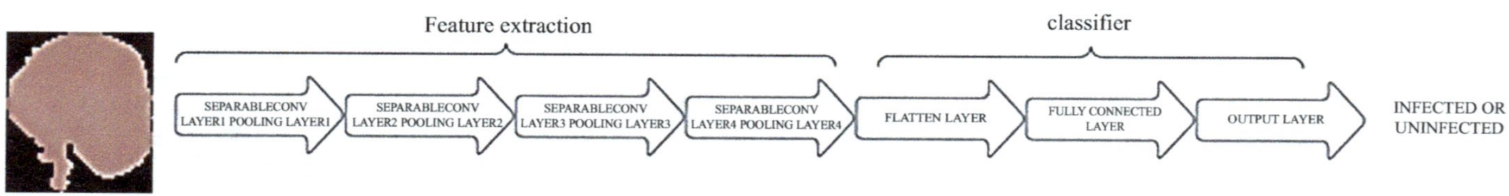

Figure 13.5 Architecture of separable convolution neural network

descent method. As an adaptive learning rate approach, Adam calculates individual learning rates for various parameters. Adam employs estimates of the first and second moments of the gradient to change the learning rate for each weight of the neural network, which is how it gets its name, adaptive moment estimation.

One measure of accuracy for a classification model that takes into account both anticipated and actual values is binary accuracy. The percentage of projected values for binary labels that match the actual values is calculated as binary accuracy. Given that the label is binary, the anticipated value is the likelihood that the predictions will come true, which in this case is 1. Calculating binary accuracy involves dividing the number of records that were correctly predicted by the total number of records.

Both the training and test data are adapted into the architecture, and the test data serve as validation data for the loss calculation. Every epoch shows the training loss for models that used training data, as well as the validation loss for models that used test data instead of validation data. Since the model was compliant with the metrics "binary accuracy," binary accuracy will be presented for training data and validation data at each epoch. Using a single picture prediction as input, the model was assessed.

13.5 Result

Malaria, a potentially fatal blood illness, is spread by mosquitoes. Fever, exhaustion, nauseousness, and headaches are typical malaria symptoms. As a result, test data are used to evaluate the models as shown in figure 13.6. In contrast to normal neural networks with convolution layers, which train and validate the model with a greater accuracy loss, pointwise and depthwise neural networks experienced a much lower accuracy loss as shown in Figure 13.7.

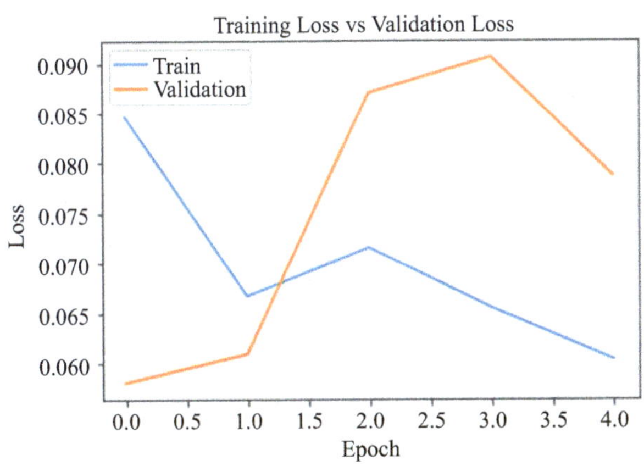

Figure 13.6 Neural network with pointwise and depthwise

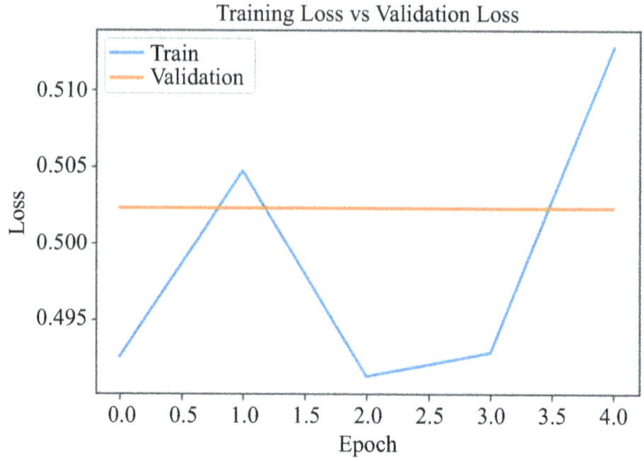

Figure 13.7 Normal CNN

13.6 Conclusion

The most troublesome sickness for people of all ages is malaria, we may infer from this. Even if it is uncommon in areas with a moderate environment, malaria is nevertheless common in countries that are tropical or subtropical. Therefore, this research concludes that a microscopic red blood cell picture may be classed as an uninfected cell or parameterized cell, therefore providing a judgment as to whether the person was afflicted or unaffected by the malaria sickness. The key message is to demonstrate how a separable CNN performs better than a conventional convolution neural network that was built.

References

[1] Arunkumar, T. R. and Jayanna, H. S. (2022). A novel light-weight approach for the classification of different types of psoriasis disease using depth wise separable convolution neural networks. *Indian Journal of Science and Technology*, 15(13), 561–569.

[2] Zhang, Y., Wang, H., Xu. R., Yang, X., Wang, Y., and Liu, Y. High-precision seedling detection model based on multi-activation layer and depth-separable convolution using images acquired by drones. *Drones.* 2022; 6(6):152. https://doi.org/10.3390/drones6060152

[3] Hassan, E. and Lekshmi, V. L. Scene text detection using attention with depthwise separable convolutions. *Applied Sciences.* 2022; 12(13):6425. https://doi.org/10.3390/app12136425

[4] Zhu, Z., Wang, S., and Zhang, Y. (2022). ROENet: a ResNet-based output ensemble for malaria parasite classification. *Electronics*, 11(13), 2040.

[5] Sengar, N., Burget, R., and Dutta, M. K. (2022). A vision transformer based approach for analysis of plasmodium vivax life cycle for malaria prediction

using thin blood smear microscopic images. *Computer Methods and Programs in Biomedicine*, 224, 106996.

[6] Agarwal, D., Sashanka, K., Madan, S., Kumar, A., Nagrath, P., and Jain, R. (2022). Malaria cell image classification using convolutional neural networks (CNNs). In *Proceedings of Data Analytics and Management* (pp. 21–36). Springer, Singapore.

[7] Jabbar, M. A. and Radhi, A. M. (2022). Diagnosis of malaria infected blood cell digital images using deep convolutional neural networks. *Iraqi Journal of Science*, 63, 380–396.

[8] Manning, K., Zhai, X., and Yu, W. (2022). Image analysis and machine learning-based malaria assessment system. *Digital Communications and Networks*, 8(2), 132–142.

Chapter 14

Analysis of preperimetric glaucoma using a deep learning classifier and CNN layer-automated perimetry

Dhinakaran Sakthipriya[1], *Thangavel Chandrakumar*[1], *B. Johnson*[1], *J. B. Prem Kumar*[1] *and K. Ajay Karthick*[1]

Glaucoma is an eye condition that, in its later stages, can cause blindness. It is caused by a damaged optic nerve and has few early signs. A glaucomatous eye can be diagnosed using perimetry, tonometry, and ophthalmoscopy. The fundamental criterion for pre-primary glaucoma (PPG) is the presence of a glaucomatous eye image, fundus, or either in the presence of an apparently normal visual field (VF). The most common way for defining an aberrant VF using conventional-automated perimetry is Anderson and Patella's criterion. This study describes a deep learning technique for analyzing fundus images for glaucoma that is generic. The research design is focused on various conditions on several samples and architectural designs, unlike previous studies. The results show that the model is either the same as or better than what has been done before. The suggested prediction models exhibit precision, sensitivity, and specificity in distinguishing glaucomatous eyes from healthy eyes. Clinicians can utilize the prediction results to make more informed recommendations. We may combine various learning models to enhance the precision of our predictions. The CNN model includes decision rules for making predictions. It can be used to describe the reasons for specific predictions.

14.1 Introduction

Glaucoma is frequently associated with an increase in stress within the sight. Ophthalmology seems to be a family trait, and it is typically not diagnosed until late adulthood. The retina, which delivers visual data to the brain, can be harmed by increased eye stress. In a few years, glaucoma can cause irreparable vision loss or perhaps total blindness if the disease continues. The majority of glaucoma patients do not experience early pain or symptoms. Regular visits to an ophthalmologist are necessary so that glaucoma can be diagnosed and treated before an irreversible

[1]Thiagarajar College of Engineering, India

visual loss occurs. A person's vision cannot be restored once it is lost. However, reducing his eye pressure will help him maintain his vision. The majority of glaucoma patients who adhere to their medication regimen and get routine eye exams are able to maintain their vision. Every human has both an optic disk and a cup, but glaucomatous eyes have an abnormally broad cup compared to the optic disk. Generally, glaucoma is diagnosed by an ophthalmologist analyzing the patient's photos and identifying any irregularities. Due to image noise and other factors that make precise analysis difficult, this technique is very time-consuming and not always accurate. In addition, if a machine is taught to conduct analysis, it eventually gets more efficient than human analysis.

Our eyes are quite often employed if the person's body has multiple senses. Visual processing requires a considerable portion of the intellect. Glaucoma, which is frequently due to a rise in hypertension, is an important responsibility of permanent loss of sight globally. Early glaucoma perception is challenging, but it is treatable [1]. Globally, glaucoma is the leading reason for permanent blindness and has a progressive effect on the optic nerve [2]. Analysis of glaucoma is determined by the healthcare history of the person, intraocular pressure, the width of the layer of visual nerve impulses, and modifications to the structure of the optic disk, especially length across, size, and region. In 2013, there were 64.3 million cases of glaucoma among those aged 40–80 years, according to a survey [3]. Currently, it is detected using four tests: (1) identification of high intraocular pressure, (2) evaluation of optic disk injury using the optic neuropathies ratio, (3) estimation of choroidal thickness, and (4) identification of typical line of sight abnormalities. Combining diagnostic organization and function techniques such as non-invasive diagnostic and field of vision evaluation, glaucoma can be diagnosed [4]. Deep learning algorithms have enhanced computer vision in recent years and are now a part of our life [5]. The methods for machine learning are suitable for glaucoma diagnosis. Parallelization and functional approaches are the two most used techniques for diagnosing glaucoma. Glaucoma is diagnosed utilizing digitally acquired fundus images. In recent chapters, the researchers proposed a plan for computerized ophthalmology diagnosis and classification by extracting features from cup segmentation [6]. For the computer-aided system, segmenting the blind spot and optic neuropathies regions is a difficult process. Combination of enhanced image techniques and field of study is required to identify the attributes with the highest degree of bias. Methods for diagnosing images of the fundus of the eye are established on the edge detection of vascular structures and the optic disk. Using the textural characteristics of digital fundus images, nerve fiber layer damage is detected [7].

Developing a computerized approach for detecting glaucoma by analyzing samples is the purpose of this project. This framework includes the gathering of a visual image dataset, pre-processing to decrease image noise, feature routine process, and the grouping of images as glaucomatous or not. For learning inputs, a neural network architecture based on convolutions will be responsible. Various performance measures and receiver operating attributes/areas under the curve true positive rate are frequently applied as evaluation criteria for diagnostic systems.

A database containing retinal fundus images from patients at a medical center will be utilized to evaluate our suggested framework.

14.2 Literature survey

In order to distinguish between healthy eyes and those with glaucoma, it is necessary to examine the retina. Mookiah *et al.* [8] devised a computer-based diagnosis method that employs discrete wavelet transform features. Using energy characteristics learned from many wavelet filters, Dua *et al.* [9] developed classifiers. Yadav *et al.* [10] developed a neuron layers analysis for fluid pressure on the retina analysis based on the textural qualities of the eye surrounding part of the optic disk. For their convolutional neural network (CNN), Chen *et al.* [11] proposed a six-layer structure with four various convolutional layers and two fully linked layers. This study employs ORIGA and SCES. An AUC of 0.831 was achieved by randomly selecting 99 photos from the ORIGA database to be used for training and the remaining 551. When using 1,676 depictions from the SCES repository for testing and 650 depictions from the ORIGA repository for training, the area under the curve is 0.887%.

A support vector machine was proposed for classification by Acharya *et al.* [12], while the Gabor transform was proposed for identifying subtle shifts in the backdrop. It was the Kasturba Medical College of Manipal, India's private database, and it contained 510 images. A total of 90% of the images were used for instruction, while the other 10% were put to use in evaluations. There was a 93.10% rate of accuracy, an 89.75% rate of sensitivity, and a 96.2% rate of specificity. To achieve automatic glaucoma recognition, Raghavendra *et al.* [6] proposed employing a CNN with 18 layers. In this study, a conventional CNN was used, complete with convolution layers, max pooling, and a fully linked layer for classification. The initial procedure involves using 70% of the instances for training and 30% for assessment. From an internal database, we pulled 589 depictions of healthy eyes and 837 depictions of glaucomatous ones. The process was done 50 times with different training and test sets each time. Results showed unique metrics ranges.

Zilly *et al.* [13] presented a technique for estimating the cup-to-disc ratio in order to diagnose glaucoma by first isolating the blind spot from depictions of the light-sensitive layer to implement the transfer method with a convolution layer. There are still underexploited characteristics that can further enhance diagnosis, despite the fact that this type of study extracts various medical characteristics. Moreover, deep learning is only employed for segmentation and not for diagnosis. In Ref. [10], a CNN strategy for automatic dispersion syndrome detection was developed. In order to classify depictions of healthy and unhealthy eyes, this network employed an n-tier architecture. The efficacy of this method was measured independently over a wide range of datasets.

In order to assess the benefit of layer analysis-based deep learning models in glaucoma diagnosis, we conduct experiments on hospital-collected datasets that include examples of various eye illnesses related to our issue statement.

14.3 Methodology

Required methodological sections describe the suggestion of deep CNN procedures for identifying and classifying low-tension eye problems that cause problems for the optic nerve. The current state of ocular glaucoma detection using AI and algorithms is limited in filtering options and is laborious to implement. Image classification using deep neural networks has been offered as a viable method. An in-depth CNN was trained for this research with a focus on classification. We use image collections to investigate the condition of the optical fundus.

The proposed study builds on prior work by creating and deploying a deep CNN to detect and categorize glaucomatous eye diseases. It is composed of various layer components of CNN. It is implemented in accordance with the seven phases of layering as shown in Figure 14.1. The images-associated categorization label is

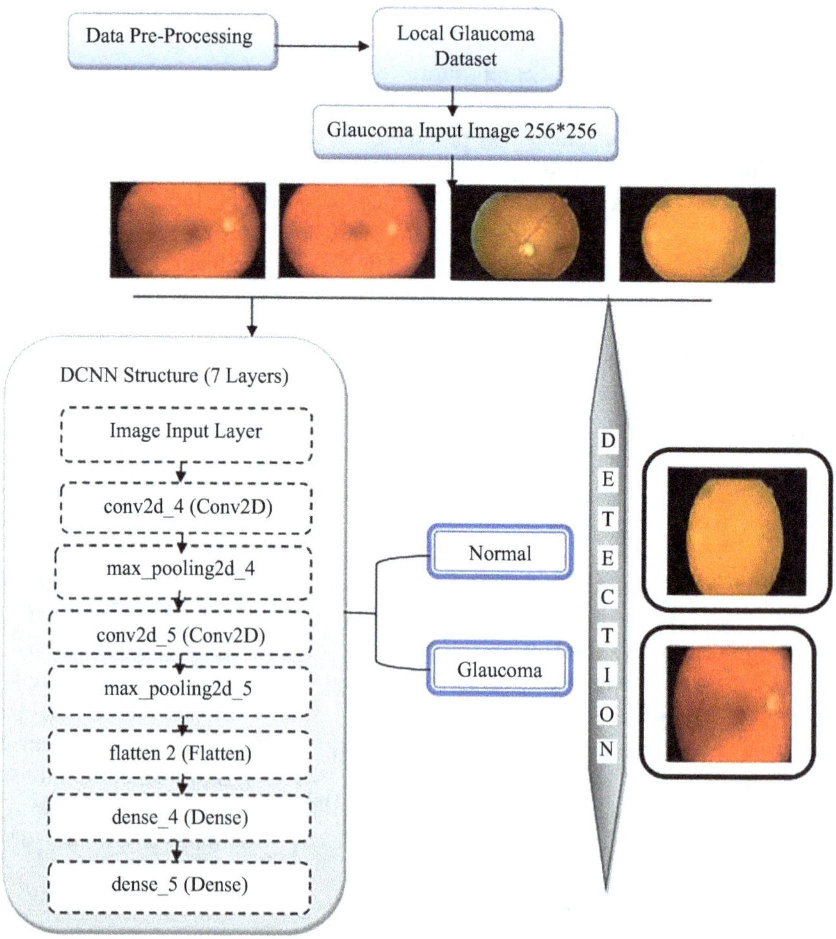

Figure 14.1 Proposed framework for eye detection

generated at the conclusion of the layer split-up analysis to help with the prediction. The subsequent CNN network uses this categorization as an input to determine ocular pictures (normal or glaucoma).

14.3.1 Procedure for eye detection

Algorithm 14.1 Proposed glaucoma prognosis

Input: No. of eye images with two various class labels where a ∊ n.
Outputs: Classification of each image and identification of glaucoma for each image sequence.

1. Glaucoma detection estimation—CNN Layer
2. Pre-Process = Input (n)
3. Partition input into sets for training and testing
4. Eye diseases (Layer Spilt Analysis with Accuracy)
5. **if** finding ordinary
6. **stop**
7. **else** eye-illness
8. **end if**

According to the research, it was possible to differentiate between two key approaches for glaucoma prognosis: the general technique applying eye detections (as shown in Algorithm 14.1) and the generic method employing a deep convolutional network with several layers (represented in Figure 14.1). In the studies that used a generic process to forecast glaucoma eye disease, it was feasible to describe a pipeline with four layers separated up based on the analytic method, such as Filter size, Kernel shape, input shape, and Activation, as depicted in Figure 14.2. To determine if a diagnostic test accurately predicts whether the eye is normal or affected by glaucoma. In Figure 14.1, we see the typical implementation of a depth CNN model, which employs a number of processing layers that are taught to represent data at different levels of abstraction. Visual image extraction may not be essential if the model accountable for this behavior includes a series of processing steps that replicate biological processes. Therefore, this paradigm uses a function to translate input into output. In Figure 14.2, the convolutional input layer is shown in detail; it is this layer's job to make predictions due to the nerve damage it causes, this eye disease glaucoma can cause total and permanent blindness if left untreated.

14.3.2 Deep CNN architecture

Figure 14.2, which represents the layer split-up analysis, is an example that is a deep CNN (DCNN). It has many of the same properties as a standard neural

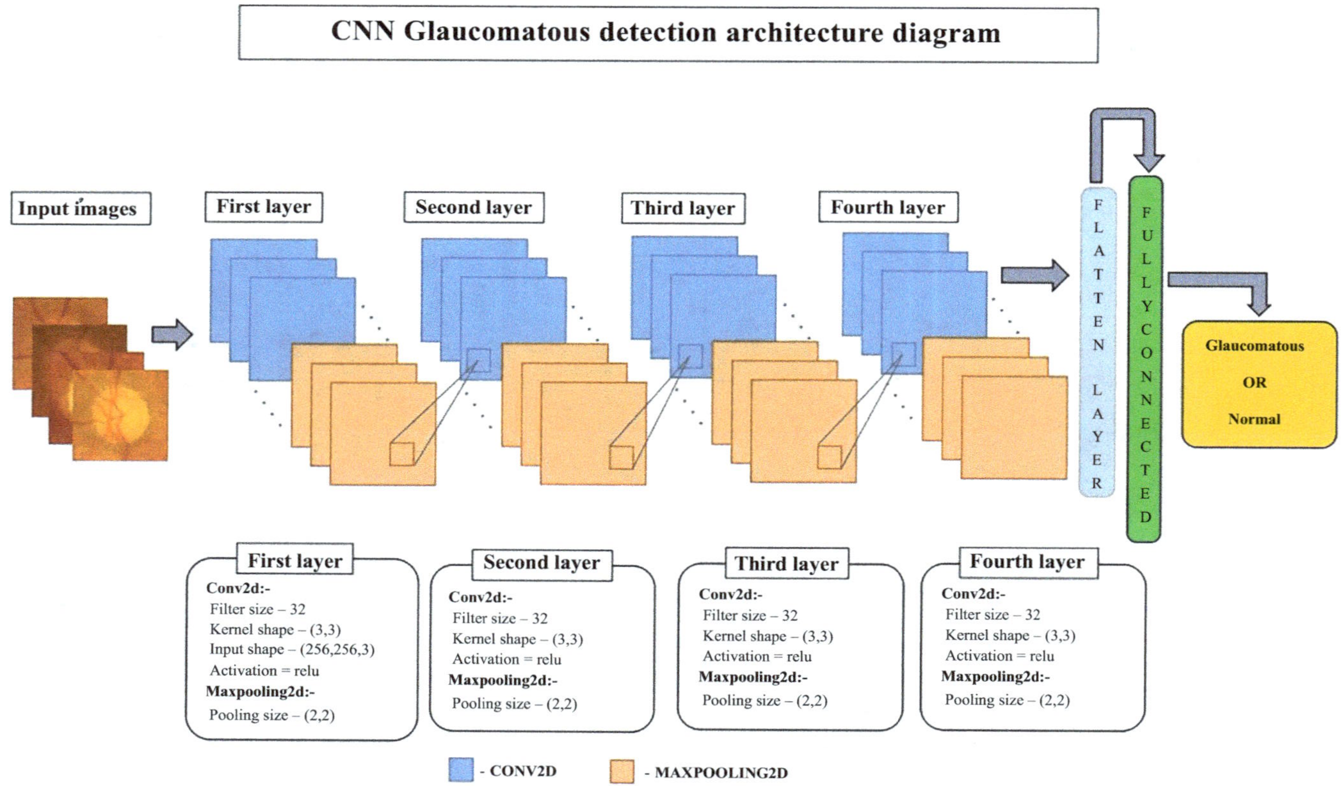

Figure 14.2 Layer-wise representation of the architecture of DCNN

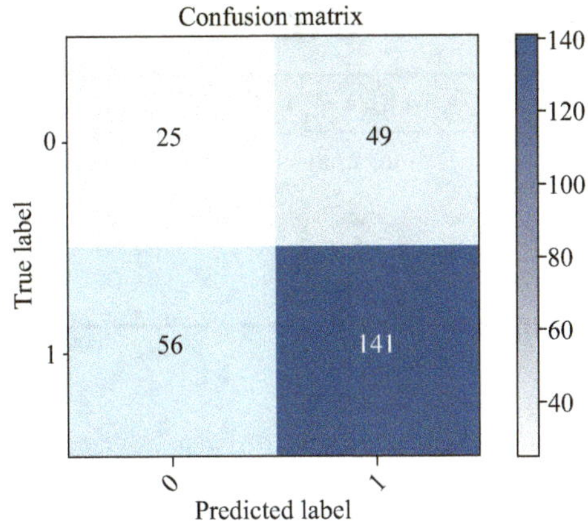

Figure 14.3 Confusion matrix (0 – glaucoma and 1 – normal)

network, such as numerous layers of neurons, different learning rates, and so on. As indicated in Figure 14.2, for network selection in this study, we employed four distinct CNN layer approaches. Figure 14.1 illustrates how the proposed DCNN operates. This part contains the implementation of CNN's layers, as mentioned below.

In this section of the DCNN architecture, four CNN layers are employed to classify the glaucoma illness level. Deep CNN is its title (DCNN). This network detects glaucoma-affected pictures using classed images from the DCNN network. The illness degree of glaucoma is classified into two phases: normal and glaucoma, with an early stage defining the start of the disease. Consequently, four distinct CNN layer architectures were developed; one for each tier of glaucoma detection in the deep classification-net phase. The design of DCNN is depicted in Figure 14.3; we utilized four distinct layers and four phases: filter size, kernel shape, input shape, and activation. The dimensionality of the variance in a deep convolutional neural network's layers is shown in Figure 14.2.

14.4 Experiment analysis and discussion

Applying Python for this research suggests that the very in-depth CNN was evaluated on the following system capacity with a processor of Intel Core i5, CPU capacity: 2.50 GHz, and 8GB Ram, the Jupiter Notebook, a number of statistical values were computed.

14.4.1 Pre-processing

The regional visual glaucoma image collection eliminates image noise with adaptive histogram equalization. Images of localized retinal glaucoma are gathered from

Table 14.1 Comparative analysis for eye detection

Period	References	Specifics	Techniques	Output
2018	[6]	Retinal images	CNN model	Glaucoma or not
2020	[14]			
2019	[15]			
2019	[16]			
2019	[17]			

Table 14.2 Dataset description of images of various groups and subsets

Class	Total	Train	Validation	Test
Normal	1,420	710	355	355
Glaucoma	1,640	820	410	410
Total	**3,060**	**1,530**	**765**	**765**

a variety of commercial and public medical centers. The collection includes 3,060 visual images in total. Each image falls into one of two categories, namely, normal and glaucoma. 54% of photos belong to one class, while 46% belong to the other. The distribution of the dataset's training, validation, and testing subsets are presented in Table 14.1. The 710 and 820 pictures are separated into normal and glaucoma categories images for evaluation purposes on the supplied dataset. The collection contains test, training, and validation datasets. Three professional clinical assistants were tasked with distinguishing between the two stages of glaucoma eye illness outlined in Tables 14.1–14.3. Where experts disagreed, the majority vote was used to label the photos.

14.4.2 Performance analysis

By using the following statistical equations, the suggested DCNN yields the following results: Sensitivity = a/(a + c), Specificity = a/(a + b), Accuracy = (a + d)/(a + c + d + b), and Precision = a/(a + b).

Here, TP denotes true positives that are accurately discovered among glaucoma photographs, whereas TN denotes true negatives that correctly identify mistakenly categorized images. As illustrated in Figure 14.3, false acceptance and false rejection denote classes that were correctly and incorrectly selected.

14.4.3 CNN layer split-up analysis

We used DCNN layer (L) validation with L = 1,2,3,4 to analyze the effectiveness of the proposed system for detecting glaucoma disease using eyes datasets. Eyes images comprise 3,060 retinal fundus images for graph analysis segmentation. These images were taken during a medical examination of the patients at India's Hospital. In addition, the layer validation results are reported in Figure 14.4.

Table 14.3 Calculated percentages of statistical measures

Eye disease glaucoma detection					
No.	**Types**	**Sensitivity (%)**	**Specificity (%)**	**Accuracy (%)**	**Precision (%)**
1	Glaucoma	72.63	69.23	72.85	66.65
2	Normal	92.00	95.62	88.90	93.45
	Average	82.31	82.42	80.87	80.03

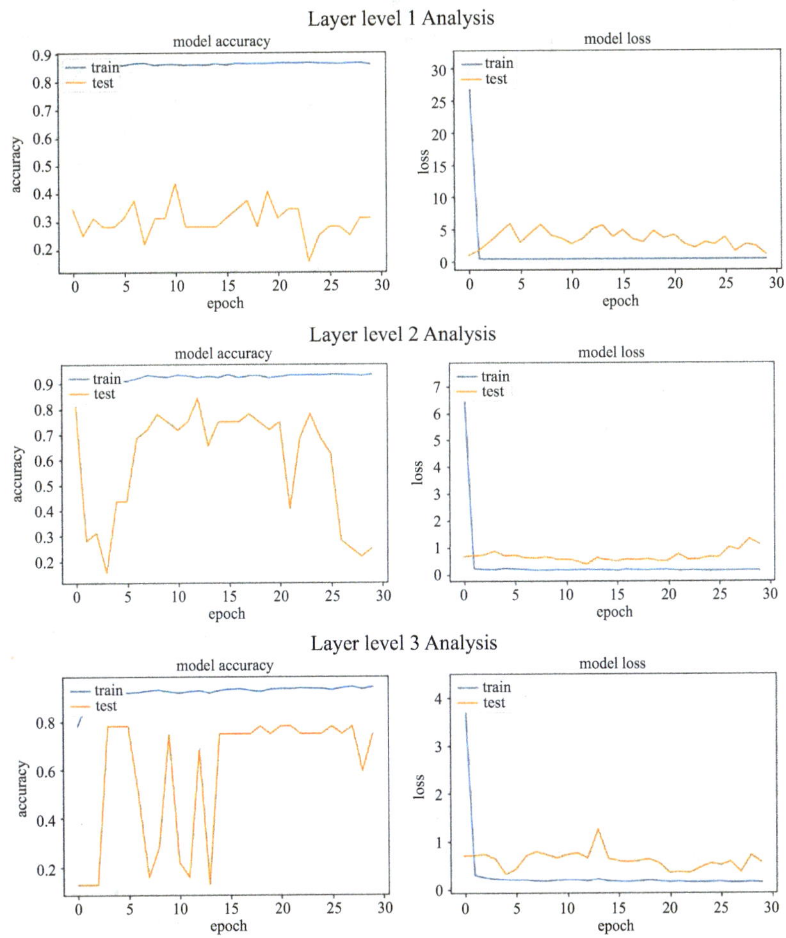

Figure 14.4 CNN layer split-up statistical analysis

- The layer level 1 analysis graph yielded a poor level of accuracy, i.e., 36.45%.
- The layer level 2 analysis graph yielded a moderate level of accuracy, between 80% and 70%.
- The layer level 3 analysis graph yielded the highest accuracy, i.e., 80.87%.

14.5 Conclusion

In this study, we apply the robust learning algorithms of deep-learning approaches to analyze images of evaluation of the visual retina for pressurized eye changes and normal. Implementing the CNN architecture to 3,060 photographs means extracting characteristics from unprocessed pixel data images using a multilayer helps construct a depth level of CNN layer analysis based to predict the result of healthy and unhealthy eyes. The depth level of convolutional neural network technique is integrated with a seven-layer parameter to detect two distinct glaucoma classifications. The DCNN model is rated on an aggregate of the statistical measures of specificity, sensitivity, accuracy, and precision of 82.31% SE, 82.42% SP, 80.87% ACC, and 80.03% PRC, respectively. This research suggests deep CNN glaucoma model produced statistically unique results for the ordinary and glaucoma groups. The successful results are comparable to those of cutting-edge technologies and were competitive in treating challenging glaucoma eye disease problems. The suggested DCNN technique performs well; however, in the future, this model will be utilized to forecast various eye disorders, with a focus on layer-splitting-based analysis due to its superior performance in predicting this type of disease.

References

[1] Abbas, Q. (2017). Glaucoma-deep: detection of glaucoma eye disease on retinal fundus images using deep learning. *International Journal of Advanced Computer Science and Applications*, 8(6), 41–45.

[2] Shaikh, Y., Yu, F., and Coleman, A. L. (2014). Burden of undetected and untreated glaucoma in the United States. *American Journal of Ophthalmology*, 158(6), 1121–1129.

[3] Tham, Y. C., Li, X., Wong, T. Y., Quigley, H. A., Aung, T., and Cheng, C. Y. (2014). Global prevalence of glaucoma and projections of glaucoma burden through 2040: a systematic review and meta-analysis. *Ophthalmology*, 121(11), 2081–2090.

[4] Taketani, Y., Murata, H., Fujino, Y., Mayama, C., and Asaoka, R. (2015). How many visual fields are required to precisely predict future test results in glaucoma patients when using different trend analyses?. *Investigative Ophthalmology & Visual Science*, 56(6), 4076–4082.

[5] Aamir, M., Irfan, M., Ali, T., *et al.* (2020). An adoptive threshold-based multi-level deep convolutional neural network for glaucoma eye disease detection and classification. *Diagnostics*, 10(8), 602.

[6] Raghavendra, U., Fujita, H., Bhandary, S. V., Gudigar, A., Tan, J. H., and Acharya, U. R. (2018). Deep convolution neural network for accurate diagnosis of glaucoma using digital fundus images. *Information Sciences*, 441, 41–49.

[7] Mookiah, M. R. K., Acharya, U. R., Lim, C. M., Petznick, A., and Suri, J. S. (2012). Data mining technique for automated diagnosis of glaucoma using higher order spectra and wavelet energy features. *Knowledge-Based Systems*, 33, 73–82.

[8] Dua, S., Acharya, U. R., Chowriappa, P., and Sree, S. V. (2011). Wavelet-based energy features for glaucomatous image classification. *IEEE Transactions on Information Technology in Biomedicine*, 16(1), 80–87.

[9] Yadav, D., Sarathi, M. P., and Dutta, M. K. (2014, August). Classification of glaucoma based on texture features using neural networks. In *2014 Seventh International Conference on Contemporary Computing (IC3)* (pp. 109–112). IEEE.

[10] Chen, X., Xu, Y., Wong, D. W. K., Wong, T. Y., and Liu, J. (2015, August). Glaucoma detection based on deep convolutional neural network. In *2015 37th Annual International Conference of the IEEE Engineering in Medicine and Biology Society (EMBC)* (pp. 715–718). IEEE.

[11] Devalla, S. K., Chin, K. S., Mari, J. M., *et al.* (2018). A deep learning approach to digitally stain optical coherence tomography images of the optic nerve head. *Investigative Ophthalmology & Visual Science*, 59(1), 63–74.

[12] Acharya, U. R., Ng, E. Y. K., Eugene, L. W. J., *et al.* (2015). Decision support system for the glaucoma using Gabor transformation. *Biomedical Signal Processing and Control*, 15, 18–26.

[13] Zilly, J., Buhmann, J. M., and Mahapatra, D. (2017). Glaucoma detection using entropy sampling and ensemble learning for automatic optic cup and disc segmentation. *Computerized Medical Imaging and Graphics*, 55, 28–41.

[14] Chai, Y., Liu, H., and Xu, J. (2020). A new convolutional neural network model for peripapillary atrophy area segmentation from retinal fundus images. *Applied Soft Computing*, 86, 105890.

[15] Li, L., Xu, M., Wang, X., Jiang, L., and Liu, H. (2019). Attention based glaucoma detection: a large-scale database and CNN model. In *Proceedings of the IEEE/CVF Conference on Computer Vision and Pattern Recognition* (pp. 10571–10580).

[16] Liu, H., Li, L., Wormstone, I. M., *et al.* (2019). Development and validation of a deep learning system to detect glaucomatous optic neuropathy using fundus photographs. *JAMA Ophthalmology*, 137(12), 1353–1360.

[17] Bajwa, M. N., Malik, M. I., Siddiqui, S. A., *et al.* (2019). Two-stage framework for optic disc localization and glaucoma classification in retinal fundus images using deep learning. *BMC Medical Informatics and Decision Making*, 19(1), 1–16.

Chapter 15

Deep learning applications in ophthalmology—computer-aided diagnosis

M. Suguna[1] *and Priya Thiagarajan*[1]

Artificial intelligence (AI) is proving to be a fast, versatile, and accurate tool to aid and support healthcare professionals in diagnosing and screening for a multitude of diseases and disorders. Several specialties have successfully incorporated AI into their healthcare services. The eye care specialty of ophthalmology has several successful applications of AI in disease detection. The applications of AI to analyze images, mainly the retinal fundus image (RFI) in ophthalmology, are proving to be very effective tools, not only for ophthalmologists but also for other specialists including neurologists, nephrologists, and cardiologists. The diseases that are diagnosable using AI are discussed in detail as an essential guide for AI designers working in the medical imaging domain in ophthalmology. The challenges and future trends including the use of multi-disease detection systems and smartphone RFI cameras are studied. This would be a game changer in screening programs and rural health centers and remote locations. Intelligent systems work as an effective and efficient tool to analyze RFI and assist healthcare specialists in diagnosing, triaging, and screening for a variety of diseases. More testing and better models need to be introduced to enhance the performance metrics further. More medical image datasets need to be available in the public domain to encourage further research. Though intelligent systems can never replace healthcare specialists, they can potentially be life-saving and cost-effective, especially in rural and remote locations.

15.1 Introduction

Ophthalmology is the field of medicine which has made significant strides in employing Artificial Intelligence (AI) to analyze images to detect diseases and disorders. In a first of its kind, the United States Food and Drug Administration (US FDA) has approved a device that uses AI to detect diabetic retinopathy (DR) in adult diabetics [1].

[1]Department of Computer Science and Engineering, Thiagarajar College of Engineering, India

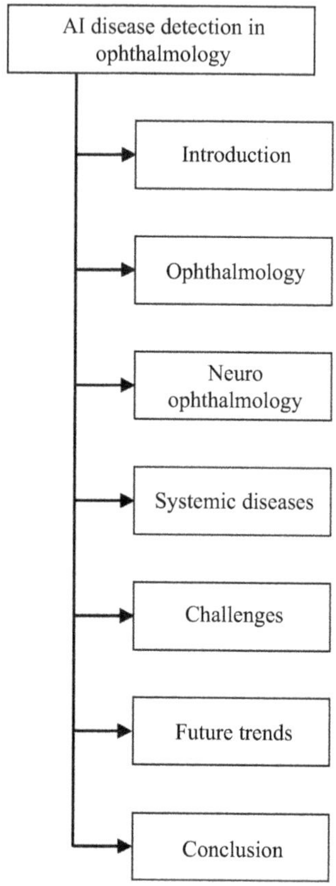

Figure 15.1 Structure of this chapter

The eye, often referred to as the window of the soul, is now providing us a window to our systemic health too.

This chapter deals with the medical applications of AI, more specifically deep learning (DL) and neural networks (NN) for image analysis in ophthalmology. It is divided into the following sections (Figure 15.1):

- Ophthalmology
- Neuro-ophthalmology
- Systemic disease detection in ophthalmology
- Challenges
- Future trends

The main image which opens up several diagnostic avenues in ophthalmology is the *retinal fundus image* (RFI).

The *Ophthalmology* section starts with a brief description of the human eye, the location, and also the parts of the retinal fundus. The process of retinal fundus image capture with retinal fundus cameras is also described.

Then we present a brief introduction to ocular diseases and evidence for the successful use of AI for the detection and screening of the following diseases:

- Diabetic retinopathy (DR)
- Age-related macular degeneration (ARMD or AMD)
- Glaucoma
- Cataract

In the *Neuro-ophthalmology* section, we discuss the following diseases and the current use of DL for the detection of the following diseases from retinal images:

- Papilledema/pseudopapilledema
- Alzheimer's disease (AD)

In the *Systemic disease detection in ophthalmology* section, we discuss how the same retinal fundus images can be used to detect and monitor even renal diseases like chronic kidney disease (CKD) and cardiovascular diseases (CVD) by just visualizing the microvascular structures in the retinal fundus.

Also, several epidemiologic studies suggest that DR and diabetic nephropathy (DN) usually progress in parallel and share a close relationship. Monitoring DR gives a good indication of the status of DN too. So ophthalmic imaging is also found to play a major role in screening for and early detection of systemic illnesses.

The *Challenges* in using intelligent systems for image analysis and classification are also discussed briefly.

In the last section of this chapter, *Future trends*, we present two new areas which have exciting applications and show good results in recent studies, especially in screening program:

- Smartphone capture of retinal fundus images (with a lens assembly)
- Multi-disease detection using a single retinal fundus image

15.2 Ophthalmology

Ophthalmology is a specialty in medicine that deals with the diseases and disorders of the eye. Ophthalmologists are doctors who have specialized in ophthalmology. Ophthalmology is one of the main specialties to apply AI in healthcare. With the first US FDA-approved AI medical device, ophthalmology can be considered a pioneer in AI disease detection research [1].

Here, we will focus on the applications of AI in image analysis and classification for disease detection. The following two images are mainly used in ophthalmology:

- Retinal fundus image (RFI)
- Optical coherence tomography (OCT)

Though OCT is proving to be very useful in studying various layers of the retina, we consider only retinal fundus imaging in this chapter. Retinal fundus imaging is widely used and cost-effective, thus making it more suitable for use in remote and rural health centers.

The retina in our eye is very important for vision. The lens focuses light from images on the retina. This is converted by the retina into neural signals and sent to the brain through the optic nerve. Basically, the retina consists of light-sensitive or photoreceptor cells, which detect characteristics of the light such as color and intensity. This information is used by the brain to visualize the whole image.

The photoreceptors in the retina are of two types: rods and cones. The rods are responsible for scotopic vision (low light conditions). They have low spatial acuity. Cones are responsible for photopic vision (higher levels of light). They provide color vision and have high spatial acuity.

The rods are mainly concentrated in the outer regions of the retina. They are useful for peripheral vision. Cones are mainly concentrated on the central region of the retina and are responsible for our color vision in bright light.

There are three types of cones based on the wavelengths to which they are sensitive. They are long, middle, and short wavelength-sensitive cones. The brain perceives the images based on all the information collected and transmitted by these rods and cones.

The inner surface of the eyeball which includes the retina, the optic disk, and the macula is called as retinal fundus. A normal retinal fundus is shown in Figure 15.2. This portion of the inner eye is what is visible to the healthcare professional by looking through the pupil.

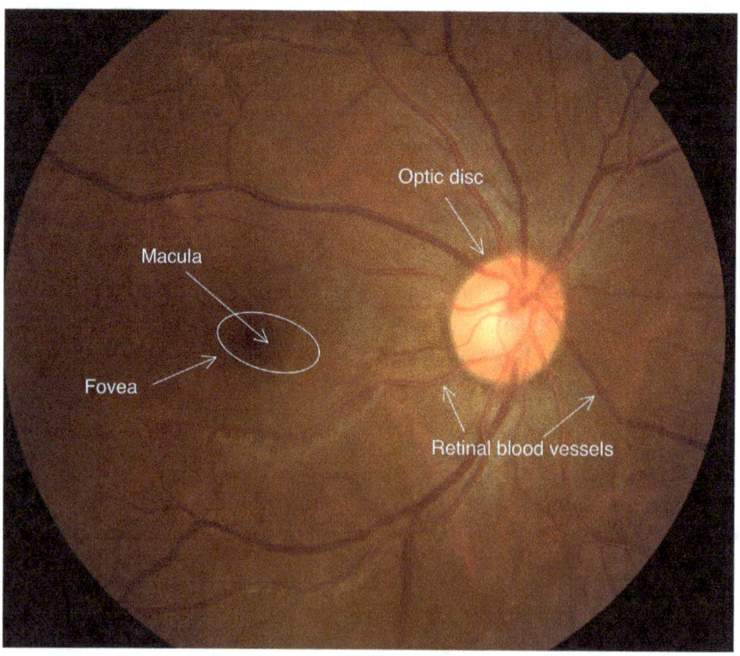

Figure 15.2 A normal retinal fundus image

The retinal fundus or the ocular fundus can be seen using an ophthalmoscope or photographed using a fundus camera. A fundus camera is a specialized camera with a low-power microscope.

The retina, the retinal blood vessels, and the optic nerve head or the optic disk can be visualized by fundus examination (Figure 15.3). The retinal fundus camera is a medical imaging device. It usually has a different set of specialized lenses and a multi-focal microscope attached to a digital camera. The digitized images can also be displayed on a monitor in addition to recording (Figure 15.4).

AI is proving to be a big boon to ophthalmologists and patients in screening, diagnosing, assessing, and staging various diseases of the eye. This has reduced waiting times for patients and unnecessary referrals to ophthalmologists. Intelligent systems in rural health centers, general practitioners' offices, and emergency departments can help with quicker diagnosis and expedite the treatment of vision and even life-threatening diseases.

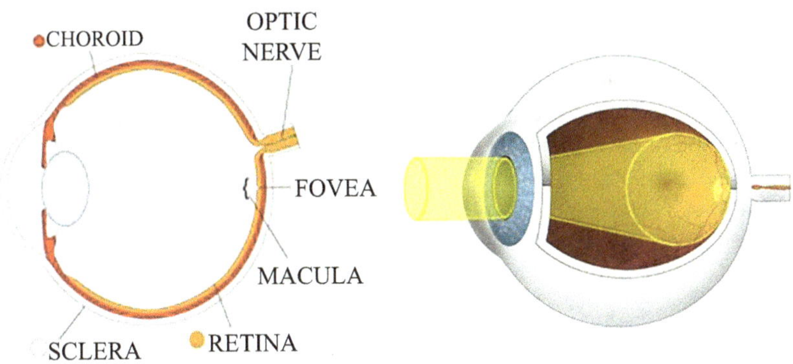

Figure 15.3 Area visualized by the fundus camera. Source: [2].

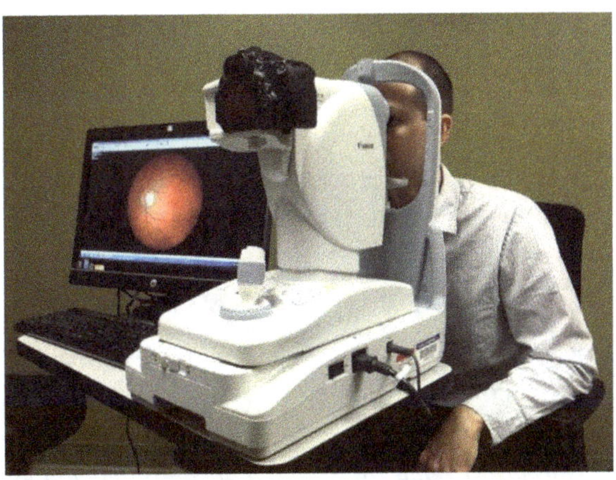

Figure 15.4 Retinal fundus image capture. Source: [3].

The retinal fundus image reveals several diseases of the eye. An ophthalmologist viewing the retinal fundus or the captured image of the retinal fundus can diagnose several diseases or disorders of the eye. With a suitable number of training images labeled by an ophthalmologist, intelligent systems can be trained to analyze the retinal fundus image and capture the characteristics to help in the decision process to diagnose the disease.

15.2.1 Diabetic retinopathy

Diabetes mellitus is a metabolic disorder which affects the way the body processes blood sugar. Diabetes mellitus causes prolonged elevated levels of blood glucose. This happens when the pancreas is not able to produce enough insulin or when the body cannot effectively use the insulin produced. This is a chronic condition.

If diabetes is not diagnosed and treated, it may lead to serious damage to nerves and blood vessels. There has been a steady increase in the incidence of diabetes and diabetic mortality over the past few years. Uncontrolled diabetes can lead to diabetic retinopathy which is caused by the damage of the blood vessels in the retina [4].

Diabetic retinopathy can cause blurry vision, floating spots in vision and may even lead to blindness. It can also cause other serious conditions like diabetic macular edema or neovascular glaucoma. Early diabetic retinopathy does not have any symptoms. But early diagnosis can help protect vision by controlling blood sugar with lifestyle changes or medications.

There are four stages in diabetic retinopathy ranging from mild non-proliferative to the proliferative stage. An RFI with features of DR is shown in Figure 15.5.

AI to detect diabetic retinopathy from retinal fundus images has achieved very high accuracy levels. An AI-based disease detection system has been approved by the USFDA to detect diabetic retinopathy in adults and is currently used successfully.

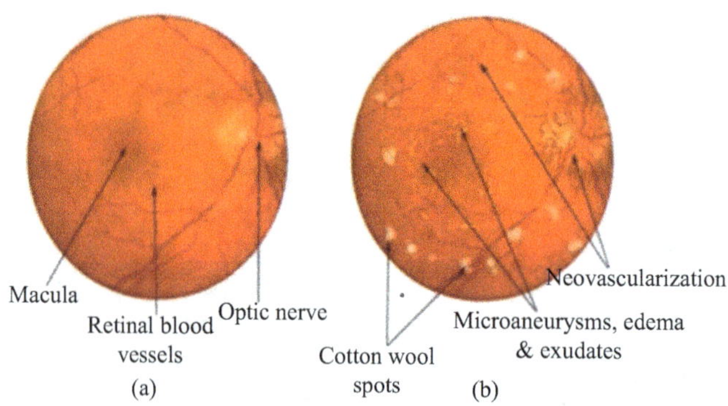

Figure 15.5 (a) Normal RFI and (b) RFI in diabetic retinopathy with neovascularization and microaneurysms. Source: [3].

15.2.2 Age-related macular degeneration

Age-related macular degeneration is an eye disease caused by aging. When aging causes the macula to degenerate, blurry or wavy areas may appear in the central vision region. Vision loss may not be noticeable in early AMD. Anyone who is 55 years or older is at risk for AMD. The risk increases with family history, history of smoking, and older age. Early diagnosis and intervention are essential to preserve vision. Loss of central vision makes reading or driving very difficult.

Early and late AMD images are shown in Figure 15.6, in comparison to a normal RFI. AI is employed for detecting and monitoring the progress of age-related macular degeneration.

15.2.3 Glaucoma

Glaucoma is usually caused by an abnormal fluid buildup and hence increased intraocular pressure in the eye. This causes damage to the optic nerve which may lead to visual losses. The excess fluid may be caused by any abnormality in the drainage system of the eye. It can cause hazy or blurred vision, eye pain, eye redness, and colored bright circles around light. A healthy optic disk and a glaucomatous disk are shown in Figure 15.7.

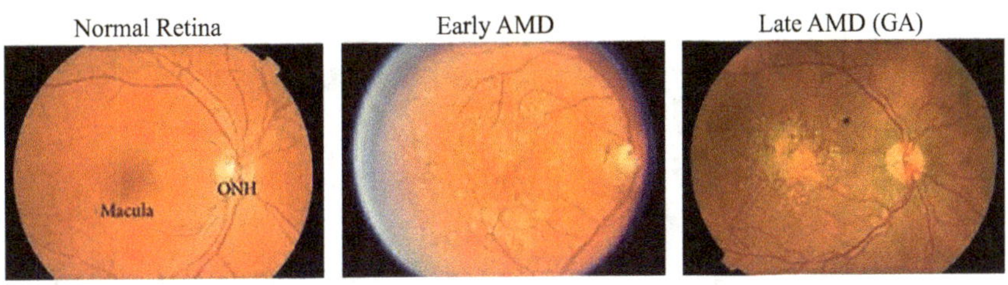

Figure 15.6 Normal retina in comparison with early and late AMD. Early AMD with extra-cellular drusen deposits around the macula. Late AMD with hyperpigmentation around the drusen. Source: [5].

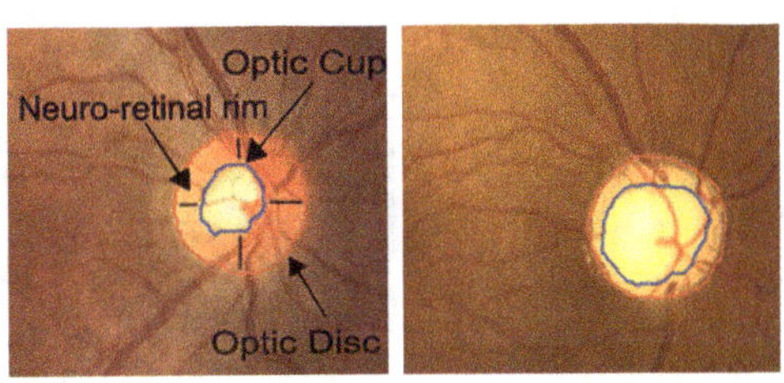

Figure 15.7 Healthy optic disk and glaucomatous optic disk with cupping (increase in optic cup size and cup–disk ratio). Source: [6].

Treatment for glaucoma involves lowering the intraocular pressure. Uncontrolled glaucoma may lead to blindness. Therefore, early detection and intervention are essential. Retinal fundus images analyzed by AI can be used to detect glaucoma. This can aid in early diagnosis and intervention.

15.2.4 Cataract

Globally, cataract is a leading cause of blindness. It can be treated and blindness prevented by timely diagnosis and surgical intervention. A cataract is defined as opacity in any part of the lens in the eye. This opacity is usually caused by protein breakdown in the lens. When the lens has increased opacity, focusing images on the retina is not done efficiently, and this may lead to blurry vision and loss of sight. The progression of this disease is slow. Early diagnosis and timely surgical intervention can save vision.

RFI in various stages of cataracts is shown in Figure 15.8. Cataract-related AI systems are still under development [8]. Studies are going on for disease detection and also for calculating pre-cataract surgery intraocular lens power. In addition to retinal fundus images, slit lamp images are also used with AI for cataract detection.

Table 15.1 lists the existing literature on the use of AI in ophthalmology. The dataset(s) used and the successful models along with significant results are also tabulated.

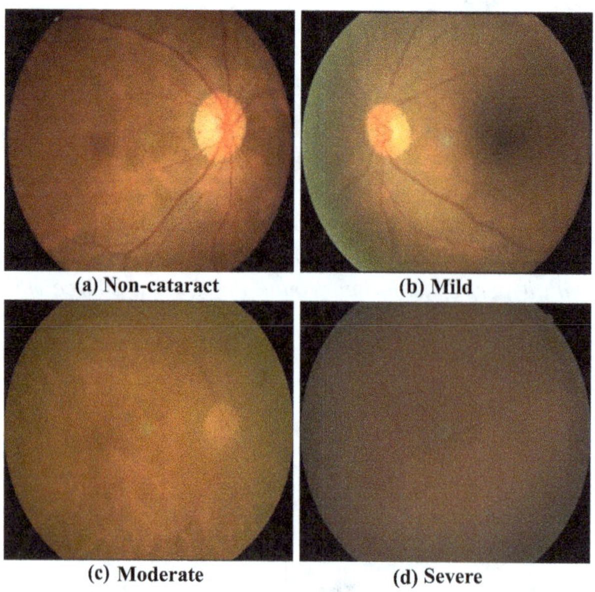

Figure 15.8 Comparison of normal RFI with various stages of cataract images showing blurriness due to lens opacity. Source: [7].

Table 15.1 AI for disease detection in ophthalmology

Ref. no.	Disease detected	Dataset	AI model used	Significant results
[9]	Diabetic retinopathy	Dataset created with images from Kaggle DR dataset + images from three hospitals in China	CNN-based LesionNet (with Inception V3 and FCN 32)	AUC 0.943, sensitivity 90.6%, specificity 80.7%
[10]	Diabetic retinopathy	APTOS dataset	AlexNet and Resnet101	Accuracy: 93%
[11]	Age-related macular degeneration	iChallenge-AMD and ARIA datasets	DCNN with 10-fold cross-validation	Classification accuracy of up to 99.45 with iChallenge-AMD and up to 99.55 with ARIA
[12]	Age-related macular degeneration	AMD lesions, ADAM, ARIA, STARE	CNN with custom-built architecture	AUC-ROC: 97.14%
[13]	Glaucoma	DRISHTI-DB and DRIONS-DB datasets	Support Vector Machine (SVM)	Specificity: 96.77% and 97.5%; sensitivity: 100% and 95%
[14]	Glaucoma	OHTS study images and five external datasets	ResNet-50, transformer model, DeiT	AUC-ROCResNet50: 0.79DeiT: 0.88
[15]	Cataract	Images collected from several open access datasets	Hybrid pre-trained CNN (AlexNet, VGGNet, ResNet) with TL to extract features and SVM for classification	Classification accuracy: 96.25%
[16]	Cataract	Training: Singapore MalayEye Study (SIMES)Testing: SINDI, SCES, BES	ResNet-50 (pre-trained on ImageNet) for feature extraction and XGBoost for classification	AUROC Training: 96.6% Testing: 91.6–96.5%

15.3 Neuro-ophthalmology

Neuro-ophthalmology is a highly specialized field which merges neurology and ophthalmology. It is usually connected with visual symptoms arising from brain diseases or disorders. The main areas in neuro-ophthalmology which currently have several research studies going on are papilledema detection and detection of Alzheimer's disease.

15.3.1 Papilledema

Papilledema is caused by an increase in the intracranial pressure of the brain. This causes the swelling of the optic nerve which is visible as a swelling of the optic disk in

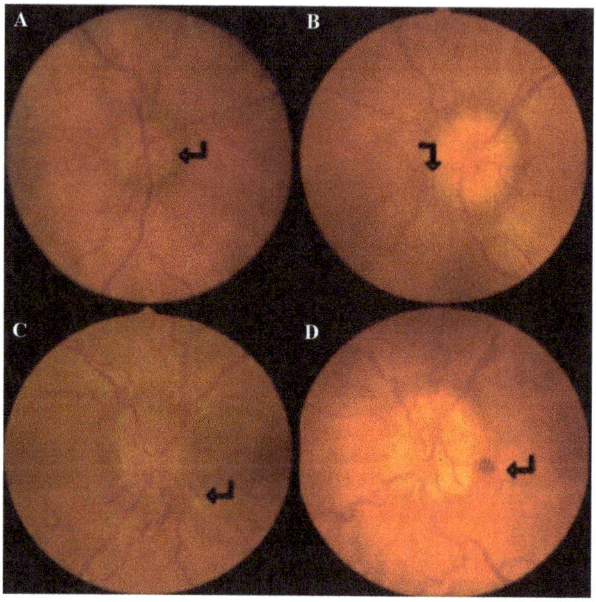

Figure 15.9 RFI in papilledema showing various grades of optic disk swelling (A - mild, B - moderate, C&D - severe). Source: [17].

retinal fundus images (Figure 15.9). This is a dangerous condition and if left undiagnosed and untreated, can lead to blindness or in some cases may even lead to death. Symptoms may include blurry vision, loss of vision, headaches, nausea, and vomiting.

The increase in intracranial pressure may be caused by space-occupying lesions or infections or hydrocephalus and sometimes idiopathic intracranial hypertension[18]. The treatment for papilledema is to treat the underlying cause which will bring down the intracranial pressure to normal levels.

Swelling of the optic disk due to non-brain-related conditions is termed pseudopapilledema though it is not as dangerous as papilledema, it still needs further evaluation. A timely and accurate diagnosis helps identify papilledema earlier and avoids unnecessary referrals and further invasive tests.

15.3.2 Alzheimer's disease

Alzheimer's is a neurological disease which is caused by brain atrophy where brain cells begin to die. It is a progressive disorder leading to a decline in language skills, thinking ability, behavioral, and social skills. This affects a person's ability to function and live independently. Alzheimer's disease can also lead to depression, social withdrawal, irritability, and aggressiveness and delusions. An early confirmed diagnosis can help with treatment and lifestyle changes. Unfortunately, there is no treatment available now which will completely cure Alzheimer's disease.

Simple definitive diagnostic tests are not available for detecting early Alzheimer's. So a non-invasive retinal fundus image analysis using AI can be quite useful. RFI in Alzheimer's disease is shown in Figure 15.10. Research is in its early stages but the results obtained are promising.

Table 15.2 lists the existing literature for AI disease detection in neurology using RFI, along with significant results.

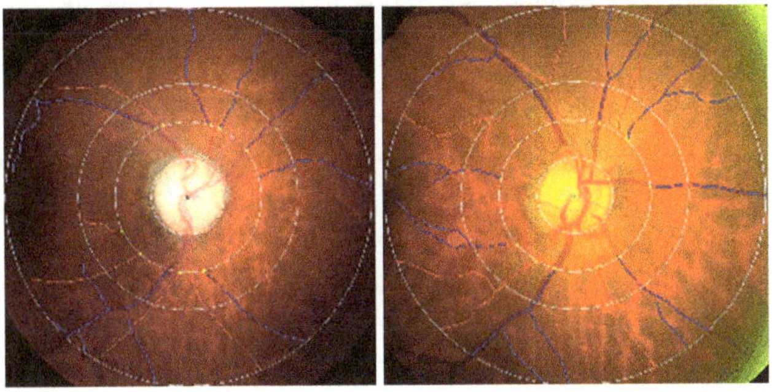

Figure 15.10 RFI in Alzheimer's disease shows reduced retinal vascular fractal dimension and increased retinal vascular tortuosity. Source: [19].

Table 15.2 AI for disease detection in neurology (using RFI)

Ref. no.	Disease detected	Dataset	AI model used	Significant results
[20]	Papilledema	100 retinal fundus images from the STARE dataset	CNN-based UNet and DenseNet	Accuracy: 99.89%
[21]	Papilledema in pediatric patients	331 pediatric fundus images-US hospital data	CNN-based DenseNet	Accuracy: ~81% in distinguishing papilledema and pseudopapilledema
[22]	Papilledema	Training dataset created with 14,341 retinal fundus images (multi-center, multinational) Testing dataset 1505 images	CNN-based U-Net and DenseNet	Accuracy: up to 94.8%
[23]	Papilledema severity	Training dataset created with 2,103 retinal fundus images (multi-center, multinational). Testing dataset 214 images	CNN-based UNet and VGGNet	Accuracy up to 87.9% in grading the severity of the papilledema
[24]	Alzheimer's disease	12,949 RFI (648 patients with AD and 3,240 people without AD)	Efficient NetB2	Accuracy: up to 83·6% Sensitivity: 93·2% Specificity: 82·0% AUROC: 0·93
[25]	Diseases/ disorders in neuro-ophthalmology	Review paper		
[26]	Diseases/ disorders in neuro-ophthalmology	Review paper		

15.4 Systemic diseases

The retinal fundus image is unique, as it is not only used to diagnose diseases of the eye and diseases of the brain, but it is also used to diagnose nephrological diseases and cardiovascular diseases and risks. In this section, we will see the applications of AI, using retinal fundus images to diagnose renal diseases and heart diseases.

15.4.1 Chronic kidney disease

Chronic kidney disease affects about 15% of the adult population. This occurs when the kidneys are damaged and cannot filter blood properly. It is a slow-progression disease where kidney function is gradually lost. This can lead to a buildup of fluid and body waste causing electrolyte imbalances. Symptoms include fatigue and weakness, sleep problems, high blood pressure, nausea, vomiting, etc.

Chronic kidney disease is irreversible and progressive. Chronic kidney disease is associated with other conditions like anemia, cardiovascular disease, and bone disorders. Chronic kidney disease is also known to be associated with ocular fundus abnormalities. Microvascular retinopathy, macular degeneration, retinal hemorrhage, etc. manifest in the eye in CKD patients.

The relationship between renal disease and vision disorders was discovered in the early 19th century. Further studies indicate that the overall presence of ocular disorders among CKD patients was around 45% which is significantly higher than the general population. It is also found that many patients with diabetes-associated renal failure also have diabetic retinopathy [27,28].

The retinal fundus imaging allows direct visualization of the microvasculature. The retinal vascular abnormalities may reflect similar vascular changes in kidneys, heart, and other tissues [29,30]. So retinal fundus imaging provides a very good non-invasive method of assessing the vascular conditions present in the body. Deep learning models using both metadata of the patients including age, sex, height, weight, BMI, and blood pressure and retinal images lead to substantially higher performance metrics for CKD.

This is useful for screening, diagnosing, and monitoring patients at high risk. Treatment is to treat the cause and control the loss of kidney function. The main risk factors are diabetes, high blood pressure, family history, and heart disease.

15.4.2 Cardiovascular diseases

Cardiovascular disease collectively refers to many conditions of the heart or blood vessels. It is commonly caused by the buildup of fat inside the arteries and has an increased risk of blood clots. Globally, cardiovascular disease is a main cause of death and disability. It is a preventable disease which can be prevented and managed by a healthy lifestyle. It is essential to identify the risk and prevent CVD as it is the cause of about 32% of all deaths worldwide [31].

Cardiovascular diseases include angina, heart attack, heart failure, and stroke. High blood pressure, high cholesterol, diabetes, smoking, and a sedentary lifestyle are all risk factors for cardiovascular diseases. Cholesterol deposits in the heart or arteries or atherosclerosis reduce blood flow to the heart in coronary artery diseases.

Retinal vasculature along with data about other risk factors is used by intelligent systems to predict the risk of cardiovascular diseases. Studies show that they predict circulatory mortality and stroke, better than heart attacks. This is because the heart attack is more of a macrovascular event. Such an intelligent system is used to triage and identify people at medium to high risk for further assessment and suitable intervention and treatment.

Table 15.3 lists the existing literature regarding the usage of AI to analyze RFI to predict nephrological and cardiovascular diseases.

Table 15.3 AI for disease detection in nephrology and cardiology (using RFI)

Ref. no.	Disease detected	Dataset	AI model used	Significant results
[32]	Chronic kidney disease	CC-FII dataset: 86,312 RFI from 43,156 participants. Cohort validation: 8,059 participants. Validation with smartphone-captured RFI	CNN: ResNet-50, RF, MLP	AUC: Internal test set: 0.864 External test set: 0.848 Smartphone captured images: 0.897
[33]	Renal function impairment	25,706 RFI: from 6,212 patients	CNN: VGG19	AUC: 0.81 AUC: up to 0.87 in subgroup stratified with HbA1c
[34]	Chronic kidney disease	SEED dataset training: 5,188 patients; validation: 1,297 patients External testing: SP2 dataset: 3,735 patients BES dataset: 1,538 patients	Deep learning algorithm (DLA): not specified	AUC: up to 0·911 for image DLA, 0·916 for risk factors, and 0·938 for hybrid DLA
[35]	Cardiovascular disease risk prediction	Dataset created with RFI from both eyes of 411,518 individuals and the BRAVE dataset for validation	Inception-Resnet-v2	AUC: Internal validation: up to 0.976 External validation: up to 0.876
[36]	Cardiovascular risk prediction	216,152 RFI from five datasets (South Korea, Singapore, and the UK)	CNN-based RetiCAC	AUROC: 0·742, 95% CI 0·732–0·753)
[37]	Cardiovascular mortality and funduscopic atherosclerosis score	Training: 15,408 RFI from Seoul National University-Hospital Cohort study: 32,000+ images from Korean population	DL: FASXception model with transfer learning from ImageNet	AUROC: 0.713, AUPRC: 0.569
[38]	Coronary artery disease	Prospective study of 145 patients	GCNN	Sensitivity: 0.649 Specificity: 0.75 Accuracy: 0.724 AUC: 0.753F1-score: 0.603 Precision: 0.471

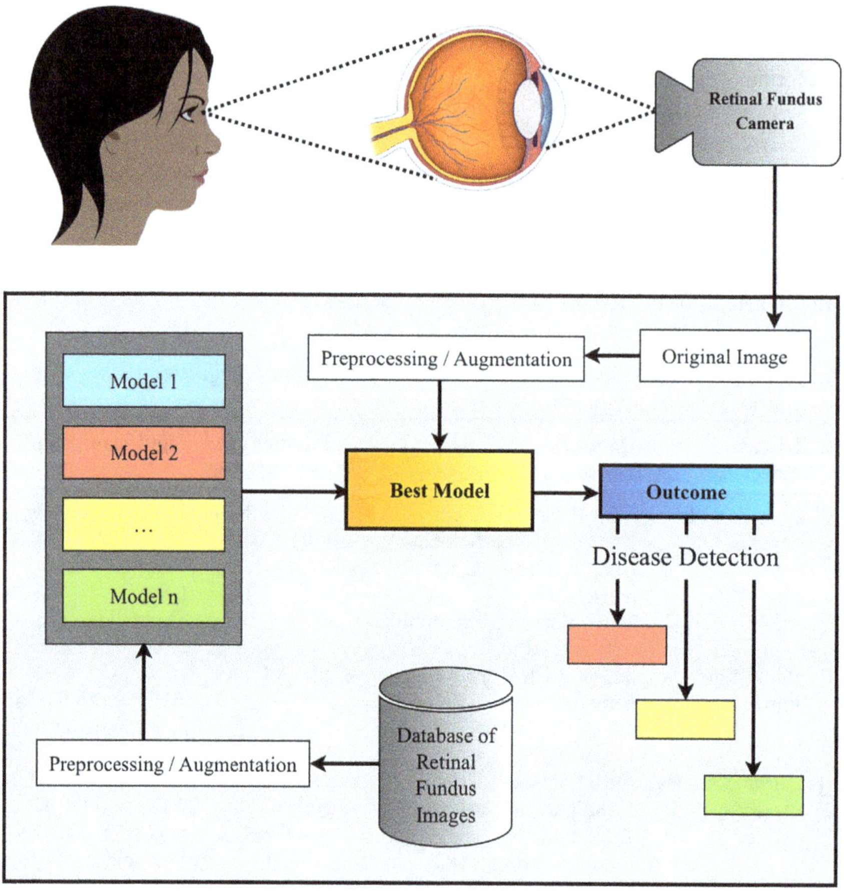

Figure 15.11 A schematic diagram of intelligent disease detection with RFI

Figure 15.11 shows a schematic diagram of an intelligent disease detection system. The process of image capture, choosing the best model, training, and image classification/decision are explained in the diagram.

15.5 Challenges and opportunities

Several challenges are present in building and using deep learning decision-making systems in healthcare applications. A few are discussed below.

- **Availability of data**—Deep learning systems need large volumes of data for effective training and testing. Not much data is available, especially in the public domain. Privacy and legal issues need to be addressed and data made available for researchers. This would be hugely beneficial for future research.

Bias-free data, covering all edge cases will lead to highly accurate disease detection systems.

- **Training and education of healthcare professionals**—Continuous training and education of healthcare professionals in using intelligent systems will help them integrate them quickly and efficiently in their healthcare practice.
- **Collaborative research**—Collaborative research ventures between system designers and medical experts will help in the creation of newer models catering to the needs of doctors and patients.

15.6 Future trends

Two promising trends are worth mentioning in the field of AI assistance in ophthalmology. They are

- Smartphone capture of retinal fundus images (with a lens assembly)
- Multi-disease detection using a single retinal fundus image

15.6.1 Smartphone image capture

Instead of patients having to visit tertiary care specialty centers, it would be beneficial if the RFI capture system is portable and easily available. Recent studies employ low-cost lens arrays with smartphones (Figure 15.12) as image capture devices and have shown high performance metrics. The existing works in this field are discussed below.

Chalam *et al.* (2022) have described using a low-cost lens complex with a smartphone to capture retinal fundus images with ease even in primary care and emergency room settings. This can capture high-quality images comparable to actual retinal fundus cameras for screening purposes [40].

Shah *et al.* (2021) studied the use of smartphone-assisted direct ophthalmoscope imaging for screening for DR and DME in general practice. They found the

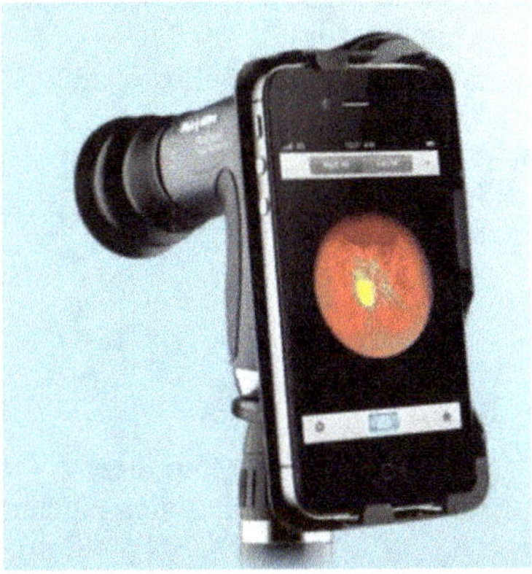

Figure 15.12 Smartphone-based RFI capture using lens assembly. Source: [39].

diagnosis made using the camera was in substantial agreement with the clinical diagnosis for DR and in moderate agreement for DME [41].

Gupta *et al.* (2022) have proposed a DIY low-cost smartphone-enabled camera which can be assembled locally to provide images which can then be analyzed using CNN-based deep learning models. They achieved high accuracy with a hybrid ML classifier [42].

Nakahara *et al.* (2022) studied a deep learning algorithm for glaucoma screening and concluded it had a high diagnostic ability, especially if the disease was advanced [43].

Mrad *et al.* (2022) have used data from retinal fundus images acquired from smartphone cameras and achieved high accuracy in detecting glaucoma. This can be a cost-effective and efficient solution for screening programs and telemedicine programs using retinal fundus images [13].

The concept of using smartphone-based cameras for image capture is a significant development in screening program and also a huge boon for remote and rural areas. A centrally located intelligent system can then be used with these images for assessing, triaging, and assisting medical experts.

15.7 Multi-disease detection using a single retinal fundus image

In community screening programs, it becomes essential to have any abnormalities detected when present. It is not sufficient or efficient to just look for just one condition like DR or papilledema. Studies show this can be implemented with suitably trained models or a combination of models. A study of existing literature is given in Table 15.4.

Table 15.4 AI for multi-disease/disorder detection (using RFI)

Ref. no.	Disease/disorder	Dataset	AI model used	Significant results
[44]	12 major fundus diseases including diabetic retinopathy, retinal vein occlusion, retinal detachment, age-related macular degeneration, possible glaucomatous optic neuropathy and papilledema	Training data: 56,738 images Testing data: 8,176 images (one internal and two external sets)	DL CNN: SeResNext50	Significantly higher sensitivity as compared to human doctors, but lower specificity
[45]	46 conditions in 29 classes	RFiMD dataset (3,200 images)	Multi disease detection pipeline: DCNN pre-trained with ImageNet and transfer learning, ensemble model	AUROC: 0.95 for disease risk classification 0.7 for multi-label scoring

(Continues)

Table 15.4 (Continued)

Ref. no.	Disease/disorder	Dataset	AI model used	Significant results
[46]	39 retinal fundus conditions	249,620 fundus images from heterogeneous sources	2-level hierarchical system with 3 groups of CNN and Mask RCNN	F1 score: 0.923, Sensitivity: 0.978, Specificity: 0.996 and (AUROC): 0.9984
[47]	Glaucoma, maculopathy, pathological myopia, and retinitis pigmentosa	Dataset with 250 RFI	MobileNetV2 and transfer learning	Accuracy: 96.2% Sensitivity: 90.4% Specificity: 97.6%

15.8 Conclusion

AI systems are proving to be a big boon for doctors and patients alike. It is a very useful tool for medical experts as it can save them a lot of time by triaging the patient's needs and also alerting the doctors if immediate medical care is indicated by the AI findings.

The RFI is a non-invasive, cost-effective imaging tool which finds applications in disease detection and monitoring systems in several specialties. Research to find new applications of AI for retinal diseases and to improve the performance of current intelligent systems is ongoing and has seen a huge surge since 2017 (Figure 15.13) [48]. Collaboration between the medical experts to provide domain knowledge and the AI experts will lead to the development of better systems.

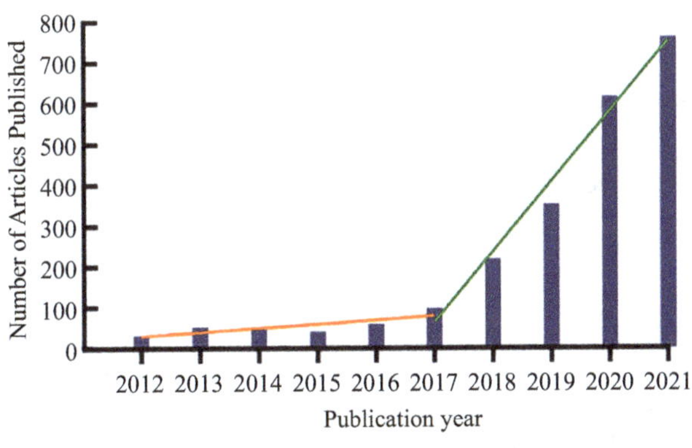

Figure 15.13 Science citation index (SCI) papers published between 2012 and 2021 on AI to study various retinal diseases. Source: [48].

AI	Artificial Intelligence	DN	Diabetic Nephropathy
AMD	Age Related Macular Degeneration	DR	Diabetic Retinopathy
ARMD	Age Related Macular Degeneration	GCNN	Graph Convolutional Neural Network
AUC	Area Under Curve	ICP	Intracranial Pressure
AUPRC	Area Under Precision Recall Curve	LCBO	Life choice-based optimizer
AUROC	Area under receiver operating curve	MLP	Multi Layer Perceptron
CAC	Coronary Artery Calcium	NN	Neural Network
CAD-RADS	Coronary Artery Disease Reporting and Data System	RCNN	Region based Convolutional Neural Network
CKD	Chronic Kidney Disease	RF	Random Forest
CNN	Convolutional Neural Network	SSD	Social Ski-Driver
CVD	Cardiovascular Disease	SVM	Support Vector Machine
DCNN	Deep Convolutional Neural Network	TL	Transfer Learning
DL	Deep Learning	US FDA	United States Food and Drug Administration
DME	Diabetic Macular Edema	WHO	World Health Organisation

Figure 15.14 Common abbreviations used

Further work suggested includes trying models other than CNN-based variants to see if performance can be enhanced with the usage of fewer resources. Also, curation of more public datasets, especially for rarer diseases and conditions is essential for further research. The smartphone-based RFI capture needs to be studied further as it can revolutionize screening programs at higher performance and lower costs. High-performance metrics and reliability will also improve the confidence of the doctors and patients in AI-based healthcare systems.

15.9 Abbreviations used

A few commonly used abbreviations in this chapter are listed in Figure 15.14.

References

[1] https://www.fda.gov/news-events/press-announcements/fda-permits-marketing-artificial-intelligence-based-device-detect-certain-diabetes-related-eye retrieved on 10.01.2023.

[2] https://ophthalmology.med.ubc.ca/patient-care/ophthalmic-photography/color-fundus-photography/ retrieved on 10.01.2023.

[3] Paradisa, R. H., Bustamam, A., Mangunwardoyo, W., Victor, A. A., Yudantha, A. R., and Anki, P. (2021). Deep feature vectors concatenation for eye disease detection using fundus image. *Electronics*, 11(1), 23. https://doi.org/10.3390/electronics11010023

[4] https://www.who.int/news-room/fact-sheets/detail/diabetes retrieved on 10.01.2023.

[5] Gao, J., Liu, R., Cao, S., *et al.* (2015). NLRP3 inflammasome: activation and regulation in age-related macular degeneration. *Mediators of Inflammation*. 2015, 11 pages. 10.1155/2015/690243.

[6] Diaz-Pinto, A., Colomer, A., Naranjo, V., Morales, S., Xu, Y., and Frangi, A. (2018). *Retinal Image Synthesis for Glaucoma Assessment Using DCGAN and VAE Models: 19th International Conference*, Madrid, Spain, November 21–23, 2018, Proceedings, Part I. 10.1007/978-3-030-03493-1_24.

[7] Xu, X., Guan, Y., Li, J., Zerui, M., Zhang, L., and Li, L. (2021). Automatic glaucoma detection based on transfer induced attention network. *BioMedical Engineering* OnLine, 20. 10.1186/s12938-021-00877-5.

[8] Goh, J. H. L., Lim, Z. W., Fang, X., *et al.* (2020). Artificial intelligence for cataract detection and management. *The Asia-Pacific Journal of Ophthalmology*, 9(2), 88–95.

[9] Wang, Y., Yu, M., Hu, B., *et al.* (2021). Deep learning-based detection and stage grading for optimising diagnosis of diabetic retinopathy. *Diabetes/Metabolism Research and Reviews*, 37(4), e3445.

[10] Faiyaz, A. M., Sharif, M. I., Azam, S., Karim, A., and El-Den, J. (2023). Analysis of diabetic retinopathy (DR) based on the deep learning. *Information*, 14(1), 30.

[11] Chakraborty, R. and Pramanik, A. (2022). DCNN-based prediction model for detection of age-related macular degeneration from color fundus images. *Medical & Biological Engineering & Computing*, 60(5), 1431–1448.

[12] Morano, J., Hervella, Á. S., Rouco, J., Novo, J., Fernández-Vigo, J. I., and Ortega, M. (2023). Weakly-supervised detection of AMD-related lesions in color fundus images using explainable deep learning. *Computer Methods and Programs in Biomedicine*, 229, 107296.

[13] Mrad, Y., Elloumi, Y., Akil, M., and Bedoui, M. H. (2022). A fast and accurate method for glaucoma screening from smartphone-captured fundus images. *IRBM*, 43(4), 279–289.

[14] Fan, R., Alipour, K., Bowd, C., *et al.* (2023). Detecting glaucoma from fundus photographs using deep learning without convolutions: transformer for improved generalization. *Ophthalmology Science*, 3(1), 100233.

[15] Yadav, J. K. P. S. and Yadav, S. (2022). Computer-aided diagnosis of cataract severity using retinal fundus images and deep learning. *Computational Intelligence* 38(4), 1450–1473.

[16] Tham, Y. C., Goh, J. H. L., Anees, A., *et al.* (2022). Detecting visually significant cataract using retinal photograph-based deep learning. *Nature Aging*, 2(3), 264–271.

[17] Mollan, S., Markey, K., Benzimra, J., *et al.* (2014). A practical approach to, diagnosis, assessment and management of idiopathic intracranial hypertension. *Practical Neurology*, 14, 380–390. 10.1136/practneurol-2014-000821.

[18] Guarnizo, A., Albreiki, D., Cruz, J. P., Létourneau-Guillon, L., Iancu, D., and Torres, C. (2022). Papilledema: a review of the pathophysiology, imaging findings, and mimics. *Canadian Association of Radiologists Journal,* 73(3), 557–567. doi:10.1177/08465371211061660.

[19] Liao, H., Zhu, Z., and Peng, Y. (2018). Potential utility of retinal imaging for Alzheimer's disease: a review. *Frontiers in Aging Neuroscience*, 10, 188. 10.3389/fnagi.2018.00188.

[20] Saba, T., Akbar, S., Kolivand, H., and Ali Bahaj, S. (2021). Automatic detection of papilledema through fundus retinal images using deep learning. *Microscopy Research and Technique*, 84(12), 3066–3077.

[21] Avramidis, K., Rostami, M., Chang, M., and Narayanan, S. (2022, October). Automating detection of Papilledema in pediatric fundus images with explainable machine learning. In *2022 IEEE International Conference on Image Processing (ICIP)* (pp. 3973–3977). IEEE.

[22] Milea, D., Najjar, R. P., Jiang, Z., *et al.* (2020). Artificial intelligence to detect papilledema from ocular fundus photographs. *New England Journal of Medicine*, 382, 1687–1695. doi:10.1056/NEJMoa1917130.

[23] Vasseneix, C., Najjar, R. P., Xu, X., *et al.* (2021). Accuracy of a deep learning system for classification of papilledema severity on ocular fundus photographs. *Neurology*, 97(4), e369–e377.

[24] Cheung, C. Y., Ran, A. R., Wang, S., *et al.* (2022). A deep learning model for detection of Alzheimer's disease based on retinal photographs: a retrospective, multicentre case-control study. *The Lancet Digital Health*, 4(11), e806–e815.

[25] Leong, Y. Y., Vasseneix, C., Finkelstein, M. T., Milea, D., and Najjar, R. P. (2022). Artificial intelligence meets neuro-ophthalmology. *Asia-Pacific Journal of Ophthalmology (Phila)*, 11(2), 111–125. doi:10.1097/APO.0000000000000512. PMID: 35533331.

[26] Mortensen, P. W., Wong, T. Y., Milea, D., and Lee, A. G. (2022). The eye is a window to systemic and neuro-ophthalmic diseases. *Asia-Pacific Journal of Ophthalmology (Phila)*, 11(2), 91–93. doi:10.1097/APO.00000 00000000531. PMID: 35533329.

[27] Ahsan, M., Alam, M., Khanam, A., *et al.* (2019). Ocular fundus abnormalities in pre-dialytic chronic kidney disease patients. *Journal of Biosciences and Medicines*, 7, 20–35. doi:10.4236/jbm.2019.711003.

[28] Mitani, A., Hammel, N., and Liu, Y. (2021). Retinal detection of kidney disease and diabetes. *Nature Biomedical Engineering*, 5, 487–489. https://doi.org/10.1038/ s41551-021-00747-4.

[29] Farrah, T. E., Dhillon, B., Keane, P. A., Webb, D. J., and Dhaun, N. (2020). The eye, the kidney, and cardiovascular disease: old concepts, better tools, and new horizons. *Kidney International*, 98(2), 323–342.

[30] Gupta, K. and Reddy, S. (2021). Heart, eye, and artificial intelligence: a review. *Cardiology Research*, 12(3), 132–139. doi:10.14740/cr1179.

[31] https://www.who.int/news-room/fact-sheets/detail/cardiovascular-diseases-(cvds) retrieved on 10.01.2023

[32] Zhang, K., Liu, X., Xu, J., *et al.* (2021). Deep-learning models for the detection and incidence prediction of chronic kidney disease and type 2 diabetes from retinal fundus images. *Nature Biomedical Engineering*, 5(6), 533–545.

[33] Kang, E. Y. C., Hsieh, Y. T., Li, C. H., *et al.* (2020). Deep learning–based detection of early renal function impairment using retinal fundus images: model development and validation. *JMIR Medical Informatics*, 8(11), e23472.

[34] Sabanayagam, C., Xu, D., Ting, D. S., *et al.* (2020). A deep learning algorithm to detect chronic kidney disease from retinal photographs in community-based populations. *The Lancet Digital Health*, 2(6), e295–e302.

[35] Ma, Y., Xiong, J., Zhu, Y., *et al.* (2021). Development and validation of a deep learning algorithm using fundus photographs to predict 10-year risk of ischemic cardiovascular diseases among Chinese population. *medRxiv*.

[36] Rim, T. H., Lee, C. J., Tham, Y. C., *et al.* (2021). Deep-learning-based cardiovascular risk stratification using coronary artery calcium scores predicted from retinal photographs. *The Lancet Digital Health*, 3(5), e306–e316.

[37] Chang, J., Ko, A., Park, S. M., *et al.* (2020). Association of cardiovascular mortality and deep learning-funduscopic atherosclerosis score derived from retinal fundus images. *American Journal of Ophthalmology*, 217, 121–130.

[38] Huang, F., Lian, J., Ng, K. S., Shih, K., and Vardhanabhuti, V. (2022). Predicting CT-based coronary artery disease using vascular biomarkers derived from fundus photographs with a graph convolutional neural network. *Diagnostics*, 12(6), 1390.

[39] Karakaya, M. and Hacisoftaoglu, R. (2020). Comparison of smartphone-based retinal imaging systems for diabetic retinopathy detection using deep learning. *BMC Bioinformatics*, 21, 259. 10.1186/s12859-020-03587-2.

[40] Chalam, K. V., Chamchikh, J., and Gasparian, S. (2022). Optics and utility of low-cost smartphone-based portable digital fundus camera system for screening of retinal diseases. *Diagnostics*, 12(6), 1499.

[41] Shah, D., Dewan, L., Singh, A., *et al.* (2021). Utility of a smartphone assisted direct ophthalmoscope camera for a general practitioner in screening of diabetic retinopathy at a primary health care center. *Indian Journal of Ophthalmology*, 69(11), 3144.

[42] Gupta, S., Thakur, S., and Gupta, A. (2022). Optimized hybrid machine learning approach for smartphone based diabetic retinopathy detection. *Multimedia Tools and Applications*, 81(10), 14475–14501.

[43] Nakahara, K., Asaoka, R., Tanito, M., *et al.* (2022). Deep learning-assisted (automatic) diagnosis of glaucoma using a smartphone. *British Journal of Ophthalmology*, 106(4), 587–592.

[44] Li, B., Chen, H., Zhang, B., *et al.* (2022). Development and evaluation of a deep learning model for the detection of multiple fundus diseases based on colour fundus photography. *British Journal of Ophthalmology*, 106(8), 1079–1086.

[45] Müller, D., Soto-Rey, I., and Kramer, F. (2021). Multi-disease detection in retinal imaging based on ensembling heterogeneous deep learning models. In *German Medical Data Sciences 2021: Digital Medicine: Recognize–Understand–Heal* (pp. 23–31). IOS Press.

[46] Cen, L. P., Ji, J., Lin, J. W., *et al.* (2021). Automatic detection of 39 fundus diseases and conditions in retinal photographs using deep neural networks. *Nature Communications*, 12(1), 1–13.

[47] Guo, C., Yu, M., and Li, J. (2021). Prediction of different eye diseases based on fundus photography via deep transfer learning. *Journal of Clinical Medicine*, 10(23), 5481.

[48] Zhao, J., Lu, Y., Qian, Y., Luo, Y., and Yang, W. (2022). Emerging trends and research Foci in artificial intelligence for retinal diseases: bibliometric and visualization study. *Journal of Medical Internet Research*, 24(6), e37532. doi:10.2196/37532. PMID: 35700021; PMCID: PMC9240965.

Chapter 16

Brain tumor analyses adopting a deep learning classifier based on glioma, meningioma, and pituitary parameters

Dhinakaran Sakthipriya[1], *Thangavel Chandrakumar*[1], *S. Hirthick*[1], *M. Shyam Sundar*[1] *and M. Saravana Kumar*[1]

Brain tumors are one of the major causes of death. Due to the aforementioned, a brain tumor may be seen using a variety of procedures. Early discovery of a brain tumor is crucial for enabling therapy. Magnetic resonance imaging is one such method. In contrast, current methods such as deep learning, neural networks, and machine learning have been used to handle a number of classification-related challenges in medical imaging in recent years. convolutional neural network (CNN) reports that magnetic resonance imaging was utilized in this study to classify three separate types of brain cancer: glioma, meningioma, and pituitary gland. This study's data set includes 3,064 contrast-enhanced T1 scans from 233 individuals. This research compares the proposed model to other models to demonstrate that our technique is superior. Pre-data and post-data preparation and enhancement outcomes were investigated.

16.1 Introduction

Our brain is composed of billions of cells; the brain is one of the body's most complex organs. When these cells in or near the brain multiply uncontrollably, brain tumors occur. This population of cells that divide uncontrolled can impair the brain's and more functional cell functions. These tumors of the brain can be classified as benign (low grade) or malignant (high grade) depending on their location, form, and texture [1–4]. For clinicians to construct cancer treatments, early cancer detection and automated tumor classification are required [5].

Imaging modalities like CT and magnetic resonance imaging (MRI) can help find brain cancers. MRI is one of the most popular therapies, because it can produce high-quality images in two dimensions (D) and three dimensions (3D) without causing the patient any pain or exposing them to radiation [6]. Moreover, MRI is

[1]Thiagarajar College of Engineering, India

regarded as the most effective and extensively used method for the identification and categorization of brain tumors [7] due to its ability to produce high-quality images of brain tissue. However, it requires a great deal of time and effort for specialists to manually examine several MR pictures simultaneously in order to discover problems. Recent years have seen a rise in the importance of Artificial Intelligence (AI) technology as a means of preventing this catastrophe. Computer-aided diagnostic (CAD) technologies are increasingly used in concert with advances in AI technology. Several diseases, including brain tumors and cancer, can be identified with speed and precision using CAD technology. The first phase of a typical CAD system is to detect and segment lesions from images, the second is to analyze these segmented tumors with numerical parameters to extract their features, and the third is to use the proper machine learning (ML) approach to predict abnormality categorization [8].

Applications for smart systems based on ML have recently been employed in many additional industries. For these systems to work effectively, useful characteristics must be found or extracted. Deep learning is a very effective subcategory of retraining machine algorithms. Its architecture comprises a number of nonlinear layers, each of which collects characteristics with greater skill by using the result of the prior layer as input [9]. The most modern machine learning technology, convolutional neural network (CNN) algorithms, is used to diagnose diseases from MRI scans. They have also been employed in many other areas of medicine, including image processing [10–12]. CNN is commonly used to categorize and grade medical pictures because preprocessing and feature extraction are not necessary before the training phase. By first classifying MR pictures as normal or abnormal and then recognizing aberrant brain MR images in accordance with various types of brain malignancies [13,14], ML- and DL-based techniques for brain tumor identification can be broken down into two main categories.

In this regard, some contemporary literary works are listed. Three distinct CNN deep learning architectures for classifying several tumor kinds (pituitary gland tumors, glioma tumors, and meningioma tumors) using brain MRI data sets (GoogleNet, AlexNet, and VGGNet). Using the VGG16 architecture, they were able to attain 98.69% accuracy [15]. To present a capsule network for categorizing brain tumors (CapsNet). To improve accuracy performance, they additionally compiled CapsNet feature maps from several convolution layers. We were able to accurately classify 86.50% of data, according to the final tally [16]. A variation of a CNN called AlexNet was used to develop a method for diagnosing glioma brain tumors. Using whole-brain MR imaging, they achieved a respectable 91.16% accuracy [17]. They proposed a technique based on deep CNN (DCNN) for finding and categorizing brain tumors. Fuzzy C-Means (FCM) is the suggested method for brain segmentation. The application's accuracy rate was 97.5% according to the final data [18]. An approach that uses both DWT and DL techniques was proposed the addition, the fuzzy k-mean approach and principal component analysis (PCA) were used to segment the brain tumor in an effort to streamline the analysis. In the end, they were successful with a 96.97% rate of accuracy [19]. An approach for classifying brain tumors was developed by using the CNN architecture and the gray-level conformation matrix (GLCM). They looked at each picture from four different angles and picked out four features: energy, correlation, contrast, and homogeneity (0, 45, 90, and 135 degrees). A total of 82.27% of the study's hypotheses were correct [20].

The objective of this project is to create a computer-aided method for detecting tumors by analyzing materials. In this framework, brain tumor images are collected, pre-processed to reduce noise, subjected to a feature routine, and then categorized according to tumor grade. A CNN architecture will be in charge of taking in data for training purposes. True positive rate is one of many performance metrics and receiver operating characteristics/areas under the curve that are used to assess diagnostic systems. To test our proposed architecture, we will use a database of retinal fundus images collected from patients at a medical facility.

16.2 Literature survey

Siar and Teshnehlab (2019) [21] analyzed a CNN that has been trained to recognize tumors using images from brain MRI scans. The first to utilize visuals was CNN. In terms of categorization, the softmax fully connected layer achieved a remarkable 98% accuracy. It's worth noting that while the radial basis function classifier has a 97.34% success rate, the decision tree (DT) classifier only manages a 94.24% success rate. We use accuracy standards, as well as sensitivity, specificity, and precision benchmarks to measure the efficacy of our networks. Komarasamy and Archana (2023) [22], a variety of specialists, have created a number of efficient methods for classifying and identifying brain tumors. The detection time, accuracy, and tumor size challenges that currently exist for existing methods are numerous. Brain tumor early diagnosis increases treatment choices and patient survival rates. It is difficult and time-consuming to manually segregate brain tumors from a large volume of MRI data for brain tumor diagnosis.

Correctly, diagnosing a brain tumor is vital for improving treatment outcomes and patient survival rates (Kumar, 2023) [23]. However, manually analyzing the numerous MRI images generated in a medical facility may be challenging (Alyami *et al.*, 2023) [24]. To classify brain tumors from brain MRI images, the authors of this research use a deep convolutional network and the salp swarm method to create a powerful deep learning-based system. The Kaggle dataset on brain tumors is used for all tests. Preprocessing and data augmentation procedures are developed, such as ideas for skewed data, to improve the classification success rate (Asad *et al.*, 2023) [25]. Using a series of cascading U-Nets, it was intended to identify tumors. DCNN was also created for patch-based segmentation of tumor cells. Prior to segmentation, this model was utilized to pinpoint the location of brain tumors. The "BraTS-2017" challenge database, consisting of 285 trained participants, 146 testing subjects, and 46 validation subjects, was used as the dataset for the proposed model.

Ramtekkar *et al.* (2023) [26] proposed a fresh, upgraded, and accurate method for detecting brain tumors. The system uses a number of methods, such as preprocessing, segmentation, feature extraction, optimization, and detection. A filter made up of Gaussian, mean, and median filters is used in the preprocessing system. The threshold and histogram techniques are used for image segmentation. Extraction of features is performed using a co-occurrence matrix of gray-level occurrences (Saladi *et al.*, 2023) [27]. Brain tumor detection remains a difficult task in medical image processing. The purpose of this research is to describe a more precise and accurate method for detecting brain cancers in neonatal brains. In certain ways, the brain of an infant differs from that of an adult, and adequate preprocessing techniques are advantageous for avoiding errors in results.

The extraction of pertinent characteristics is an essential first step in order to accomplish appropriate categorization (Doshi *et al.*, 2023) [28]. In order to refine the segmentation process, this research makes use of the probabilistic FCM approach. This research provides a framework for lowering the dimensionality of the MRI brain picture and allows for the differentiation of the regions of interest for the brain's MRI scan to be disclosed (Panigrahi & Subasi, 2023) [29]. Early identification of brain tumors is the essential need for the treatment of the patient. Brain tumor manual detection is a highly dangerous and intrusive procedure. As a result, improvements in medical imaging methods, such as magnetic resonance imaging, have emerged as a key tool in the early diagnosis of brain cancers.

Chen (2022) [30] analyses brain disorders, such as brain tumors, which are serious health issues for humans. As a result, finding brain tumors is now a difficult and demanding process. In this research, a pre-trained ResNeXt50(324d) and an interpretable approach are suggested to use past knowledge of MRI pictures for brain tumor identification. Youssef *et al.* (2022) [31] developed an ensemble classifier model for the early identification of many types of patient infections associated with brain tumors that combine data augmentation with the VGG16 deep-learning feature extraction model. On a dataset with four different classifications (glioma tumor, meningioma tumor, no tumor, and pituitary tumor), we do the BT classification using the suggested model. This will determine the kind of tumor if it is present in the MRI. The proposed approach yields a 96.8% accuracy for our model (ZainEldin *et al.*, 2022) [32]. It takes a while to identify a brain tumor, and the radiologist's skills and expertise are crucial. As the number of patients has expanded, the amount of data that must be processed has greatly increased, making outdated techniques both expensive and useless [40] (Kandimalla *et al.*, 2023) [33]. The major goal is to provide a feasible method for using MRIs to identify brain tumors so that choices about the patients' situations may be made quickly, effectively, and precisely. On the Kaggle dataset, collected from BRATS 2015 for brain tumor diagnosis using MRI scans, including 3,700 MRI brain pictures, with 3,300 revealing tumors, our proposed technique is tested.

16.3 Methodology

The DCNN approaches recommended for finding and categorizing various forms of tumors that create difficulties for the brain are described in the required methodological parts. Deep neural networks have been proposed as a workable solution for image categorization. For this study, a CNN that specializes in classification was trained.

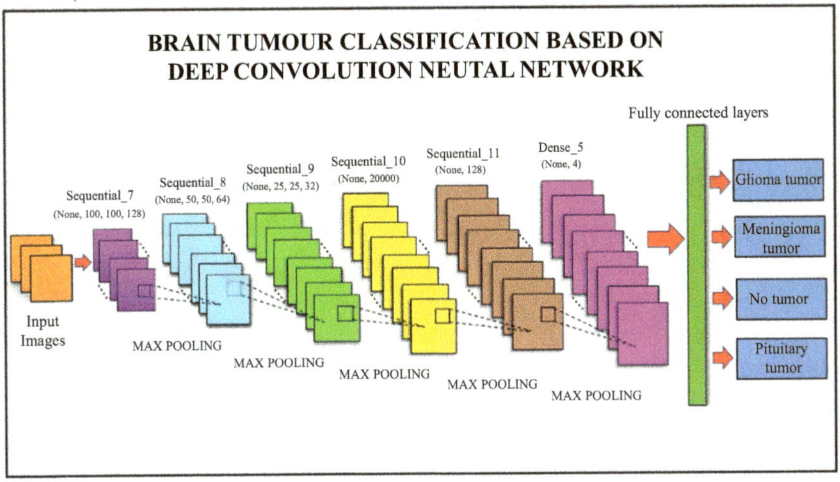

Figure 16.1 Layer-wise representation of the architecture of DCNN

A dataset for brain tumors would likely include medical imaging such as MRI along with patient information such as age, sex, and medical history. The data may also include labels or annotations indicating the location and type of tumor present in the images. The dataset could be used for tasks such as training machine learning models to detect and classify brain tumors, or for research on the characteristics of different types of brain tumors. Preprocessing of brain tumor images typically includes steps such as image registration, intensity normalization, and noise reduction. Image registration aligns multiple images of the same patient acquired at different times or with different modalities to a common coordinate system. The CNN would be trained using this data to learn to recognize the features of a brain tumor. A testing set would also consist of medical imaging data, but this data would not be used during the training process. A CNN with seven layers could be used for brain tumor detection. A large dataset with labeled brain tumors would be needed. Once trained, the network could be used to identify brain tumors in new images. Performance analysis in brain tumors typically involves evaluating various treatment options and determining which ones are most effective at treating the specific type of brain tumor. Factors that are commonly considered in performance analysis include overall survival rates, progression-free survival rates, and the side effects associated with each treatment. Additionally, imaging techniques such as MRI are often used to evaluate the size and progression of the tumor over time.

By developing and applying a DCNN to identify and classify various forms of brain tumors, the suggested study advances previous research. It is made up of different CNN layer components. It is carried out in line with the seven layering processes. Particularly, the naming and classification of brain tumors. The recommended method represents a positive advancement in the field of medical analysis. Additionally, radiologists are predicted to gain from this applied research activity. Obtaining a second opinion will help radiologists determine the kind, severity, and size of tumors much more quickly and easily. When brain tumors are found early, professionals can create more efficient treatment plans that will benefit the patient's health. At the end of the layer split-up analysis, a categorization label for the picture is created to aid with prediction.

16.3.1 Procedure for brain tumor detection

Algorithm: Proposed brain tumor prognosis

Input: The first step in the algorithm is to collect a large dataset of brain MRI images. This dataset should include both normal and abnormal images, such as those with brain tumors.

Outputs: Classification of each image and identification of Brain Tumor for each image sequence.

1. Brain Tumor detection estimation – CNN Layer
2. Pre-Process = The next step is to preprocess the images by removing noise and enhancing the quality of the images. This can be done using techniques such as image denoising and image enhancement.
3. Partition input into sets for training and testing.
4. Brain Tumor diseases **(Layer Spilt Analysis with Accuracy)**
5. **if finding** ordinary
6. stop
7. else Brain Tumor
8. end if

According to the research, it was possible to differentiate brain tumors for prognosis: The CNN is trained on a dataset of labeled MRI scans, with the tumors annotated. During inference, the CNN is used to analyze an MRI scan and predict the presence and location of tumors. Another approach is to use a 2D CNN to analyze computed tomography (CT) scans of the brain, which can be useful for detecting and segmenting tumors in the brain. This can be done using techniques such as texture analysis, shape analysis, and intensity analysis. The extracted features will be used as input to the classifier. This can be done by calculating performance metrics such as accuracy, precision, and recall, the final step is to localize the tumor within the brain. This can be done using techniques such as region growing, active contours, or level sets.

16.3.2 Deep CNN (DCNN) architecture

Figure 16.1, which represents the layer split-up analysis, is an example that is a deep CNN (DCNN). It has many of the same properties as a standard neural network, such as numerous layers of neurons, different learning rates, and so on. As indicated in Figure 16.1, for network selection in this study, we employed four

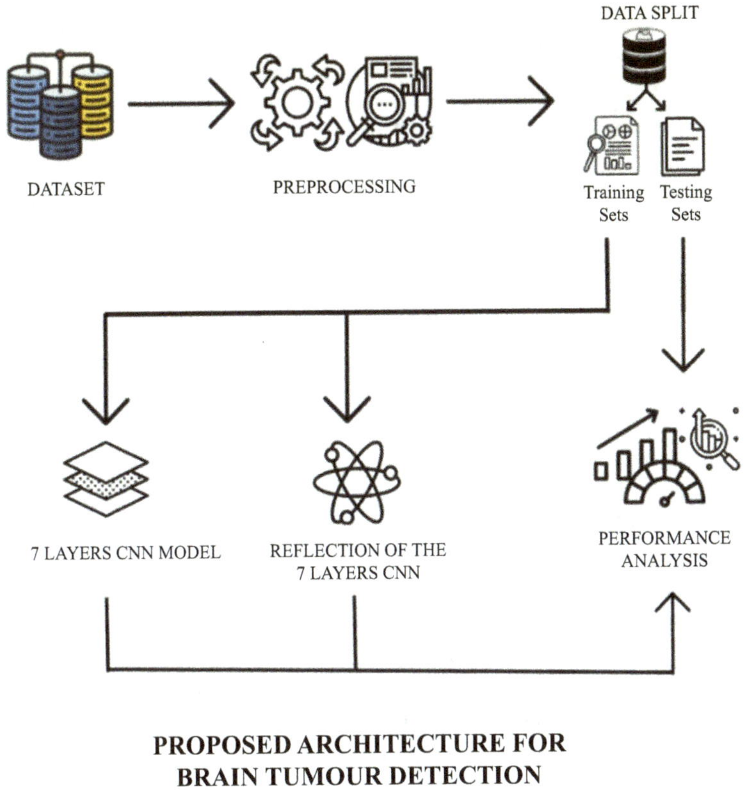

Figure 16.2 Proposed framework for brain tumor detection

distinct CNN layer approaches. Figure 16.2 illustrates how the proposed DCNN operates. This part contains the implementation of CNN's layers, as mentioned below.

Four CNNs are utilized in this section of the ML-CNN architecture to classify the level of brain tumor illness. It goes by the moniker classification-net CNN (CN-CNN). To identify photos impacted by brain tumors (pituitary tumor, meningioma tumor, and glioma tumor) and four categories of images, this network employs classed images from the DN-CNN network. The progression of a brain tumor is broken down into four stages: advanced, early, moderate, and normal. Early refers to the glaucoma illness's onset, moderate refers to the disease's medium value, advanced refers to the peak value, and normally refers to the no tumor disease value. We constructed one CNN method for each stage of brain tumor identification in the classification-net phase, using a total of four CNN architectures in this section.

CN-internal CNN's structure. We employed 7 layers, 40 epochs in size, and a learning rate of 0.001. In this table, the input picture has a size of 128×128 and a filter size of 33, there are 6 filters, and the first convolutional layer's stride is 1. The second convolutional layer has a smaller size (64 64), but the stride and number of filters remain the same (16 filters). The size is 3,232 with 25 filters in the third convolutional layer, and the filter size and stride are also constant.

16.4 Experiment analysis and discussion

Python was used to analyze the very in-depth convolutional neural network, which was tested on the Jupiter Notebook's Intel Core i5 processor with a 2.50 GHz clock speed and 8GB of RAM. A variety of statistical results were calculated.

16.4.1 Preprocessing

As such, preprocessing serves primarily to enhance the input image and build it in a highly efficient human or machine vision system. Preprocessing also aids in increasing the SNR, removing noisy artifacts, smoothing the image from the inside out, and preserving the image's edges, which is very important when dealing with human subjects. The raw male image can be seen more clearly by increasing the SNR settings. To prepare an image for analysis by a human or machine vision system, pre-processing is essential. Pre-processing also aids in boosting SNR, removing noisy artifacts, smoothing the image from the inside out, and conserving the image's edges, which is very important when dealing with human subjects. In order to improve the SNR values and, by extension, the clarity of raw human photographs, it is usual practice to employ adjective differentiation improvement-assisted modified sigmoid processes (Tables 16.1 and 16.2).

Table 16.1 Dataset description of images of various groups and subsets

Class	Total	Train	Validation	Test
Pituitary	1,204	602	301	301
Glioma	1,180	820	410	410
Meningioma	1,044	522	261	261
No tumor	1,136	568	284	284
Total	**4,564**	**2,282**	**1,141**	**1,141**

16.4.2 Performance analysis

Table 16.2 Calculated percentages of statistical measures

No.	Types	Sensitivity (%)	Specificity (%)	Accuracy (%)	Precision (%)
1	Pituitary	93.03	88.48	93.25	94.08
2	Glioma	88.56	83.61	88.95	89.30
3	Meningioma	82.00	85.62	78.90	83.45
4	No tumor	79.67	82.59	76.54	84.28
	Average	85.81	85.07	84.39	67.78

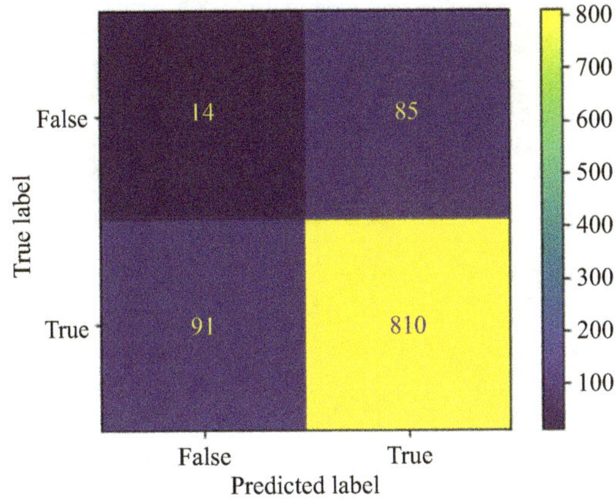

Figure 16.3 Confusion matrix

16.4.3 Brain tumor deduction

Here, TP denotes true positives that are accurately discovered among glaucoma photographs, whereas TN denotes true negatives that correctly identify mistakenly categorized images. As illustrated in Figure 16.3, false acceptance and false rejection denote classes that were correctly and incorrectly selected.

16.4.4 CNN layer split-up analysis

The efficiency of the suggested technique for diagnosing brain tumor illness utilizing brain tumor datasets was examined using DCNN layer (L) validation with L = 1. For the purpose of segmenting graphs, brain tumor photos include 3,060 images with and without tumors. The patients at India's Hospital were being examined when these pictures were shot. Layer validation results are presented in Figure 16.4.

- The segmentation graph shows how the layer has been divided up into accuracy vs. epoch and loss vs. epoch.
- An accuracy of 78.17% was obtained from the layer level 1 analysis graph.

16.5 Conclusion

Deep learning is a branch of machine learning that involves training artificial neural networks to perform tasks such as image or speech recognition. In the medical field, deep learning algorithms have been used to assist in the detection and diagnosis of brain tumors. These algorithms can analyze medical images, such as MRI, and identify regions of the brain that may contain a tumor. However, it's

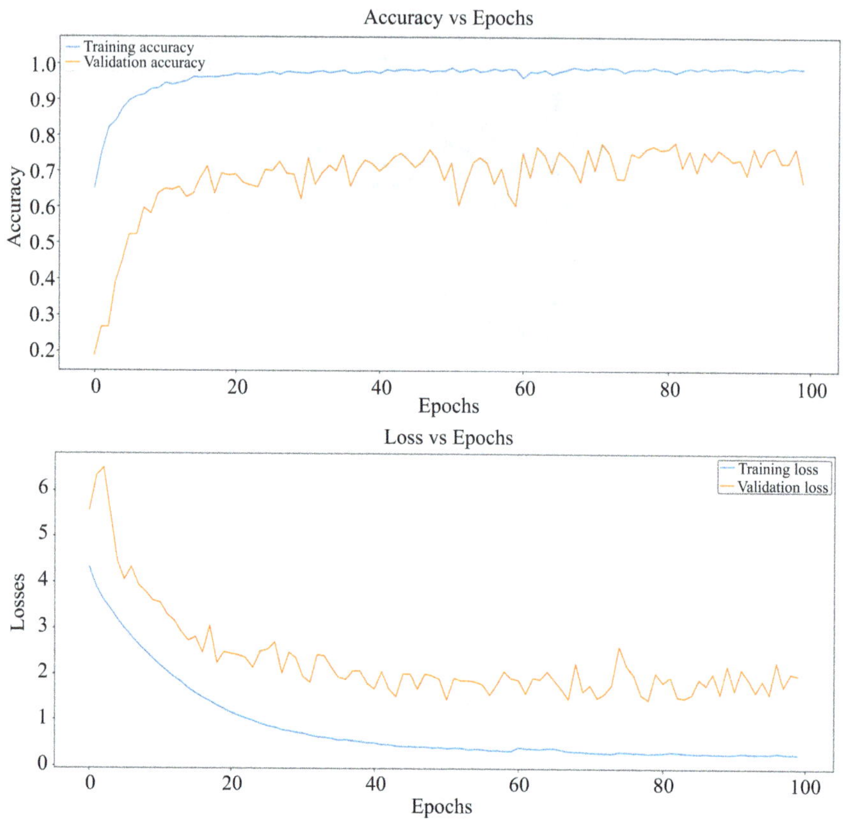

Figure 16.4 CNN layer split-up statistical analysis

important to note that deep learning should be used in conjunction with a radiologist's expertise and other medical diagnostic tools to make a definitive diagnosis. A brain tumor is an abnormal growth of cells within the brain or the skull. Symptoms of a brain tumor can include headaches, seizures, vision or speech problems, and changes in personality or cognitive function. Treatment options for a brain tumor can include surgery, radiation therapy, and chemotherapy, and the choice of treatment depends on the type and location of the tumor, as well as the patient's overall health. A conclusion of a brain tumor using a CNN would involve analyzing medical imaging data, such as MRI or CT scans, using the CNN to identify any potential tumors. The CNN would be trained on a dataset of labeled images to learn the features that indicate a tumor. Once the CNN has been trained, it can then be used to analyze new images and make predictions about the presence of a tumor. The accuracy of the predictions will depend on the quality of the training dataset and the specific architecture of the CNN.

References

[1] Mohsen, H., El-Dahshan, E. S. A., El-Horbaty, E. S. M., and Salem, A. B. M. (2018). Classification using deep learning neural networks for brain tumors. *Future Computing and Informatics Journal*, 3(1), 68–71.

[2] Khambhata, K. G. and Panchal, S. R. (2016). Multiclass classification of brain tumor in MR images. *International Journal of Innovative Research in Computer and Communication Engineering*, 4(5), 8982–8992.

[3] Das, V. and Rajan, J. (2016). Techniques for MRI brain tumor detection: a survey. *International Journal of Research in Computer Applications & Information Technology*, 4(3), 53e6.

[4] Litjens, G., Kooi, T., Bejnordi, B. E., *et al.* (2017). A survey on deep learning in medical image analysis. *Medical Image Analysis*, 42, 60–88.

[5] Pereira, S., Meier, R., Alves, V., Reyes, M., and Silva, C. A. (2018). Automatic brain tumor grading from MRI data using convolutional neural networks and quality assessment. In *Understanding and Interpreting Machine Learning in Medical Image Computing Applications: First International Workshops, MLCN 2018, DLF 2018, and iMIMIC 2018, Held in Conjunction with MICCAI 2018, Granada, Spain, September 16–20, 2018, Proceedings 1* (pp. 106–114). Springer International Publishing.

[6] Le, Q. V. (2015). A tutorial on deep learning. Part 1: nonlinear classifiers and the backpropagation algorithm. Google Brain, Google Inc. Retrieved from https://cs.stanford.edu/~quocle/tutorial1.pdf

[7] Kumar, S., Dabas, C., and Godara, S. (2017). Classification of brain MRI tumor images: a hybrid approach. *Procedia Computer Science*, 122, 510–517.

[8] Vidyarthi, A. and Mittal, N. (2015, December). Performance analysis of Gabor-Wavelet based features in classification of high grade malignant brain tumors. In *2015 39th National Systems Conference (NSC)* (pp. 1–6). IEEE.

[9] Deng, L. and Yu, D. (2014). Deep learning: methods and applications. *Foundations and Trends® in Signal Processing*, 7(3–4), 197–387.

[10] Zikic, D., Glocker, B., Konukoglu, E., *et al.* (2012, October). Decision forests for tissue-specific segmentation of high-grade gliomas in multi-channel MR. In *MICCAI (3)* (pp. 369–376).

[11] Pereira, S., Pinto, A., Alves, V., and Silva, C. A. (2016). Brain tumor segmentation using convolutional neural networks in MRI images. *IEEE Transactions on Medical Imaging*, 35(5), 1240–1251.

[12] Alam, M. S., Rahman, M. M., Hossain, M. A., *et al.* (2019). Automatic human brain tumor detection in MRI image using template-based K means and improved fuzzy C means clustering algorithm. *Big Data and Cognitive Computing*, 3(2), 27.

[13] Tharani, S. and Yamini, C. (2016). Classification using convolutional neural network for heart and diabetics datasets. *International Journal of Advanced Research in Computer and Communication Engineering*, 5(12), 417–422.

[14] Ravì, D., Wong, C., Deligianni, F., *et al.* (2016). Deep learning for health informatics. *IEEE Journal of Biomedical and Health Informatics*, 21(1), 4–21.

[15] Rehman, A., Naz, S., Razzak, M. I., Akram, F., and Imran, M. (2020). A deep learning-based framework for automatic brain tumors classification using transfer learning. *Circuits, Systems, and Signal Processing*, 39, 757–775.

[16] Afshar, P., Mohammadi, A., and Plataniotis, K. N. (2018, October). Brain tumor type classification via capsule networks. In *2018 25th IEEE International Conference on Image Processing (ICIP)* (pp. 3129–3133). IEEE.

[17] Khawaldeh, S., Pervaiz, U., Rafiq, A., and Alkhawaldeh, R. S. (2017). Noninvasive grading of glioma tumor using magnetic resonance imaging with convolutional neural networks. *Applied Sciences*, 8(1), 27.

[18] Abiwinanda, N., Hanif, M., Hesaputra, S. T., Handayani, A., and Mengko, T. R. (2019). Brain tumor classification using convolutional neural network. In *World Congress on Medical Physics and Biomedical Engineering 2018: June 3–8, 2018, Prague, Czech Republic (Vol. 1)* (pp. 183–189). Singapore: Springer.

[19] Anaraki, A. K., Ayati, M., and Kazemi, F. (2019). Magnetic resonance imaging-based brain tumor grades classification and grading via convolutional neural networks and genetic algorithms. *Biocybernetics and Biomedical Engineering*, 39(1), 63–74.

[20] Widhiarso, W., Yohannes, Y., and Prakarsah, C. (2018). Brain tumor classification using gray level co-occurrence matrix and convolutional neural network. *IJEIS (Indonesian Journal of Electronics and Instrumentation Systems)*, 8(2), 179–190.

[21] Siar, M. and Teshnehlab, M. (2019, October). Brain tumor detection using deep neural network and machine learning algorithm. In *2019 9th International Conference on Computer and Knowledge Engineering (ICCKE)* (pp. 363–368). IEEE.

[22] Komarasamy, G. and Archana, K. V. (2023). A novel deep learning-based brain tumor detection using the Bagging ensemble with K-nearest neighbor. *Journal of Intelligent Systems*, 32.

[23] Kumar, K. S., Bansal, A., and Singh, N. P. (2023, January). Brain tumor classification using deep learning techniques. In *Machine Learning, Image Processing, Network Security and Data Sciences: 4th International Conference, MIND 2022, Virtual Event, January 19–20, 2023, Proceedings, Part II* (pp. 68–81). Cham: Springer Nature Switzerland.

[24] Alyami, J., Rehman, A., Almutairi, F., *et al.* (2023). Tumor localization and classification from MRI of brain using deep convolution neural network and salp swarm algorithm. *Cognitive Computation*. https://doi.org/10.1007/s12559-022-10096-2.

[25] Asad, R., Imran, A., Li, J., Almuhaimeed, A., and Alzahrani, A. (2023). Computer-aided early melanoma brain-tumor detection using deep-learning approach. *Biomedicines*, 11(1), 184.

[26] Ramtekkar, P. K., Pandey, A., and Pawar, M. K. (2023). Innovative brain tumor detection using optimized deep learning techniques. *International*

Journal of System Assurance Engineering and Management, 14, 459–473. https://doi.org/10.1007/s13198-022-01819-7

[27] Saladi, S., Karuna, Y., Koppu, S., *et al.* (2023). Segmentation and analysis emphasizing neonatal MRI brain images using machine learning techniques. *Mathematics*, 11(2), 285.

[28] Doshi, R., Hiran, K. K., Prakash, B., and Vyas, A. K. (2023). Deep belief network-based image processing for local directional segmentation in brain tumor detection. *Journal of Electronic Imaging*, 32(6), 062502.

[29] Panigrahi, A. and Subasi, A. (2023). Magnetic resonance imagining-based automated brain tumor detection using deep learning techniques. In *Applications of Artificial Intelligence in Medical Imaging* (pp. 75–107). Academic Press.

[30] Chen, S. (2022, December). An application of prior knowledge on detection of brain tumors in magnetic resonance imaging images. In *2022 6th International Seminar on Education, Management and Social Sciences (ISEMSS 2022)* (pp. 3087–3094). Atlantis Press.

[31] Youssef, S. M., Gaber, J. A., and Kamal, Y. A. (2022, December). A computer-aided brain tumor detection integrating ensemble classifiers with data augmentation and VGG16 feature extraction. In *2022 5th International Conference on Communications, Signal Processing, and their Applications (ICCSPA)* (pp. 1–5). IEEE.

[32] ZainEldin, H., Gamel, S. A., El-Kenawy, E. S. M., *et al.* (2022). Brain tumor detection and classification using deep learning and sine-cosine fitness grey wolf optimization. *Bioengineering*, 10(1), 18.

[33] Kandimalla, S. Y., Vamsi, D. M., Bhavani, S., and VM, M. (2023). Recent methods and challenges in brain tumor detection using medical image processing. *Recent Patents on Engineering*, 17(5), 8–23.

Chapter 17

Deep learning method on X-ray image super-resolution based on residual mode encoder–decoder network

Khan Irfana Begum[1], *G.S. Narayana*[1], *Ch. Chulika*[1] *and Ch. Yashwanth*[1]

Deep learning aims to improve the resolution of bicubically damaged images. These existing approaches do not work well for actual single super-resolution. To encode incredibly effective features and to regenerate high-quality images, we introduce an encoder-decoder residual network (EDRN) for a real single image super-resolution (SISR).

17.1 Introduction

High-quality medical resonance (MR) images are difficult to capture due to prolonged scan time, low spatial coverage, and signal-to-noise ratio. Super-resolution (SR) helps to resolve this by converting low-resolution MRI images to high-quality MRI images. SR is a process of merging low-resolution images to achieve high-resolution images. SR is categorized into two types, namely, multi-image SR (MISR) and single image SR (SISR). MISR reconstructs a high-resolution image from multiple degraded images. However, MISR is rarely employed in practice, due to the unavailability of multiple frames of a scene. On the contrary, a high-resolution image is intended to be produced using SISR from a single low-resolution image.

SISR is categorized into non-learning-based methods and learning-based methods. Interpolation and wavelet methods fall under the category of non-learning techniques. Interpolation methods re-sample an image to suit transmission channel requirements and reconstruct the final image.

The commonly used techniques for interpolation are nearest neighbor, bi-cubic and bi-linear up-scaling. Bi-linear and bi-cubic interpolations calculate

[1]Electronics and Communications Engineering, Velagapudi Ramakrishna Siddhartha Engineering College, India

the distance-weighted average of 4 and 16 closely located pixels. The nearest neighbor algorithm considers only one neighbor pixel to compute missing pixels. In general, interpolation methods produce jagged artifacts due to the simple weighted average phenomenon. wavelet improves resolution uniformly in all directions on the same plane. It is used to withdraw information from images but has a higher computational cost.

Data hierarchy representations became proficient using a machine learning method known as deep learning (DL). In many areas of artificial intelligence, including computer vision, and natural processing, DL has a major advantage over traditional machine learning techniques. In general, the development of computer technology and the advancement of complex algorithms are responsible for the strong ability of DL to handle unstructured data. We employ an encoder decoder structure for real SISR aiming at high quality image in Figure 17.1 from its low quality version in Figure 17.2.

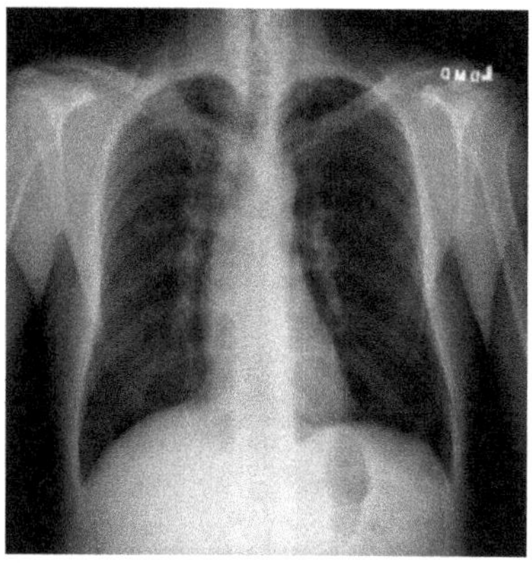

Figure 17.1 Positive pneumonia X-rays high quality

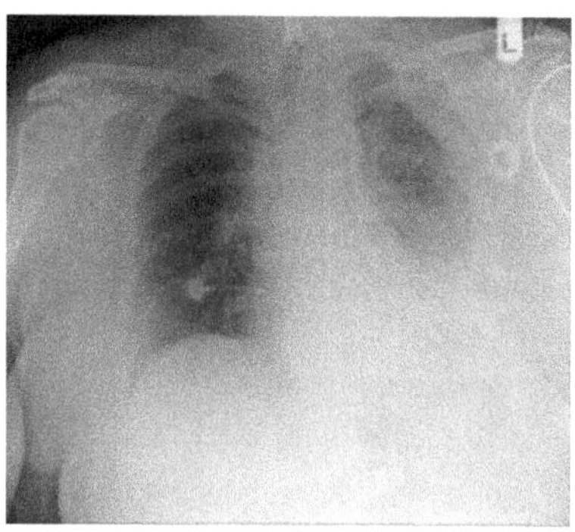

Figure 17.2 Positive pneumonia X-rays low quality

17.2 Preliminaries

17.2.1 Encoder–decoder residual network

Encoder–decoder structure enhances the context information of input shallow features. We used a coarse-to-fine method in the network to recover missed data and eliminate noise. The coarse-to-fine method firstly rebuilds the coarse information by small features and further recreates the finer step by step, in order to reduce the impact of noise, batch normalization is employed to scale down/up convolution layers.

We introduced an encoder–decoder residual network (EDRN) for restoring missed data and to decrease noise. The encoder–decoder was developed to capture connections among large-range pixels. With additional data, the structure can encode the data. The EDRN is divided into four sections: the network of feature encoder (FE), the network of large-scale residual restoration (L-SRR), the network of middle-scale residual restoration (M-SRR), and the network of small-scale residual restoration (S-SRR). A network of full convolution (FCN) has been proposed for image semantic segmentation and object recognition.

After removing the completely linked layers, FCN is made up of convolution and de-convolution processes, which are commonly referred to as encoder and decoder. Convolution is always followed by pooling in FCNs, whereas de-convolution is always followed by un-pooling, although image restoration operations in FCNs result in the loss of image data.

The M-SRR and L-SRR algorithms were employed to improve the quality of blurred images. Because of its light-weight structure, the SRCNN model has become a framework for image super resolution. However, acknowledges that deeper and more complicated networks can lead to higher SR performance, raising the complexity of network training drastically. Our network's decoder structure is made up of L-SRR, M-SRR, and S-SRR.

17.3 Coarse-to-fine approach

A coarse-to-fine approach gradually reconstructs high-quality images. By using residual learning, we can represent lost information and noise reduction at each scale. Batch normalization is applied to the down-scaling and up-scaling convolution layers in our experiments. Furthermore, we distinguish our work with and without batch normalization implementation on all convolution layers. It suggests that using batch normalization in part can lead to improved restoration performance.

Zhang *et al.* proposed a residual in residual structure (RIRBs) made up of several residual channel-wise attention blocks (RCAB). Unlike commonly used residuals, the FEN uses a convolution layer.

$$\mathrm{P}_0 = \mathrm{K}_0(\mathrm{P}_{\mathrm{LR}}), \tag{17.1}$$

P_0 stands for outermost skip connection and extracts low-level features which will be used for encoding, where 64 features from the RGB channels are extracted by the convolution layer K_0.

$$P_1 = K_{e1}(P_0), \tag{17.2}$$

A convolution layer with stride 2, rectified linear units (ReLU), and batch normalization (BN) are the three processes that make up the down-scaling process K_{e1}. K_{e1} reduces the spatial dimension of input. P_1 characterizes first downscale features, by using the second skip connection and second down-scaling process.

$$P_2 = K_{e2}(P_1), \tag{17.3}$$

where K_{e1} and K_{e2} are similar and reduce the spatial dimension of the input features by half and extract K_{e2} 256 features. P_2 stands for the innermost skip connection. The L-SRR $P_{L,mc}$ output can be written as

$$P_{L,mc} = K_{L,mc}(K_{L,4}(\cdots(K_{L,1}(P_2))\cdots)) + P_2, \tag{17.4}$$

where $P_{L,mc}$ stands for the last convolution layer of the LSRR and $K_{L,1}$, $K_{L,2}$, $K_{L,3}$, and $K_{L,4}$ stand for the RIRBs. We merge the first down-scaled features with the coarse large-scale residual features after and then we send the resulting data to the M-SRR for refinement.

The large-scale features are present in the input to M-SRR. The M-SRR objective is to recover missed data and reduce noise at a finer level. Additionally, the finer features are added to the first down-scaled features using a

$$PM = K_{M,mc}(K_{M,2}(K_{M,1}(K_{jn1}(P_{L,mc})))) + P_1, \tag{17.5}$$

where $K_{M,mc}$ stands for the final convolution layer of the M-SRRN, $K_{M,1}$ and $K_{M,2}$ for the RIRBs, ReLU layer, and BN layer is the de-convolution layer with stride 1, or K_{jn1}. Both K_{jn1} and M-convolution SRRN's layer extract 128 features. PM stands for block, and RCAB incorporates an adaptive channel-wise attention mechanism to detect the channel-wise relevance. As a result, rather than evaluating all features fairly, RCAB re-scales the extracted residual features based on channel significance. We inherit this by introducing RIRB. To maintain shallow information flow, our RIRB layers have multiple RCABs, one convolution layer, and one skip connection.

17.4 Residual in residual block

Zhang *et al.* suggested RIRBs, which are composed of several Residual Channel-wise Attention Blocks (RCAB). Unlike commonly used residual blocks, RCAB incorporates an adaptive channel-wise attention mechanism to detect the channel-wise relevance. As a result, RCAB re-scales the extracted residual features based on channel relevance rather than evaluating all features equally. We inherit this by introducing RIRB. To maintain shallow information flow, our RIRB layers have multiple RCABs, one convolution layer, and one skip connection.

17.5 Proposed method

In this section, we discuss our proposed EDRN method considering the block diagram as outlined in Figure 17.3.

17.5.1 EDRN

PLR stands for the input image, and PSR for the comparable output image. For both lost information and interfering noise, we use a scale of 3. First, we extract low-level features. SRRN's layer extracts 128 features. PM stands for features that have been recovered from large- and medium-scale data loss. In the S-SRR network, to recover lost data and eliminate noise at the most granular level, we use a RIRB and a convolution layer.

$$\mathrm{PS} = \mathrm{K}_{\mathrm{S,mc}}(\mathrm{K}_{\mathrm{S,1}}(\mathrm{K}_{\mathrm{jn2}}(\mathrm{PM})) + \mathrm{P}_0, \tag{17.6}$$

where $\mathrm{K}_{\mathrm{S,mc}}$ stands for the SSRRN's final convolution layer, $\mathrm{K}_{\mathrm{S,1}}$ for the RIRB, and $\mathrm{K}_{\mathrm{jn2}}$ for deconvolution layer. $\mathrm{K}_{\mathrm{S,1}}$ convolution layer and $\mathrm{K}_{\mathrm{jn2}}$ extract 64 features. When mapping onto the RGB color space, PS stands for the qualities that have been restored, for each of the three lost information scales. To map the returning features to a super-resolved high-resolution image, we use a convolution layer. PSR = FEDRN (P_{LR}), where the term FEDRN refers to EDRN's whole architecture.

The RIRB's jth result PR, j can be written as

$$\mathrm{PR,j} = \mathrm{K}_{\mathrm{R,j}}(\mathrm{P}_{\mathrm{R,j-1}}) = \mathrm{P}_{\mathrm{R,j}}(\mathrm{K}_{\mathrm{R,j-1}}(\cdots(\mathrm{K}_{\mathrm{R,1}}(\mathrm{P}_{\mathrm{R,0}}))\cdots)), \tag{17.7}$$

where $\mathrm{K}_{\mathrm{R,j}}$ stands for the jth RCAB and $\mathrm{P}_{\mathrm{R,0}}$ for the input of the RIRB. As a result, the output of the RIRB can be written as

$$\mathrm{PR} = \mathrm{K}_{\mathrm{R,mc}}(\mathrm{K}_{\mathrm{R,J}}(\cdots(\mathrm{K}_{\mathrm{R,1}}(\mathrm{P}_{\mathrm{R,0}}))\cdots)) + \mathrm{P}_{\mathrm{R,0}}, \tag{17.8}$$

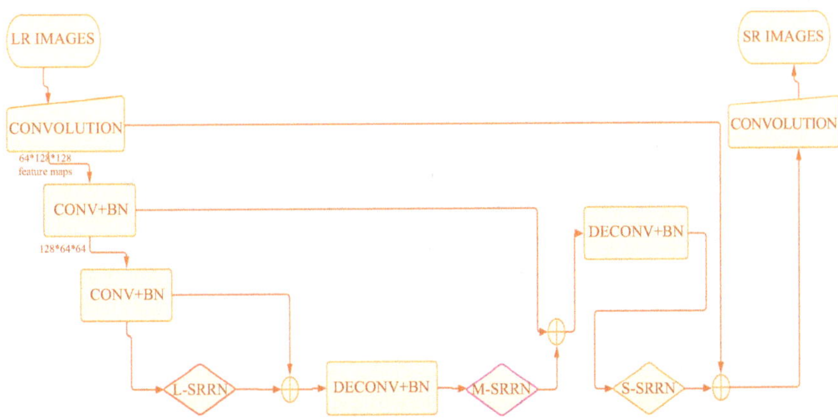

Figure 17.3 Encoder–decoder residual network

where $K_{R,mc}$ stands for the final convolution layer of the RIRB. The skip connection keeps the previous network's information. It improves network resilience.

17.6 Experiments and results

17.6.1 Datasets and metrics

A dataset is obtained from the NHICC [1] in indoor settings for the real super-resolution challenge. The dataset contains 10,000 images in that we consider 650 training photos, 200 validation images, and 150 test images. Each image has a minimum pixel resolution of 1,000 × 1,000. Since the test dataset ground truth images from the selected images and use the validation dataset to compare and illustrate our results. On NHICC [1], we also trained the single image super-resolution, and on chest X-ray [2], and also compared with the state-of-the-art approaches for ×2, ×3, and ×4. Peak signal-to-noise ratio and structural similarity are used as the evaluation metrics throughout all experiments. On the Y channel of the YCbCr space, PSNR and SSIM calculations are performed.

17.6.2 Training settings

For data augmentation, we rotated the training images 90, 180, and 270 degrees randomly and flip them horizontally. We enter 16 (RGB) low-resolution (LR) patches in each batch that subtract the dataset RGB mean. We load the learning rate to 1×10^{-4} and optimize our network using the Adam optimizer ($\alpha 1 = 0.9$, $\alpha 2 = 0.999$, $\varepsilon = 10^{-8}$) with L1 loss. On the training image, we crop areas with a size of 128 × 128 for the NIHCC 2018 real super-resolution challenge. Each 5 × 104 iteration results in a halving of the initial learning rate. We create LR pictures for the single-image super-resolution by bi-cubic down-sampling the high-resolution photos. On the LR input, we crop regions that are 48 × 48 in size, and every 2 × 105 iterations, we cut the learning rate in half.

We discuss the efficiency of the encoder–decoder and the coarse-to-fine structure of our R-EDN in this section.

17.6.3 Decoder–encoder architecture

The decoder–encoder structure helps reduce the amount of unnecessary information and encrypting the essential data, In order to show the viability of the suggested decoder-encoder structure. We also study typologies with various down-scaling/up-scaling convolution layer counts to determine the best. All implementations include 7 in order to allow for fair comparisons. The training environments, including RIRBs, are uniformly the same. The size reduces when the encoder–decoder structure is taken-off. Fixed mid-halfway feature mappings, each computation costs more and takes longer to run the picture. In addition to the performance being 0.32 dB worse than the top. Additionally, down-scaling once requires more computations and has a longer runtime as a result. Three times down-scaling results in shorter runtime but 0.2 dB worse than PSNR performance. On the contrary, down-scaling/up-scaling once and three times would be 1 dB and 0.03 dB higher, respectively, when the coarse-to-fine approach is not used, as indicated in the last lines. The comparison shows that using two down-scaling/up-scaling processes is the best and highlights the efficiency of the encoder–decoder structure.

17.6.4 Coarse-to-fine approach

We go on to show how successful the coarse-to-fine architecture is. We eliminate the RIRBs in M-SRR and S-SRR and combine all seven RIRBs into L-SRR. The performance is 0.03 dB worse without the coarse-to-fine architecture when the network only downscales or upscales once. The coarse-to-fine architecture can increase by about 0.01 dB when the network has two scaling of up/down components. The performance can be greatly enhanced by the coarse-to-fine architecture when the network has three down-scaling/up-scaling components. The comparison shows clearly how beneficial the coarse-to-fine architecture.

17.6.5 Investigation of batch normalization

It has been established that batch normalization (BN) is ineffective for traditional SISR. The offered training datasets are somewhat tiny, and the input low-resolution image contains unknown noise for genuine SISR. So, in order to lessen the impact of unknown noise and alleviate the overfitting phenomena, we choose to use BN. To the best of our understanding, BN is not ideal for small batches or patches, and various formulas for test and training are not acceptable for picture super-resolution. Therefore, we carefully consider BN and balance these concepts. We compare the results while maintaining the same training parameters to show the validity of BN usage. The performance is 29.66 dB when BN is not used. While the execution time is 0.45 s longer, the positive gain is 0.3 dB. When down-scaling/up-scaling convolution, we apply BN compared to using BN on every convolution layer, 0.02 dB improvement and 0.3 s are seen faster. According to the comparison above, using BN suitable for down-scaling/up-scaling convolution layers reduces noise. However, adding BN to every convolution layer will not result in further improvement.

17.6.6 Results for classic single image X-ray super-resolution

We apply our network to conventional SISR to further illustrate the efficacy and reliability of our suggested methods. We only add an up-sample network made up of convolution layers and a pixel shuffler to our EDRN to replace the last convolution layer. We contrast our results with nine cutting-edge, traditional single-image super-resolution techniques: SRCNN [2], RED [3], VDSR [4], LapSRN [5], MemNet [6], EDSR [7], SRMDNF [8], D-DBPN [8], and RCAN [9]. We have evaluated the low resolution images in Figure 17.4 and the results are displayed in Figure 17.5. From their publication, the outcomes of the other techniques are gathered. Our results outperform SRMD [8], which is suggested for tackling the super-resolving of multi-degradation, at all scales in PSNR and SSIM. When compared to the other approaches, our EDRN performs somewhat worse than

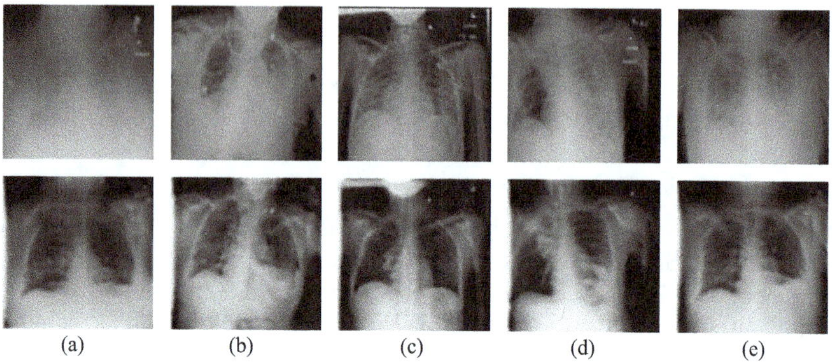

Figure 17.4 Low-resolution X-ray images on the top in that the information of the figure is inappropriate; the below images are the high-resolution images of (a), (b), (c), (d), and (e)

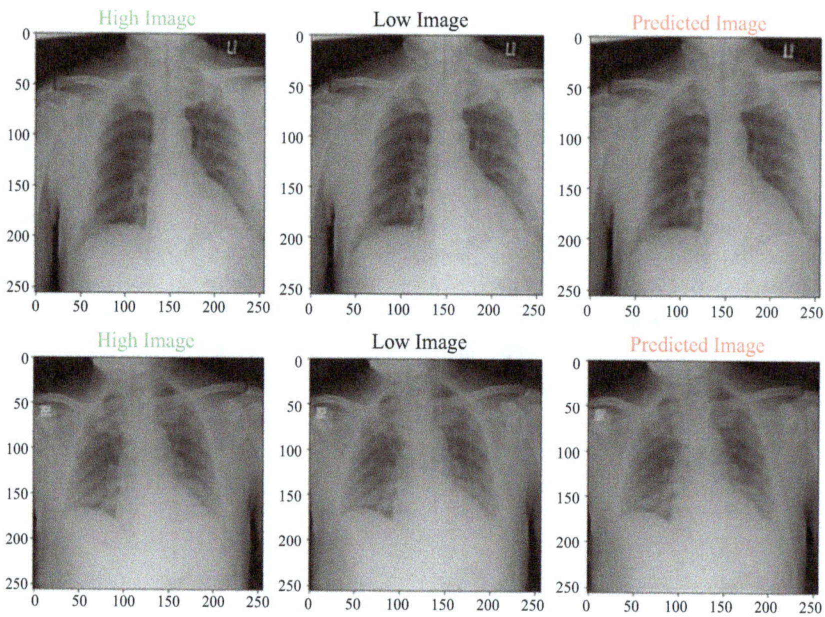

Figure 17.5 High images, low images, and predicted images. In this, we take the images of high and low from the datasets. The predicted images are from the experimental result.

RCAN [9] for scaling 2, but similarly with EDSR [7] and D-DBPN [8]. Only 74 convolution layers overall in our EDRN. With more than 400 convolution layers, RCAN [9] stacks 10 residual groups made up of 20 residual channel-wise blocks of attention. Our results cannot match the performance of the compared results.

RCAN [9], D-DBPN [8], and EDSR [7] were used for scaling $\times 3$ and $\times 4$. The usefulness and reliability of our EDRN can be further illustrated by comparing it to traditional SISR. First, due to the vast dataset and lack of noise influence, BN is not appropriate for traditional single-image SR. Second, the relationship between large-range pixels is captured by the encoder–decoder structure. When scaling 3 and 4, the input itself has a big receptive field, thus the down-scaling operation would lose a lot of details, making it more challenging to recover finer lost information. Third, our EDRN has a quicker execution time compared to EDSR [7], D-DBPN [8], and RCAN [9]. As the strictly fair comparison demonstrated, our EDRN can nevertheless produce comparable results even when using certain incorrect components and a smaller network. The usefulness and reliability of our EDRN can be further illustrated by comparing it to traditional single-picture super-resolution.

17.7 Conclusion

We presented an EDRN for actual single-image super-resolution in this chapter. Because of the bigger receptive field, the encoder–decoder structure may extract features with more context information. The coarse-to-fine structure can gradually restore lost information while reducing noise impacts. We also spoke about how to use normalization. The batch normalization provided for down-scaling/up-scaling convolution layers can minimize the effect of noise. Our EDRN can effectively recover a high-resolution image from a distorted input image and is capable of high-frequency details.

References

[1] M.H. Yeh The complex bidimensional empirical mode decomposition. *Signal Process*, 92(2), 523–541, 2012.

[2] H. Li, X. Qi, and W. Xie. Fast infrared and visible image fusion with structural decomposition. *Knowledge-Based Systems*, 204, 106182, 2020.

[3] B. Lim, S. Son, H. Kim, S. Nah, and K. M. Lee. Enhanced deep residual networks for single image super-resolution. In *2017 IEEE Conference on Computer Vision and Pattern Recognition Workshops (CVPRW)*, pp. 1132–1140, July 2017.

[4] J. Kim, J. K. Lee, and K. M. Lee. Accurate image super-resolution using very deep convolutional networks. In *2016 IEEE Conference on Computer Vision and Pattern Recognition (CVPR)*, pp. 1646–1654, June 2016.

[5] D. Kingma and J. Ba. Adam: a method for stochastic optimization. In *International Conference on Learning Representations (ICLR)* 2015, December 2015.

[6] X. Mao, C. Shen, and Y. Yang. Image restoration using very deep convolutional encoder-decoder networks with symmetric skip connections. In *Advances in Neural Information Processing Systems*, vol. 29, pp. 2802–2810, 2016. Curran Associates, Inc.

[7] W. S. Lai, J. B. Huang, N. Ahuja, and M. H. Yang. Deep laplacian pyramid networks for fast and accurate super-resolution. In *2017 IEEE Conference on Computer Vision and Pattern Recognition (CVPR)*, pp. 5835–5843, July 2017.

[8] K. Zhang, W. Zuo, and L. Zhang. Learning a single convolutional super-resolution network for multiple degradations. In *2018 IEEE/CVF Conference on Computer Vision and Pattern Recognition (CVPR)*, pp. 3262–3271, June 2018.

[9] Y. Zhang, K. Li, K. Li, L. Wang, B. Zhong, and Y. Fu. Image super-resolution using very deep residual channel attention networks. In *Computer Vision – ECCV2018*, pp. 294–310, 2018. Cham: Springer International Publishing.

[10] E. Agustsson and R. Timofte. Ntire 2017 challenge on single image super-resolution: dataset and study. In *2017 IEEE Conference on Computer Vision and Pattern Recognition Workshops (CVPRW)*, pp. 1122–1131, July 2017.

[11] C. Dong, C. C. Loy, K. He, and X. Tang. Learning a deep convolutional network for image super-resolution. In *Computer Vision – ECCV2014*, pp. 184–199, 2014. Cham: Springer International Publishing.

[12] M. Haris, G. Shakhnarovich, and N. Ukita. Deep back projection networks for super-resolution. In *2018 IEEE/CVF Conference on Computer Vision and Pattern Recognition*, pp. 1664–1673, June 2018.

[13] C. Dong, C. C. Loy, K. He, and X. Tang. Image super-resolution using deep convolutional networks. *IEEE Transactions on Pattern Analysis and Machine Intelligence*, 38(2), 295–307, 2016.

Chapter 18

Melanoma skin cancer analysis using convolutional neural networks-based deep learning classification

Balakrishnan Ramprakash[1], *Sankayya Muthuramalingam*[1], *S.V. Pragharsitha*[1] *and T. Poornisha*[1]

Melanoma, a variant of a world-threatening disease known as skin cancer, is less common but the most serious type which develops in the cells that produce melanin (gives pigmentation to the human skin). Melanoma risk appears to be rising among those under 40, particularly women. The major signs of melanoma include a variation in a normal-sized mole and the appearance of unusual pigment in the skin. Hence the detection of melanoma should be narrowed down to moles that are different in size possessing a diameter larger than 6 mm and color combinations that are not normal. Melanoma is more susceptible to spreading to many other regions of the body if it is not detected and treated early. A traditional medical way for detection of melanoma is epiluminescence microscopy or dermoscopy but unfortunately that was not much easier way as melanoma was considered a master of obscure, as in recent years the field of medicine was into great evolution scientists had invented automated diagnoses to detect these kinds of life dire disease at its very earlier stage. The dataset taken was from the skin cancer pictures from the ISIC archive. This research works on an approach for predicting skin cancer by classifying images as either malignant or benign using deep convolutional neural networks. Algorithms such as Inception v3 and MobileNetV2 were finalized among other deep learning approaches as the accuracy it provided was 99.13% and 92.46%, respectively, and it had given a minimal loss. Assessing research articles published in reputable journals on the topic of skin cancer diagnosis. For easier understanding, research findings of all the algorithms are presented in graphs, comparison tables, techniques, and frameworks. The unique stand of the chapter is that a web application for which the skin lesion or skin texture abnormality image is given as an input and predicts if the skin is either malignant or benign very accurately.

[1]Thiagarajar College of Engineering, India

18.1 Introduction

The skin is the largest organ in the human body, and skin cancer is the most prevalent worldwide health issue. The skin regulates body temperature in the human body. In general, the skin connects to other organs such as muscles, tissues, and bones, and protects us from harmful heat, ultraviolet rays, and infections. The nature of skin cancer varies depending on the weather, making it utterly unpredictable. Diepgen and Mahler (2002) have described the epidemiology of skin cancer briefly. The best and most efficient strategy to improve the survival rate of those who have been diagnosed with melanoma is to identify and treat it in its earliest stages. The advancement of dermoscopy techniques can significantly improve the accuracy of melanoma diagnosis, thereby increasing the rate of survival for patients. Dermoscopy is a methodology for monitoring the skin carefully. It is a method that employs polarized light to render the contact region translucent and display the subsurface skin structure. Manual interpretation of dermoscopy images is an arduous and longtime process to implement. Even though melanoma is early diagnosis can result in a relatively high chance of survival. Skin cancer has become the most prevalent medical condition globally. In general, the skin of the human body connects to other organs such as muscles, tissues, and bones, and protects us from harmful heat, ultraviolet rays, and infections. The nature of skin cancer varies depending on the weather, making it utterly unpredictable. Diepgen and Mahler (2002) have described the epidemiology of skin cancer briefly.

18.2 Literature survey

Daghrir *et al.* (2020) have developed a hybrid method to examine the presence of any suspicious lesions to detect melanoma skin cancer using the prediction techniques like CNN and two different classical machine learning (ML) classifiers and have achieved higher accuracy in the process of examining the presence of suspicious lesions that may cause melanoma skin cancer. Dildar *et al.* (2021) have presented a systematic literature review based on techniques such as ANN, CNN, KNN, and GAN that have been widely utilized in the process of skin cancer early detection. Hosny *et al.* (2018) have proposed a skin lesion classification method which automatically classifies the lesion, which uses transfer learning has been used to substitute the topmost layer with a softmax, which has been used to classify three different lesions, namely, common nevus, melanoma, and atypical nevus, and is applied to AlexNet and have achieved an accuracy of 98.61%; it uses ph2 dataset for the training and testing.

Nahata and Singh (2020) have used Python with Keras and Tensorflow as the backend to develop a CNN model using data from the ISIC challenge archive, which can be employed for the timely identification of skin cancer for training and testing. Vidya and Karki (2020) extracted features to detect early skin lesions and used ML techniques to classify the skin lesion as melanoma or benign. In her work published in 2019, Vijayalakshmi developed a comprehensive model that automates the detection of skin diseases to enable early diagnosis. This model encompasses three key phases: data collection and augmentation, design and development of the model architecture, and ultimately accurate disease prediction. She has used machine learning techniques such as SVM and CNN and has augmented the model with image processing tools for a more accurate prediction, resulting in 85% accuracy for the developed model. Li *et al.* (2016) have used novel data synthesis techniques to merge the individual images of skin lesions with the fill body images and have used deep learning techniques like CNN to build a model that uses the synthesized images as the data for the model to detect malignant skin cancer with greater accuracy than the traditional detection and tracking methods.

Monika *et al.* (2020) have used ML techniques and image processing tools to classify skin cancer into various types of skin-related cancers and have used dermoscopic images as the input for the pre-processing stage; they have removed the unwanted hair particles that are present in the skin lesions using the dull razor method and have performed image smoothing using the median filter as well as the gaussian filter are both employed to remove the noise. Nawaz *et al.* (2022) have developed a fully automated method for the earlier detection of skin cancer using the techniques of RCNN and FKM (fuzzy k-means clustering) and have evaluated the developed method using three standard datasets, namely, PH2, International Skin Imaging Collaboration dataset (2017), and International Symposium on Biomedical Imaging dataset (2016), achieving an accuracy of 95.6%, 95.4%, and 93.1%, respectively. Hasan *et al.* (2019) have used ML and image processing to design an artificial skin cancer system; they have used feature extraction methods to extract the affected skin cells features from the skin images and segmented using the DL techniques have achieved 89.5% accuracy and 93.7% training accuracy for the publicly available dataset.

Hossin *et al.* (2020) have used multilayered CNN techniques in conjunction with regularization techniques such as batch normalization and dropout to classify dermoscopic images for the earlier detection of skin cancer, which helps to reduce the medical cost, which may be high if the cancer is detected at a later stage. Ansari and Sarode (2017) used SVM and image processing techniques for the rapid diagnosis of skin cancer, mammogram images were used as model input, and pre-processed input for better image enhancement and to remove noise; the thresholding method is used for the segmentation purpose and GLCM methods are utilized to extract the image's features, and support vector machine is used to identify the input image. Garg *et al.* (2018) have used image processing techniques to detect skin cancer from a digital image; the image is pre-processed to avoid the excessive noise that is present in the image, followed by segmentation and feature extraction from the pre-processed image, and implemented the ABCD rule, which assesses the dermoid cyst using a variety of criteria like color of the skin tissue, asymmetry, diameter, and border irregularity of the lesion.

Alquran *et al.* (2017) have used image processing tools for the earlier detection of melanoma skin cancer; their process of detection involves the collection of dermoscopic images, followed by segmentation and feature extraction; they have used thresholding to perform segmentation and have extracted the statistical

Table 18.1 Comparative analysis

Reference	Algorithm	Significance
[6]	SVM and CNN have added image processing tools to the model	Developed a model for the automated detection of skin diseases to detect diseases earlier
[11]	Multi-layered CNN techniques along with regularization techniques like batch normalization and dropout	To classify dermoscopic images to detect skin cancer earlier
[14]	Image processing tools: GLCM, ABCD, and PCA	For the detection of melanoma skin cancer earlier
[15]	The threshold in the histogram, k-means: SVM, FFNN, and DCNN	To detect skin cancer earlier
[16]	Image processing techniques: SVM and GLCM	For the creation of a skin cancer detection system that is efficient
[18]	MobileNetV2 network	Melanoma image classification of skin cancer into malignant and benign categories
[19]	Inception-v3 and ResNet-101	Classification system for skin cancer
[20]	A single Inception-v4 model	For the categorization of the HAM10000 dataset

features using techniques such as GLCM and ABCD; and they have used PCA for the selection of features, followed by the total dermoscopy score calculation. Jana *et al.* (2017) have developed a technology for skin cancer detection that may require four steps: the removal of hair, the removal of noise, resizing of the image, and sharpening of the image in the image pre-processing step; they have used techniques, such as threshold in the histogram, k-means, etc., for the segmentation purpose; extraction of features from the segmented images; and classification using the techniques such as SVM, FFNN, and DC. Thaajwer and Ishanka (2020) have used image processing techniques and SVM for the development of an efficient skin cancer diagnosis system; they have pre-processed the image to have a smoothed and enhanced image; they have used thresholding and morphological methods for segmentation; they have extracted the features using the GLCM methods; and the extracted features are used for classification with the SVM. Table 18.1 gives the clear view of comparative analysis for this proposed work.

18.3 Methodology

Figure 18.1 describes the process of collecting data, pre-processing it, and then evaluating the model on the trained dataset. The dataset was retrieved from the Kaggle resource with all rights bound to the ISIC-Archive. Both the training and testing classes contained an equal number of images. The aim of the melanoma project of the International Skin Imaging Collaboration is to decrease the increasing number of deaths caused to melanoma and to enhance the effectiveness of

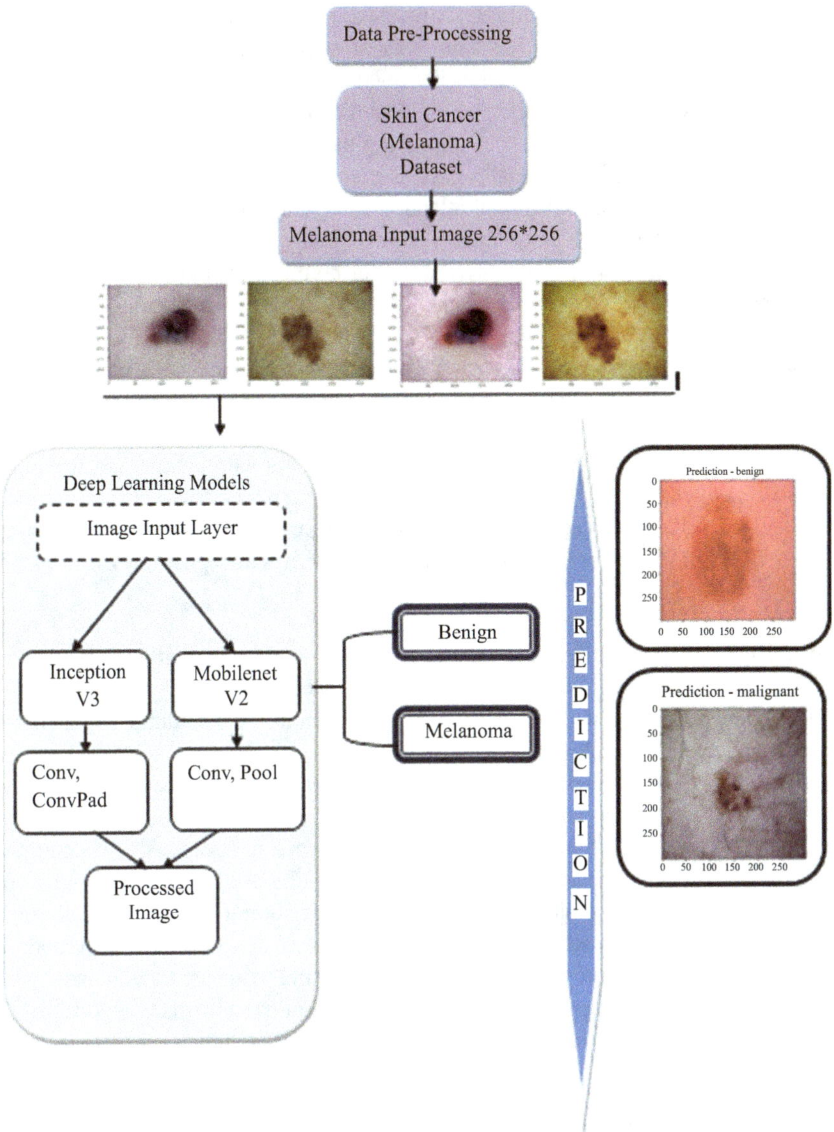

Figure 18.1 Framework proposed for melanoma detection

diagnostic testing. The dataset contains two distinct class labels, benign and malignant, denoting the less harmful and severe stages of the melanoma skin cancer disease, respectively. The primary objective of our model is to visually categorize these benign and malignant types using robust algorithms. Tensor Flow, an open-source library developed by Google that supports distributed training, immediate model iteration, simple debugging with Keras, and much more, was imported. Tensor Flow's inner computation is primarily utilized for machine learning and deep learning projects, where it contributes significantly. It consists of standard normalization and image enhancement processes. Normalization is a method used in image processing and is frequently called histogram stretching or contrast stretching. As its name implies, image enhancement is the process of improving the graphical fidelity of photographs by means of the extraction of detail which is more reliable and accurate information from them. Widely employed to call attention to or emphasize particular image elements. The sharpness and color quality of the images were improved by image enhancement, resulting in high-quality images for both benign and malignant tumors.

In addition, some data management tasks contained in data include data augmentation and methods currently used for data augmentation (black-box methods) that use deep neural networks and histogram-based methods which fall into two major categories in the problem of image classification. Additionally, subplots are required to narrow the focus on the lesion that is predicted to be cancerous or normal. Inception v3 is a deep learning model on convolution neural networks with greater and Inception v3 has deeper neural connections over Inceptions v1 and v2, however, its performance is unchanged. It is the next step in model construction, and it is a significant contributor to the success of the project. It employs adjunct classifiers as regularizers.

18.3.1 MobileNetv2

The objective of MobileNetV2 is to train a discriminative network using transfer learning to predict benign versus malignant skin cancers (Figure 18.2). This study's dataset consists of training and testing data culled from the literature. Experiments were conducted on a GPU-based cluster with the CUDA and cuDNN libraries installed. The results indicate that, except for Inception v3, the MobileNet model outperforms the other approaches in regard to precision, recall, and accuracy. The batch size is decisive that the amount of parallel processing necessary to execute the algorithm. The larger the batch size, the greater the number of parallel computations performed. This permits multiple instances of the algorithm to run concurrently on a single machine, making it efficient and quick. The batch size grows as the number of parallel computations performed increases. It is essential to ensure that the batch size is sufficient for your needs, although this technique is utilized by many algorithms. If you are running a small number of instances, a large batch size could potentially hinder your performance.

18.3.2 Inception v3

The Inception v3 model has 42 layers and a reduced margin of error than its precedents (Figure 18.3). The trained network is capable of identifying 1,000 distinct types of objects in images. The network has obtained holistic feature representations for a variety of image categories. Some of the symmetric and asymmetric building blocks of the model encompass convolutions, maximum pooling, expulsions, and fully connected layers. The model makes extensive utilization regularization, which is implemented in the activation components as well. Softmax is used to calculate the loss. MobileNetv2, a convolution neural network for image classification with 53 layers, is a further supporting algorithm. MobileNet's architecture is distinctive in that it requires minimal

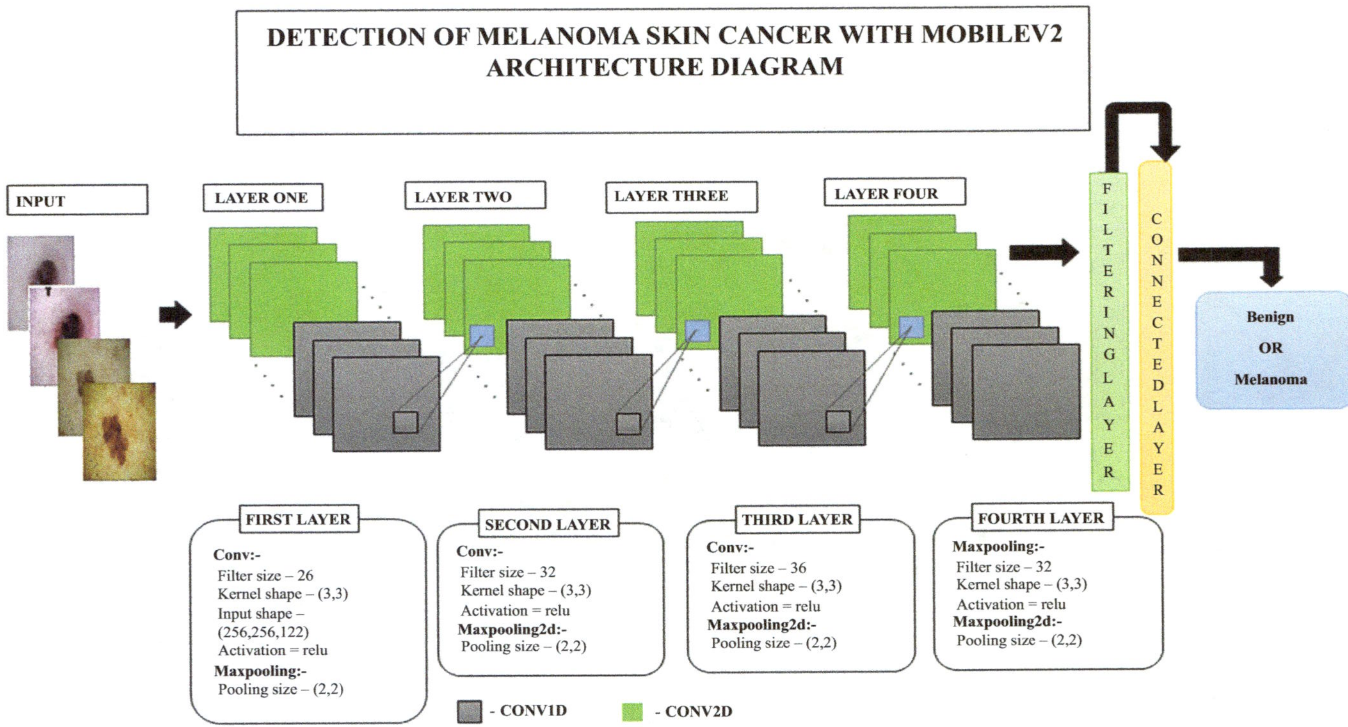

Figure 18.2 Mobilev2 architecture diagram

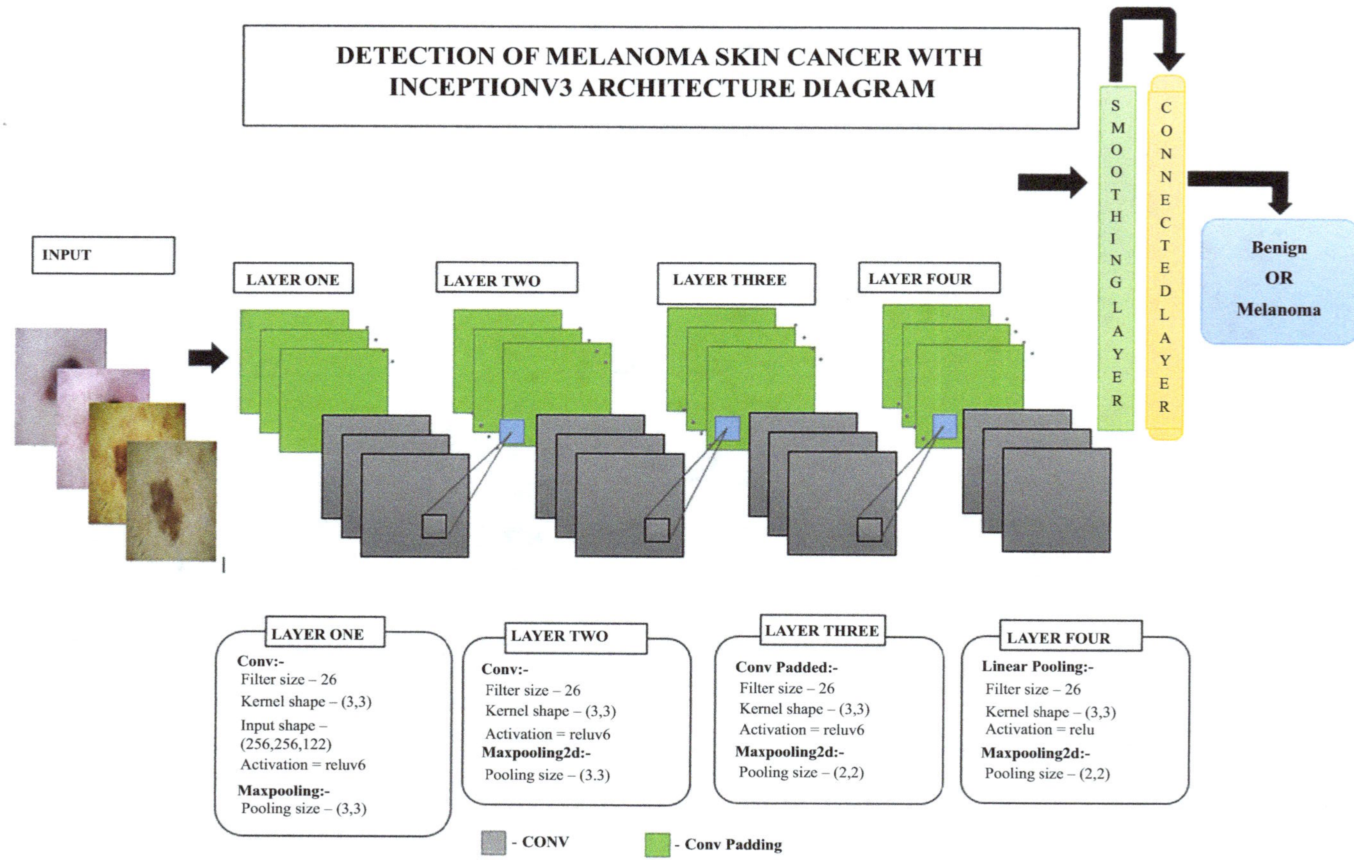

Figure 18.3 InceptionV3 architecture diagram

processing power to function. This makes it possible for computers, embedded systems, and mobile devices to operate without GPUs. MobileNetV2 contains two distinct varieties of blocks. The remaining block has a single pace. Another obstacle for shrinking by two strides. There are three levels for both kinds of blocks. It is built on a reversed residual arrangement in which residual connections connect the bottleneck layers. Lightweight fully convolutional convolutions are used as a variation source to filter intermediate expansion layer characteristics. MobileNetV2's architecture is comprised of a fully convolutional 32-filter initial layer and 19 supplemental bottleneck layers.

18.4 Results

Experimental analysis was conducted in Google Colab with 12.6GB RAM and 2.3GHz GPU: 1xTesla K80 with Python 3 Google Compute Engine backend (GPU). An experimental analysis was conducted.

18.4.1 Data pre-processing

The dataset contains an aggregate of 1,559 images for the various classes as given in Table 18.2. Each image is assigned a benign or malignant classification. Approximately half of the images contained in the dataset belong to one class, while the remaining half belong to a different class. For the purpose of separating the dataset into test and training datasets, the dataset is divided into 75%, 25%, and 5% as the test, training datasets, and validation datasets, respectively. The initial argument is that you desire to enhance your image. The second argument specifies the image enhancement function that will be employed (e.g., ImageEnhance.brightness). ImageEnhance.sharpness is an excellent image-sharpening tool for those with limited experience with image editing software. There are numerous available sharpening techniques, but this one appears to be the most effective. This tool provides complete control over the enhancement of your image. Those who desire greater control over their images also have several options. This is a good program because it allows you to customize your image so that it looks exactly how you want it to look without having to go through several steps or processes to achieve the desired result (which can be time-consuming). ImageEnhance.color is a Python library that enables a variety of image coloring techniques. It supports both RGB (and RGBA) images and 32-bit alpha channel floating point color palettes. Two libraries for deep learning, Keras and TensorFlow, are growing in popularity. As a library for neural networks, Keras encapsulates the underlying mathematics in an API. TensorFlow, a library built upon the TensorFlow framework, is utilized to develop deep

Table 18.2 Classification of ISIC images within the various classes and subsets

Subset	Melanoma	Non-melanoma	Total
Train	254	654	908
Validation	25	112	137
Test	129	385	514

learning models. ImageEnhance functions enhance the contrast of an image. It may be utilized alone or in conjunction with other commands.

18.4.2 Performance analysis

The model selected for this study is Inception v3, which achieved an accuracy of 99.13%, compared to MobileNetv2's accuracy of 92.46% (Table 18.3). Inception v3 for skin cancer image classification is a deep learning framework for detecting melanomas in human skin lesion images. Inception v3 includes methods for training deep convolutional neural networks (DCNNs) for recognizing different types of melanomas from mamma papillomatosis or contact dermatitis images and classifying benign versus malignant melanomas from histopathological slides of surgically treated patients. The Xception model consists of 71 layers of a convolutional neural network. A variant of the network trained with over one million images from the ImageNet registry. Based on image analysis, Xception forecasts whether a lesion is benign or malignant based on multiple characteristics, including skin color and texture, blood vessel density, and other characteristics. Upon inspecting the x model layers, it becomes evident that this particular layer is trainable and exhibits a specific shape. If your approach has merged on new data and is frozen, you can attempt to deactivate all or a portion of the base model and train it completely with a very small learning percentage. According to the conclusion of the research, this procedure may assist in enhancing the accuracy of the predictions. The training is conducted using the above strategy. Non-trainable parameters are parameters that cannot be modified. This may indicate that the parameter is either too new or unstable for machine learning to handle. When this occurs, it is often necessary to devise a workaround or create a new model that is better suited to the task at hand. This can be a complex ProcessPro, so it is essential to be as well-informed as possible about the parameters that will be used in your project. By recognizing which parameters are non-trainable, you can make better decisions regarding how to approach the project.

18.4.3 Statistical analysis

In the below graph of Figure 18.4 and Figure 18.5 is plotted for training and validation dataset accuracy for the Inception V3 model. The graph shows that the validation is on par with the training dataset.

Training and validation dataset losses for the Inception V3 model are plotted on this graph. The graph demonstrates that the validation dataset and the training dataset are comparable.

In the final prediction of result show the Figure 18.6 to given the model's prediction of malignant and benign skin images. The deep learning model is efficient enough to classify the malignant tissue with benign tissue.

Table 18.3 Performance metrics table for Inception V3 and Mobilenet V2

Model	Loss	Accuracy	Validation loss	Validation accuracy
Inception V3	0.0249	0.9917	0.7740	0.9176
MobileNetv2	0.0368	0.9246	0.8820	0.8788

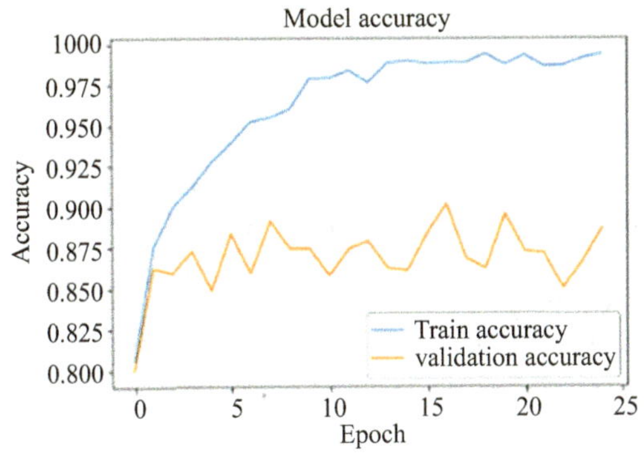

Figure 18.4 Image depicting the graph variance of the model accuracy

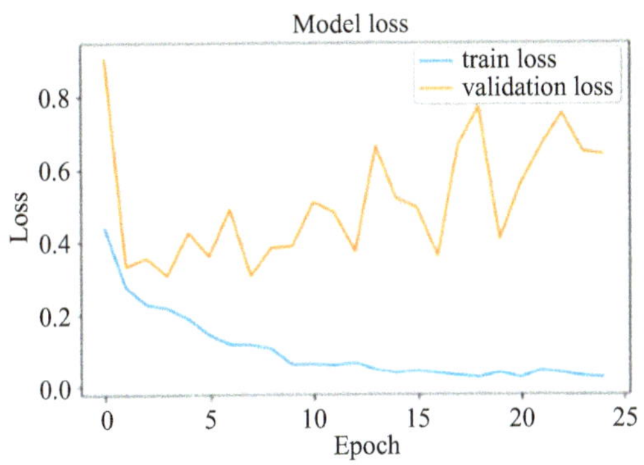

Figure 18.5 Image depicting the graph variance of the model loss

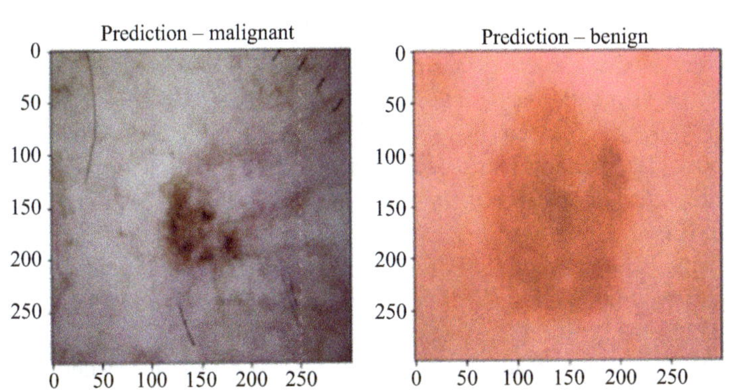

Figure 18.6 Image depicting the prediction result of the malignant type and benign type

18.5 Conclusion

Multiple deep learning methodologies are utilized in this research to identify melanoma-affected and unaffected skin cancer images. The most accurate technique for deep learning is identified. To create a multi-layer deep convolutional neural network (ML-DCNN) for glaucoma classification and detection, 1,559 raw pixel skin cancer images are prepared with MobileNetV2 and Inception v3 to extract features for the deep learning model. The deep learning model is deployed using MobileNetV2 and Inception v3: the former employs 53 layers for identifying melanoma skin cancer and the latter utilizes 48 layers for categorizing melanoma and non-melanoma skin cancers. To assess the effectiveness of deep learning models, we utilize the statistical measures of precision, validation precision, loss, and validation loss to ensure the effectiveness of our models. Inception V3 model achieves an accuracy of 99.17%, validation accuracy of 0.9176, loss of 0.0249, and validation loss of 0.7740. Evaluated by comparing to Mobile Netv2, which has an accuracy of 92.46%, validation accuracy of 0.8788, loss of 0.0368, and validation loss of 0.8820. The proposed deep learning model with InceptionV3 yielded distinct statistical values for distinct melanoma skin cancer stage categories. The obtained results are comparable to previous benchmarks, and the classification of melanoma skin cancer was made more efficient. The proposed method performs admirably; however, in the future, this model will be integrated with web applications to facilitate accessibility.

References

[1] Daghrir, J., Tlig, L., Bouchouicha, M., and Sayadi, M. (2020, September). Melanoma skin cancer detection using deep learning and classical machine learning techniques: a hybrid approach. In *2020 5th International Conference on Advanced Technologies for Signal and Image Processing (ATSIP)* (pp. 1–5). IEEE.

[2] Dildar, M., Akram, S., Irfan, M., *et al.* (2021). Skin cancer detection: a review using deep learning techniques. *International Journal of Environmental Research and Public Health*, 18(10), 5479.

[3] Hosny, K. M., Kassem, M. A., and Foaud, M. M. (2018, December). Skin cancer classification using deep learning and transfer learning. In *2018 9th Cairo International Biomedical Engineering Conference (CIBEC)* (pp. 90–93). IEEE.

[4] Nahata, H., and Singh, S. P. (2020). Deep learning solutions for skin cancer detection and diagnosis. In *Machine Learning with Health Care Perspective* (pp. 159–182). Springer, Cham.

[5] Vidya, M., and Karki, M. V. (2020, July). Skin cancer detection using machine learning techniques. In *2020 IEEE International Conference on Electronics, Computing and Communication Technologies (CONECCT)* (pp. 1–5). IEEE.

[6] Vijayalakshmi, M. M. (2019). Melanoma skin cancer detection using image processing and machine learning. *International Journal of Trend in Scientific Research and Development (IJTSRD)*, 3(4), 780–784.

[7] Li, Y., Esteva, A., Kuprel, B., Novoa, R., Ko, J., and Thrun, S. (2016). Skin cancer detection and tracking using data synthesis and deep learning. *arXiv preprint arXiv:1612.01074*.

[8] Monika, M. K., Vignesh, N. A., Kumari, C. U., Kumar, M. N. V. S. S., and Lydia, E. L. (2020). Skin cancer detection and classification using machine learning. *Materials Today: Proceedings*, 33, 4266–4270.

[9] Nawaz, M., Mehmood, Z., Nazir, T., *et al.* (2022). Skin cancer detection from dermoscopic images using deep learning and fuzzy k-means clustering. *Microscopy Research and Technique*, 85(1), 339–351.

[10] Hasan, M., Barman, S. D., Islam, S., and Reza, A. W. (2019, April). Skin cancer detection using convolutional neural network. In *Proceedings of the 2019 5th International Conference on Computing and Artificial Intelligence* (pp. 254–258).

[11] Hossin, M. A., Rupom, F. F., Mahi, H. R., Sarker, A., Ahsan, F., and Warech, S. (2020, October). Melanoma skin cancer detection using deep learning and advanced regularizer. In *2020 International Conference on Advanced Computer Science and Information Systems (ICACSIS)* (pp. 89–94). IEEE.

[12] Ansari, U. B. and Sarode, T. (2017). Skin cancer detection using image processing. *International Research Journal of Engineering and Technology*, 4(4), 2875–2881.

[13] Garg, N., Sharma, V., and Kaur, P. (2018). Melanoma skin cancer detection using image processing. In *Sensors and Image Processing* (pp. 111–119). Springer, Singapore.

[14] Alquran, H., Qasmieh, I. A., Alqudah, A. M., *et al.* (2017, October). The melanoma skin cancer detection and classification using support vector machine. In *2017 IEEE Jordan Conference on Applied Electrical Engineering and Computing Technologies (AEECT)* (pp. 1–5). IEEE.

[15] Jana, E., Subban, R., and Saraswathi, S. (2017, December). Research on skin cancer cell detection using image processing. In *2017 IEEE International Conference on Computational Intelligence and Computing Research (ICCIC)* (pp. 1–8). IEEE.

[16] Thaajwer, M. A. and Ishanka, U. P. (2020, December). Melanoma skin cancer detection using image processing and machine learning techniques. In *2020 2nd International Conference on Advancements in Computing (ICAC)* (Vol. 1, pp. 363–368). IEEE.

[17] Toğaçar, M., Cömert, Z., and Ergen, B. (2021). Intelligent skin cancer detection applying autoencoder, MobileNetV2 and spiking neural networks. *Chaos, Solitons & Fractals*, 144, 110714.

[18] Indraswari, R., Rokhana, R., and Herulambang, W. (2022). Melanoma image classification based on MobileNetV2 network. *Procedia Computer Science*, 197, 198–207.

[19] Demir, A., Yilmaz, F., and Kose, O. (2019, October). Early detection of skin cancer using deep learning architectures: resnet-101 and inception-v3. In *2019 Medical Technologies Congress (TIPTEKNO)* (pp. 1–4). IEEE.

[20] Emara, T., Afify, H. M., Ismail, F. H., and Hassanien, A. E. (2019, December). A modified inception-v4 for imbalanced skin cancer classification dataset. In *2019 14th International Conference on Computer Engineering and Systems (ICCES)* (pp. 28–33). IEEE.

[21] Yélamos, O., Braun, R. P., Liopyris, K., *et al.* (2019). Usefulness of dermoscopy to improve the clinical and histopathologic diagnosis of skin cancers. *Journal of the American Academy of Dermatology*, 80(2), 365–377.

[22] Barata, C., Celebi, M. E., and Marques, J. S. (2018). A survey of feature extraction in dermoscopy image analysis of skin cancer. *IEEE Journal of Biomedical and Health Informatics*, 23(3), 1096–1109.

[23] Leiter, U., Eigentler, T., and Garbe, C. (2014). Epidemiology of skin cancer. In Reichrath J. (ed.), *Sunlight, Vitamin D and Skin Cancer* (pp. 120–140). Springer.

[24] Argenziano, G., Puig, S., Iris, Z., *et al.* (2006). Dermoscopy improves accuracy of primary care physicians to triage lesions suggestive of skin cancer. *Journal of Clinical Oncology*, 24(12), 1877–1882.

[25] Diepgen, T. L. and Mahler, V. (2002). The epidemiology of skin cancer. *British Journal of Dermatology*, 146, 1–6.

[26] Gloster Jr, H. M. and Brodland, D. G. (1996). The epidemiology of skin cancer. *Dermatologic Surgery*, 22(3), 217–226.

[27] Armstrong, B. K. and Kricker, A. (1995). Skin cancer. *Dermatologic Clinics*, 13(3), 583–594.

Chapter 19

Deep learning applications in ophthalmology and computer-aided diagnostics

Renjith V. Ravi[1], P.K. Dutta[2], Sudipta Roy[3] and S.B. Goyal[4]

Recently, artificial intelligence (AI) that is based on deep learning has gained a lot of attention. Deep learning is a new technique that has a wide range of potential uses in ophthalmology. To identify diabetic retinopathy (DR), macular edema, glaucoma, retinopathy of prematurity, and age-related macular degeneration (AMD or ARMD), DL has been utilized in optical coherence tomography, images of fundus, and visual fields in ophthalmology. DL in ocular imaging can be used along with telemedicine as an effective way to find, diagnose, and check up on serious eye problems in people who need primary care and live in residential institutions. However, there are also possible drawbacks to the use of DL in ophthalmology, such as technical and clinical difficulties, the inexplicability of algorithm outputs, medicolegal concerns, and doctor and patient resistance to the "black box" AI algorithms. In the future, DL could completely alter how ophthalmology is performed. This chapter gives a description of the cutting-edge DL systems outlined for ocular applications, possible difficulties in clinical implementation, and future directions.

19.1 Introduction

Artificial intelligence (AI) is used in computer-aided diagnostics (CAD), which is one way to make the process of diagnosis more accurate and easier to use. "Deep learning" (DL) is the best way to use AI for many tasks, including problems with medical imaging. It has been utilized for diagnostic imaging tasks for various diseases in ophthalmology.

The fourth industrial revolution is in the development of AI. Modern AI methods known as DL have attracted a lot of attention world widen in recent years [1]. The representation-learning techniques used by DL to process the input data

[1]Department of Electronics and Communication Engineering, M.E.A. Engineering College, India
[2]Department of Engineering, Amity School of Engineering and Technology, Amity University Kolkata, India
[3]Artificial Intelligence & Data Science, Jio Institute, India
[4]City University College of Science and Technology, Malaysia

have many degrees of abstraction, eliminating the need for human feature engineering. This lets DL automatically find complicated systems in high-dimensional data by projecting those systems onto lower-dimensional manifolds. DL has achieved noticeably higher accuracy than traditional methods in several areas, including natural-language processing, machine vision, and speech synthesis [2].

In healthcare and medical technology, DL has primarily been used for medical imaging analysis, where DL systems have demonstrated strong diagnostic performance in identifying a variety of medical conditions, including malignant melanoma on skin photographs, and tuberculosis from chest X-rays [1]. Similarly, ophthalmology has benefited from DL's incorporation into the field.

An advancement in the detection, diagnosis, and treatment of eye illness is about to occur in ophthalmology. DL technology that is computer-based is driving this transformation and has the capacity to redefine ophthalmic practice [3].

Visual examination of the eye and its surrounding tissues, along with pattern recognition technology, allows ophthalmologists to diagnose diseases. Diagnostic technology in ophthalmology gives the practitioner additional information via digital images of the same structures. Because of its reliance on imagery, ophthalmology is well-positioned to gain from DL algorithms. The field of ophthalmology is starting to use DL algorithms, which have the potential to alter the core kind of work done by ophthalmologists [3]. In the next few years, computer-aided intelligence will probably play a significant role in eye disease screening and diagnosis. These technological developments may leave human resources free to concentrate on face-to-face interactions between clinicians and patients, such as discussions of diagnostic, prognostic, and available treatments. We anticipate that for the foreseeable future, a human physician will still be required to get permission and perform any necessary medical or surgical procedures. Decision-making in ophthalmology is likely to use DL algorithms sooner than many would anticipate.

19.1.1 Motivation

There is a lot of work to be done in the industrialized environment of today, utilizing a variety of electronic devices, including tablets, mobiles, laptops, and many more. Due to COVID-19's effects, most people worked mostly from home last year, utilizing a variety of internet platforms. Most individuals have vision problems as a result of these conditions. Additionally, those who have visual impairments are more susceptible to other illnesses, including diabetes, heart conditions, stroke, increased blood pressure, etc. They also have a higher risk of falling and getting depressed [4]. According to current publications, reviews, and hospital records, many people have been identified with different eye disorders such as AMD, DR, cataracts, choroidal neovascularization, glaucoma, keratoconus, Drusen, and many more. As a consequence, there is a worldwide problem that must be dealt with. According to the WHO study, medical professionals' perspectives, and researchers' theories, these eye illnesses are the main reasons why people go blind. As the world's population ages, their population will increase exponentially.

Overall, relatively few review papers that concurrently cover all DED detection methods are published in academic databases. As a result, this review of the literature is crucial for gathering research on the subject of DED detection. A detailed review of eye problems such as glaucoma, diabetic retinopathy (DR), and AMD was published by Ting *et al.* [1]. In their study, they summarized a number of studies that were chosen and published between 2016 and 2018. They provided summaries of the publications that made use of transfer learning (TL) techniques using fundus and optical coherence tomography images. They excluded the diagnosis of ocular cataract illness from their study's scope and did not include recent (2019–2020) papers that used TL techniques in their methodology. Similarly to this, Hogarty *et al.*'s [5] work applying AI in ophthalmology was lacking in comprehensive AI approaches. Mookiah *et al.* [6] evaluated research on computer-assisted DR identification, the majority of which is lesion-based DR. In [7], Ishtiaq *et al.* analyzed thorough DR detection techniques from 2013 to 2018, but papers from 2019 and 2020 were not included in their evaluation. Hagiwara *et al.* [8] examined a publication on the utilization of fundus images for computer-assisted diagnosis of GL. They spoke about computer-aided systems and optical disc segmentation systems. Numerous papers that use DL and TL approaches for GL detection but are not included in their review article exist. Reviewing publications that take into account current methods of DED diagnosis is crucial. In reality, the majority of researchers did not include in their review papers the time period of publications addressed by their investigations. Both the clinical scope (DME, DR, Gl, and Ca) and methodological scope (DL and ML) of the existing reviews were inadequate. Therefore, to cover the current techniques for DR detection developed by DL-based methods and solve the shortcomings of the aforementioned research, this paper gives a complete study of DL methodologies for automated DED identification published during the period 2014 and 2020. The government of India had launched National Programme for Control of Blindness and Visual Impairment (NPCB&VI) and has conducted a survey [9] about blindness in India. The major causes of blindness in India and the rate of blindness according to this survey are shown in Figures 19.1 and 19.2, respectively.

Despite being used to diagnose and forecast the prognosis of several ophthalmic and ocular illnesses, DL still has a lot of unrealized potentials. It would be a promising future for the healthcare sector since DL-allied approaches would radically revolutionize vision treatment, even if the elements of DL are just now beginning to be unveiled. Therefore, the use of DL to enhance ophthalmologic treatment and save healthcare costs is of particular relevance [4]. In order to keep up with ongoing advancements in the area of ophthalmology, this review delves further into researching numerous associated methods and datasets. Therefore, this study aims to open up opportunities for new scientists to understand ocular eye problems and research works in the area of ophthalmology to create a system which is completely autonomous.

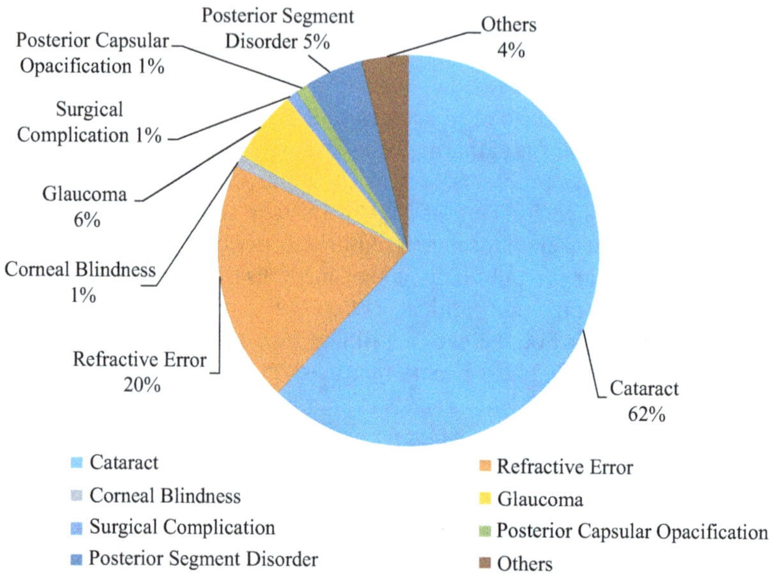

Figure 19.1 Major causes of blindness in India

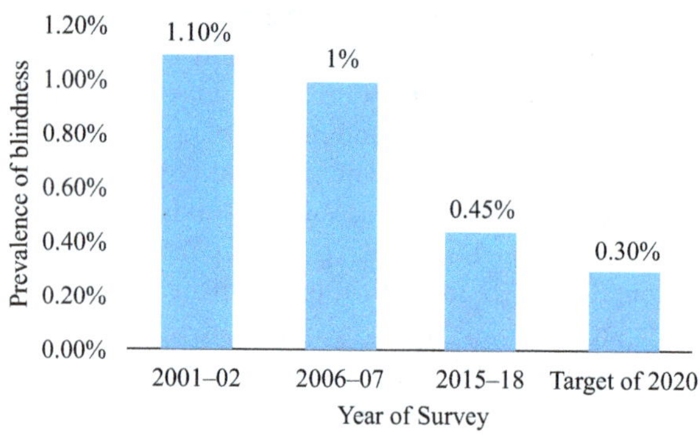

Figure 19.2 Blindness rates in India throughout time

19.2 Technical aspects of deep learning

Medical image evaluation is one of many phrases often used in connection with computer-based procedures that deal with analysis and decision-making situations. The term "computer-aided diagnosis" describes methods in which certain clinical traits linked to the illness are retrieved using image-processing filters and tools [10]. In general, any pattern categorization technique needing a training program, either supervised or unsupervised, to identify potential underlying patterns is referred to as "machine learning (ML)." Most often, the term DL belongs to

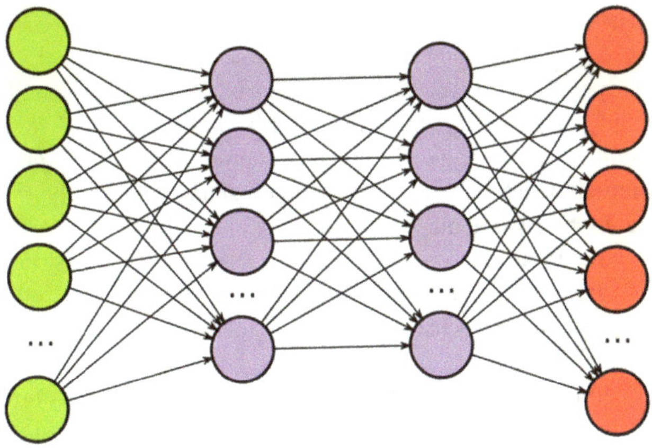

Figure 19.3 Basic structure of a deep neural network

machine learning techniques using convolutional neural networks (CNNs) [11]. The basic structure of a CNN is shown in Figure 19.3. According to the information learned from the training data, which is a crucial component of feature classification approaches, such networks use a collection of filters for image processing to retrieve different kinds of picture characteristics that the system considers suggestive of pathological indications [12]. In order to find the best image-processing filters or tools that can quantify different illness biomarkers, DL might be seen as a brute-force method. Finally, in a very broad sense, the term "artificial intelligence" (AI) denotes any system, often based on machine learning, that is able to recognize key patterns and characteristics unsupervised, without the need for outside assistance—typically from people. Whether a real AI system currently exists or not is up for debate [3].

Initially, a lack of computer capacity was the fundamental problem with CNN applications in the real world. Running CNNs with larger depths (deep learners) is now considerably more powerful and time-efficient because of the emergence of graphics processing units (GPUs), which have far exceeded the computing capability of central processing units (CPUs). These deep neural networks, sometimes termed as "DL" solutions, include neural network architectures like SegNet, GoogLeNet, and VGGNet [13]. These approaches hold great potential for both industrial applications like alternative suggestive algorithms that are akin to Netflix's reference system and for generalizing the entire content of a picture. There have been several initiatives in recent years to evaluate these systems for use in medical settings, especially biomedical image processing.

The two primary types of DL solutions for biological image analysis are:

1. Providing the DL network simply with photos and the diagnoses, labels, and stages that go along with them, sometimes known as "image-classification approaches."
2. "Semantic-segmentation approaches" refers to the process of giving the structure of image data and its accompanying ground-truth masks (black-and-white photos) wherein the pathological states associated with the illness are hand-drawn.

19.3 Anatomy of the human eye

The human body's most advanced sense organ is the eye. Compared to hearing, taste, touch, and smell put together, the brain's portion devoted to vision is far greater. The light from an image that is being captured reaches the pupil first, then travels to the retina, which turns it into an electric signal, and finally, the brain joins in to enable people to see the outside world. The anatomy of the human eye, which is shown in Figure 19.4 and consists of tissues that are light-sensitive, refractive, and supportive, allows us to see.

Refracting tissue: It focuses on light to provide us with a clean image. Spectral tissue includes

Pupil: The function of the aperture in a camera is performed by the pupil in the human eye. It enables light to reach the eye. The pupil controls the quantity of light in both bright and dark environments.

Lens: The lens is situated behind the pupil. The shape of the lens may alter depending on the circumstances.

Ciliary muscles: The accommodation process, often known as the "changing of the curvature of the lens," is the primary function of the muscle.

Cornea: The ability of the eye to focus is largely attributable to this structure. The irregularity in the cornea is the real cause of the vast majority of refractive defects in the eyes.

Light-sensitive tissue: A layer of light-sensitive tissue covers the retina. The retina functions by converting light impulses into electrical signals, which it subsequently sends to the brain for processing further and the creation of vision.

Support tissue: It contains the following:

- **Sclera:** Providing comfort to the eyeball.
- **Choroid:** The tissue that connects the retina to the sclera.

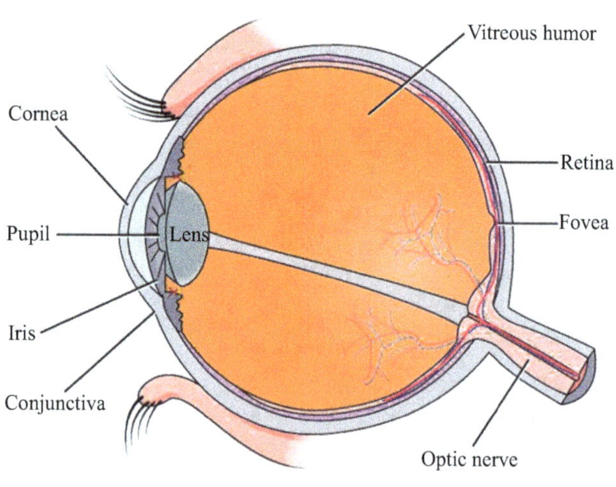

Figure 19.4 Cross-section of the human eye

As a result, as seen in Figure 19.1, vision is a tremendously sophisticated process that functions smoothly in humans. Humans' eyes enable them to observe and feel every movement in this colorful environment. Our primary connection to the surrounding world is only possible due to our eyesight.

19.4 Some of the most common eye diseases

The human eye is a truly amazing organ, and the ability to see is one of our most treasured possessions. The eye allows us to see and perceive the world that surrounds us, and this is all thanks to the eye. However, even seemingly unimportant eye problems may cause serious pain and, in severe cases, even blindness. And for that reason, it's important to keep our eyes healthy [4]. Some eye disorders have early warning symptoms, whereas others do not. People often lack the ability to recognize these signs, and whenever they do, they frequently first ignore them. The most priceless asset you have, your vision, may be preserved if you get a diagnosis early.

19.4.1 Diabetic retinopathy (DR)

Diabetes mellitus may cause damage to the retina, which is called DR (also termed diabetic eye disease). In advanced countries, it is the main cause of blindness. DR normally affects up to 80% of people having type 1 or type 2 diabetes at some point in their lives. Appropriate monitoring and treatment of the eyes might prevent the advancement of at least 90% of new instances of vision-threatening maculopathy and retinopathy to more severe forms [4]. A person's risk of developing DR increases with the length of time they have diabetes. DR causes 12% of all new occurrences of blindness each year in the United States. Additionally, it is the main contributor to blindness in adults ages 20–64 years. The damage that occurred to the human eye due to diabetic retinopathy problem is shown in Figure 19.5.

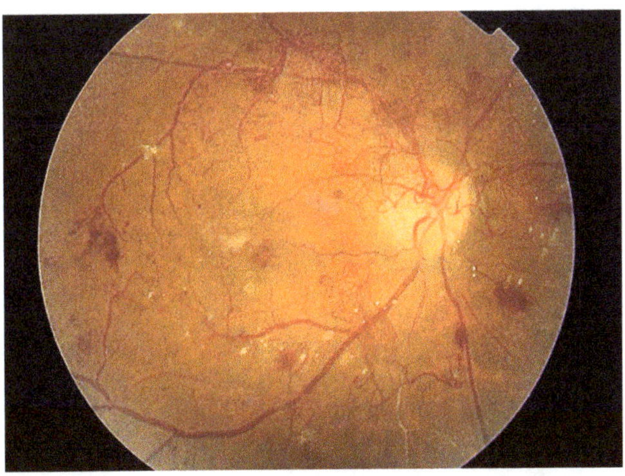

Figure 19.5 Normal eye versus diabetic retinopathy

19.4.2 Age-related macular degeneration (AMD or ARMD)

It is often called macular degeneration and is a medical disorder that may cause clouded or absent vision in the center of the visual area. In the early stages, patients often do not experience any symptoms. On the other hand, some individuals, over time, have a progressive deterioration in their eyesight, which may damage either or both eyes.

While central vision loss cannot result in total blindness, it can make it challenging to recognize people, drive, read, or carry out other everyday tasks. In addition, visual hallucinations may happen. In most cases, macular degeneration affects elderly adults. Smoking and genetic factors both contribute. The cause is damage to the retinal macula. The diagnosis is made after an extensive eye examination. There are three categories of severity: early, middle, and late. The "dry" and "wet" versions of the late type are split, with the dry type accounting for 90% of the total. The eye with AMD is depicted in Figure 19.6.

19.4.3 Glaucoma

The optic nerve of the eye may be harmed by a variety of glaucoma-related disorders, which may result in blindness and loss of vision. Glaucoma occurs as the normal fluid pressure inside the eyes progressively rises. However, recent studies show that even with normal eye pressure, glaucoma may manifest. Early therapy may often protect your eyes from severe vision loss.

The two types of glaucoma conditions are "closed angle" and "open angle" glaucoma. The open-angle glaucoma is a serious illness that develops gradually over time without the patient experiencing loss of vision until the ailment is quite far along. It is known as the "sneak thief of sight" for this reason. Ankle closure may come on quickly and hurt. Visual loss may worsen fast, but the pain and

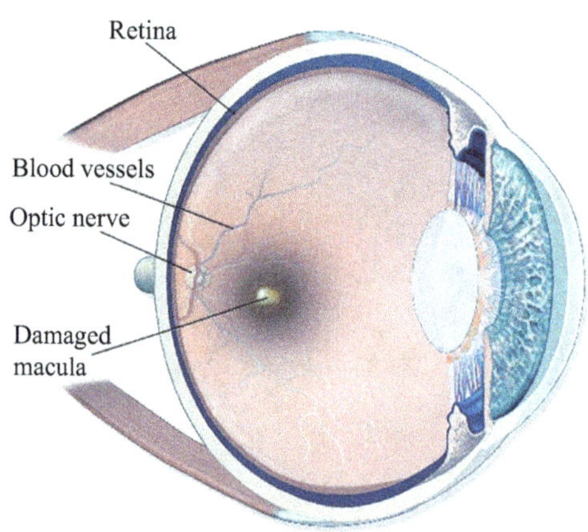

Figure 19.6 Eye with AMD

suffering prompt individuals to seek medical care before irreversible damage takes place. The normal eye versus the eye with glaucoma is shown in Figure 19.7.

19.4.4 Cataract

The cataract is a clouded region that is created on the lens of the eye, which causes a reduction in one's ability to see. The progression of cataracts is often sluggish and might impair one or both eyes. Halos around lights, fading colors, blurred or double vision, trouble with bright lights, and trouble seeing at night are just a few symptoms. As a consequence, you can have problems reading, driving, or recognizing faces [4]. Cataracts can make it hard to see, which can also make you more likely to trip and feel down. 51% of instances of vision loss and 33% of cases of vision impairment are brought on by cataracts.

The most common cause of cataracts is age, although they may also be brought on by radiation or trauma, be present at birth, or appear after eye surgery for another reason. Risk factors include diabetes, chronic corticosteroid use, smoking, prolonged sunlight exposure, and alcohol use [4]. The basic procedure comprises the lens's ability to gather protein clumps or yellow-brown pigment, which reduces light's ability to reach the retina at the back of the eye. An eye exam is used to make the diagnosis. The problem of cataracts is depicted in Figure 19.8.

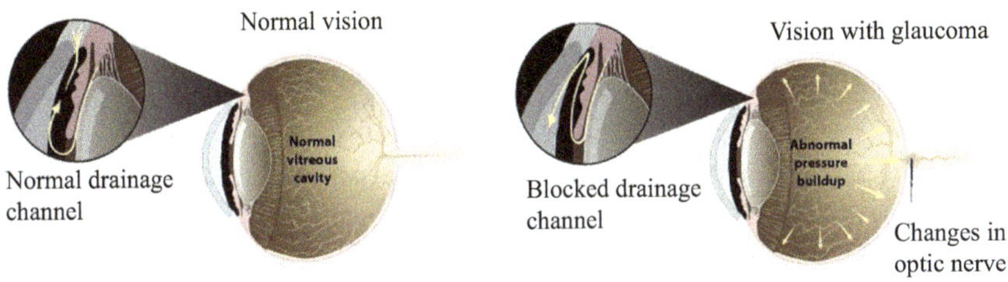

Figure 19.7 Normal eye versus glaucoma

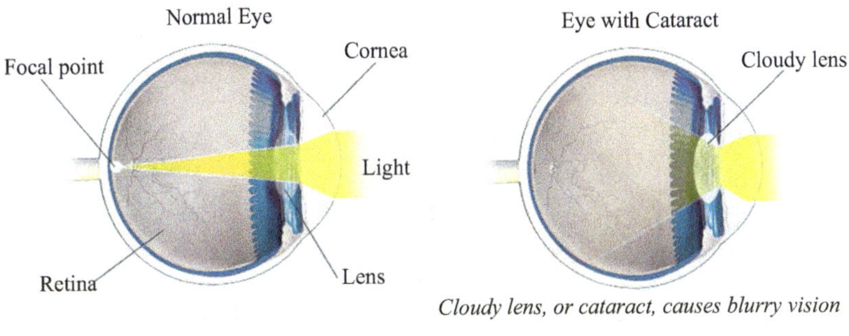

Figure 19.8 Normal eye versus cataract-affected eye

19.4.5 Macular edema

The macula is the region of the eye that aids in the perception of color, fine detail, and distant objects. It has more photoreceptors, which are light-sensitive cells, than any monitor or television we have ever seen. The bullseye of sight is the small, central portion of the retina that is most valuable. The macula is often affected by diseases such as macular edema, puckers, holes, degeneration, drusen (small yellow deposits), scarring, fibrosis, bleeding, and vitreomacular traction. Vision distortion (metamorphopsia), blank patches (scotomas), and impaired vision are typical signs of macular illness.

An abnormal build-up of fluid in the regions of the macula is known as macular edema (as shown in Figure 19.9). It seems from the side that the snake overindulged. The enlarged retina distorts pictures, making it harder to see properly, much like droplets of liquid on your computer monitor. One is more likely to have blurred, distorted, and impaired reading vision the more broad, thick, and extreme the swelling grows.

Chronic macular edema, if untreated, may result in permanent loss of vision and irreparable harm to the macula. Macular edema is often brought on by aberrant blood vessel proliferation in the deep retina or excessive leaking from injured retinal blood vessels. Neovascularization (NV) is the development of new blood vessels that do not have typical "tight junctions," which nearly always causes aberrant fluid leakage (serum from the circulation) into the retina.

19.4.6 Choroidal neovascularization

The medical name for the growing new blood vessels behind the retina of the eye is choroidal neovascularization (CNV) (subretinal) (as shown in Figure 19.10). Although it may not hurt, it can induce macular degeneration, a main reason for visual loss. Although this illness is still incurable, it may respond to therapy. An

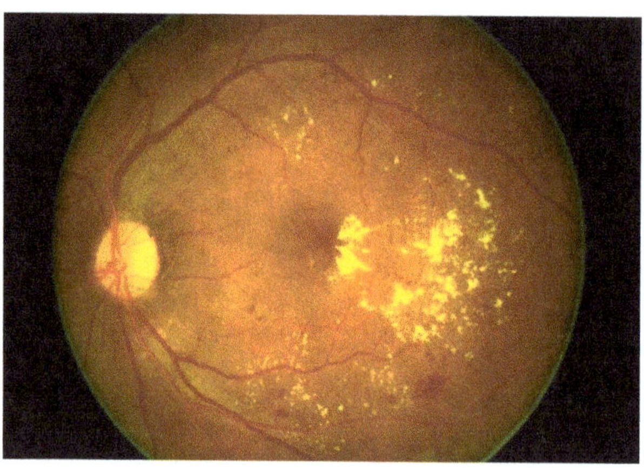

Figure 19.9 Macular edem

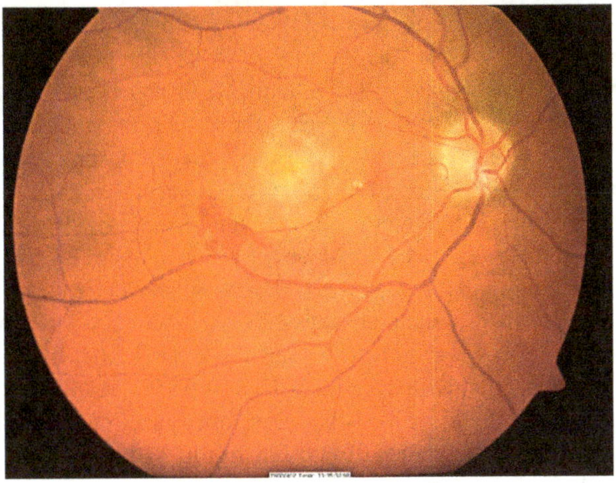

Figure 19.10 Neovascularization

ophthalmologist, a doctor who specializes in the eyes, will diagnose CNV by capturing images of your eyes using cutting-edge medical imaging technology.

19.5 Deep learning in eye disease classification

19.5.1 Diabetic retinopathy

It is a well-known method for delaying blindness, detecting DR, and triggering treatment referrals. Eye doctors, optometrists, medical specialists, diagnostic technicians, and ophthalmologic technicians are just a few of the medical experts that can perform the screening procedure. One of the most exciting uses of AI in clinical medicine at the moment is the identification and treatment of DR using fundus pictures. Recent research has shown that these algorithms can dependably equal the performance of experts and, in some situations, even outperform them while providing a much more cost-effective and comprehensive replacement for conventional screening programs.

In order to train and evaluate an NN model to discriminate the natural fundus images and the images with DR, Gargeya *et al.* [10] employed a DL approach to identify DR using a freely accessible dataset that included 75,137 number of color fundus imagery (CFP) of patients with diabetes. The model's 94% sensitivity and 98% specificity showed that screening of fundus pictures might be done reliably using an AI-based approach. Abramoff *et al.* [14] used a DL approach to screen for DR; this model has got an AUC of 0.980, specificity of 87.0%, and sensitivity of 96.8%. A clinically acceptable DRZ diagnosis method was created and evaluated by Ting *et al.* [15] based on ten datasets from the Singaporean Integrated DR Project that were conducted over a five-year period in six distinct nations or regions, such as China, Singapore, Mexico, Hong Kong, Australia, and the USA.

This model achieved accurate diagnosis in several ethnic groups with specificity, AUC, and sensitivity of 91.6%, 0.936, and 90.5%, respectively. Despite the fact that the majority of research has created reliable DL-based models for the diagnosis of DR and screening using CFP or optical coherence tomography (OCT) photos, other studies have concentrated on automatically detecting DR lesions in fundus fluorescein angiogram (FFA) images. To create an end-to-end DL framework for staging the severity of DR, non-perfusion regions, vascular discharges, and microaneurysms were categorized automatically under many labels based on DL models [16,17]. Additionally, DL-based techniques have been utilized to forecast the prevalence of DR and associated systemic cardiovascular risk factors [18] as well as predict the severity of diabetic macular edema (DME) based on the OCT from two-dimensional fundus images (sensitivity of 85%, AUC of 0.89, and specificity of 80%) [19]. Additionally, since the American Food and Drug Agency (FDA) authorized IDx-DR as the [20] first electronic AI diagnosis system and the EyRIS SELENA [21] was given clearance for medical usage in the European Union, commercial devices for DR screening have been created [22].

More recently, DL systems with good diagnostic performance were reported by Gulshan and team members [23] from Google AI Healthcare. A team of 54 US-licensed medical experts and ophthalmology fellows rated 128 and 175 retinal pictures for DR and DMO three to seven times each from May to December 2015 in order to build the DL framework. About 10,000 photos from two freely accessible datasets (EyePACS-1 and Messidor-2) were included in the test set, which at least seven US board-certified medical experts assessed with good intra-grade accuracy. For EyePACS-1 and Messidor-2, the AUC was 0.991 and 0.990, accordingly.

19.5.2 Glaucoma

If glaucoma sufferers do not get timely detection and quick treatment, they risk losing their visual fields (VF) permanently [22]. This is a clear clinical need that may benefit from using AI. AI research on glaucoma has come a long way, even though there are problems like not enough multimodal assessment and long-term natural spread. Several studies [15,24–28] have used AI to effectively identify structural abnormalities in glaucoma using retinal fundus images and OCT [29–31]. Utilizing an SVM classifier, Zangwill *et al.* [32] identified glaucoma with high accuracy. In order to diagnose glaucoma, Burgansky *et al.* [33] employed five classifiers, including a machine learning assessment of OCT image dataset.

In the diagnosis and treatment of glaucoma, VF evaluation is crucial. In order to create a DL system, VF offers a number of proven parameters. With 17 prototypes, Elze *et al.* [34] created an unsupervised method to categorize glaucomatous visual loss. VF loss in early glaucoma may be found using the unsupervised approach. DL algorithms have been employed to forecast the evolution of glaucoma in the VF. A DL model was trained by Wen *et al.* [35] that produced pointwise VF predictions up to 5.5 years in the future with an average difference of 0.410 dB and a correlation of 0.920 between both the MD of the projected and real future HVF. For an accurate assessment and appropriate care, clinical problems need thorough investigation and fine-grained grading. Huang *et al.* [36] suggested a DL method to accurately score the VF of glaucoma using information from two instruments (the Octopus and the Humphrey Field Analyzer). This tool might be used by glaucomatous patients for self-evaluation and to advance telemedicine.

Li *et al.* [37] used fundus photographs that resembled a glaucoma-like optic disc to train a machine-learning system to recognize an optical disc with a cup-to-disc proportion of 0.7 vertically. The results showed that the algorithm has a significant degree of specificity (92%), sensitivity (95.60%), and AUC for glaucoma optic neuropathy detection (0.986). In a similar study, Phene *et al.* [38] proposed a DL-based model with an AUC of 0.945 to screen for attributable glaucoma using data from over 080,000 CFPs. Furthermore, its effectiveness was shown when used with two more datasets, where the performance of AUC marginally decreased to 0.8550 and 0.8810, respectively. According to Asaoka *et al.* [39], a DL-based model trained on 4316 OCT images for the early detection of glaucoma has an AUC of 93.7%, a specificity of 93.9%, and a sensitivity of 82.5%. Xu *et al.* [40] identified gonioscopic angle closures and the primary angle closure disorder (PACD) based on a completely autonomous analysis with an AUC of 0.9280 and 0.9640 using over 4000.0 anterior regions and OCT (AS-OCT) images.

Globally, adults aged 40–80 have a 3.4% risk of developing glaucoma, and by 2040, it is anticipated that there will be roughly 112 million cases of the condition [41]. Developments in disease identification, functional and structural damage assessments throughout time, therapy optimization to avoid visual impairment and a precise long-term prognosis would be welcomed by both patients and clinicians [1].

19.5.3 Age-related macular degeneration

The primary factor behind older people losing their eyesight permanently is AMD. CFP is the most commonly used screening technique, and it can detect abnormalities such as drusen, retinal hemorrhage, geographic atrophy, and others. CFP is crucial for screening people for AMD because of its quick, non-invasive, and affordable benefits. AMD was completely diagnosed and graded with the same precision by ophthalmologists using a CFP-based DL algorithm.

The macular portion of the retina may be seen using optical coherence tomography (OCT). In 2018, Kermany *et al.* [42] used a transfer learning technique on an OCT database for choroidal neovascularization (CNV) and three other categories, using a tiny portion of the training data from conventional DL techniques. Their model met senior ophthalmologists' standards for accuracy, specificity, and sensitivity, with scores of 97.8%, 96.6%, and 97.4%. Studies on the quantitative evaluation of OCT images using AI techniques have grown in number recently. In order to automatically detect and measure intraretinal fluid (IRF) and subretinal fluid (SRF) on OCT images, Schlegl *et al.* [43] created a DL network. They found that their findings were extremely similar to expert comments. Erfurth *et al.* [44] also investigated the association between the quantity of evacuation and the visual system after intravitreal injection in AMD patients by identifying and quantifying retinal outpourings, including IRF, SRF, and pigmented epithelial detachment, using a DL algorithm. The quantitative OCT volume mode study by Moraes *et al.* [45] included biomarkers like subretinal hyperreflective material and hyperreflective foci on OCT images in addition to retinal effusion, and the results showed strong clinical applicability and were directly connected to the treatment choices of AMD patients in follow-up reviews. Yan *et al.* [46] used an attention-based DL technique to decipher CNV activity on OCT images to help a physician diagnose AMD. Zhang *et al.* [47] used a DL model for evaluating photoreceptor degradation, hyperprojection, and retinal pigment epithelium loss to quantify geographic atrophy (GA) in addition to wet AMD on OCT images. As additional indicators linked to the development of the illness, Liefers *et al.* [48] measured a number of important characteristics on OCT pictures of individuals with early and late AMD.

One of the hotspots in healthcare AI technologies is the integrated use of several modalities, which has been found to be closer to clinical research deployment. In order to diagnose AMD and polypoidal choroidal vasculopathy (PCV), Xu *et al.* [49] joined CFP and OCT images and attained 87.40% accuracy, 88.80% sensitivity, and 95.60% specificity. OCT and optical coherence tomography angiography (OCTA) pictures were used by Jin *et al.* [50] to evaluate the features of a multimodal DL model to evaluate CNV in neovascular AMD. On multimodal input data, the DL algorithm obtained an AUC of 0.97960 and an accuracy of 95.50%, which is equivalent to that of retinal experts.

In 2018, Burlina *et al.* [51] created a DL algorithm that automatically performed classification and feature extraction on more than 130,000 CFP sets of data. Compared to older ways of classifying things into two groups, their DL algorithm had more potential to be used in clinical settings. A 13-category AMD fundus imaging dataset was created by Grassmann *et al.* [52], which had 12 instances of AMD severity and one that was unable to be assessed due to low image quality. Finally, they presented an ensemble of networks on an untrained independence test set after training six sophisticated DL models. In order to detect AMD severity-related events precisely defined at the single-eye level and be able to deliver a patient-level final score paired with binocular severity, Peng *et al.* [53] employed a DeepSeeNet DL technique interconnected by three seed networks. Some AI research has centered on forecasting the probability of progression of the diseases along with AMD diagnosis based on the CFP. The focus on enhancing the DL algorithm was further expanded in 2019 by Burlina *et al.* [54], who not only investigated the clinical impact of the DL technique on the 4 and 9 classification systems of AMD severity but also observationally used a DL-based regression analysis to provide patients with a 5-year risk rating for their estimated development of the disease too advanced AMD. In a research work carried out in 2020, Yan *et al.* [55] used DL algorithms to estimate the probability of severe AMD creation by combining CFP with patients' matched genotypes linked with AMD.

19.5.4 Cataracts and other eye-related diseases

The screening and diagnosis of fundus illnesses used to get greater emphasis in ophthalmology AI research. The evidence demonstrates the promise of AI for detecting and treating a wide range of disorders, such as automated detection and criticality classifying of cataracts using slit-lamp or fundus images. In identifying various forms of cataracts [56], AI algorithms have shown good to exceptional overall diagnostic accuracy, with high AUC (0.860–1.00), accuracy (69.00%–99.50%), sensitivity (60.10%–99.50%), and specificity (63.2%–99.6%). In [57], Long *et al.* used the DL method to create an artificial intelligence platform for genetic cataracts that performs three different tasks: population-wide congenital cataract identification, risk delamination for patients with inherited cataracts, and supporting a channel of treatment methods for medical doctors. By segregating the anatomy and labeling pathological lesions, Li *et al.* [58] enhanced the efficacy of a DL algorithm for identifying anterior segment disorders in slit-lamp pictures, including keratitis, cataracts, and pterygia. In slit-lamp pictures collected by several pieces of equipment, including a smartphone using the super macro feature, another DL system performed admirably in automatically diagnosing keratitis, a normal cornea, and other corneal abnormalities (all AUCs > 0.96). AI's keratitis diagnosis sensitivity and specificity were on track with those of skilled cornea experts. Ye *et al.* [59] proposed the DL-based technique to identify and categorize myopic maculopathy in patients with severe myopia and to recommend treatment. Their model had sensitivities comparable to or superior to those of young ophthalmologists. Yoo *et al.* [60] used 10,561 eye scans and incorporated preoperative data to create a machine-learning model that could predict if a patient would be a good candidate for refractive surgery. They achieved an AUC of 0.970 and an accuracy of 93.40% in cross-validation. In order to create a set of convolutional neural networks (CNN) for recognizing malignant tumors in ocular tumors, a large-scale statistical study of demographic and clinicopathological variables was carried out in conjunction with universal health databases and multilingual clinical data. Besides having a sensitivity of 94.70% and an accuracy of 94.90%, the DL diagnostic method [61] for melanoma visualization was able to discriminate between malignant and benign tumors.

19.6 Challenges and limitations in the application of DL in ophthalmology

Despite the DL-based models' excellent degree of accuracy in many ophthalmic conditions, real-time implementation of these concepts in medical care still faces several clinical and technological hurdles (Table 19.1). These difficulties could appear at various points in both the research and therapeutic contexts. First of all, a lot of research has used training data sets from generally homogenous populations [14,23,62]. Numerous factors, including depth of field, the field of vision, picture magnifying, image resolution, and participant nationalities, are often taken into account during DL training and evaluation utilizing retinal images. To solve this problem, the data collection might be made more diverse in terms of racial composition and image-capture technology [1,15].

The low availability of extensive data for both uncommon illnesses (such as ocular tumors) and deadly diseases that are not regularly photographed in clinical

Table 19.1 The difficulties in constructing and implementing DL in ophthalmology on a clinical and technological level

Problem	Possible difficulties
Selection of data sets for training	Concerns about patient consent and privacy. Various institutional ethics committees have different requirements and rules. Limited training benchmark datasets for uncommon illnesses that are not routinely collected, such as eye tumors or common disorders (e.g., cataracts).
Testing and validating data sets	Not enough control due to an insufficient sample size. Lack of generalizability—it has not been tested in a lot of different situations or with data from a lot of different devices.
The availability of explanations for the findings	Showing the areas that DL "deemed" anomalous. Methods for producing heat maps include retinopathy tests, class activation, the fuzzy attention map, etc.
Implementation of DL systems in practice	Proper suggestions for suitable clinical deployment locations. Regulatory permission request from the health authority. Successful execution of prospective clinical studies. Medical compensation scheme and medical legislation. Challenges from an ethical standpoint.

practice, such as cataracts, has been a problem in the creation of AI models in ophthalmology. Additionally, there may be conflict and interobserver heterogeneity in the characterization of the target gene for certain conditions, like glaucoma and ROP. The algorithm picks up knowledge from the input it receives. If the training collection of images provided by the AI tool is also limited or not typical of actual patient populations, the program is unlikely to deliver reliable results [63]. More information on methods of attaining high-quality ground-truth labeling is needed for various imaging techniques.

19.6.1 Challenges in the practical implementation of DL ophthalmology

The effects of AI in ophthalmology provide unique difficulties. Making AI as practical as possible is the biggest issue facing its use in ophthalmology. It could be necessary to combine many devices with clinically acceptable performance for this objective. But even if the resolution of the photographs from regularly used devices fluctuates, these systems should be able to receive them. Other practical problems include choosing the right patients, dealing with misclassified patients like false positives and negatives, and the fact that DL systems can only be used to classify one eye-related disease at a time. Determining culpability, that is, whether the doctor or the AI practitioner is at fault, requires addressing what transpires in the misclassification situation.

19.6.2 Technology-related challenges

The necessity for sufficient training data and quality assessment is the biggest technical hurdle in the deployment of DL. In order to introduce artificial intelligence, input data must also be labeled for the training procedure, which calls for experienced practitioners. The likelihood of human error thus rises. The whole classification of datasets and system calibration procedure might take a long time, delaying the use of DL. In order to assess the effectiveness and efficiency of these systems, both existing and newly developed DL approaches in ophthalmology need expert agreement and the creation of standards and recommendations.

19.6.3 Social and cultural challenges for DL in the eyecare

Numerous socio-cultural problems come with using DL in clinical practice. Many of these difficulties are related to the typical disparities in healthcare access. Asia is a good illustration of this. There are significant inequalities throughout this huge continent, not just in terms of healthcare access but also in terms of expenditure and consumption in the healthcare sector. Numerous regions suffer from few resources, poor infrastructure, and other issues that might hinder the use and efficiency of artificial intelligence.

It takes more than simply setting up hardware and software to implement DL. Despite living in the twenty-first century, there are still a lot of important distinctions in the healthcare sector. Not only in Asia but all throughout the globe, many hospitals lack the funding necessary to adopt DL. Infrastructure issues and electrical issues are also significant obstacles.

19.6.4 Limitations

The use of DL in medical practice might come with various hazards. Some computer programs use algorithms that have a high risk of false-negative retinal disease diagnosis. Diagnostic mistakes might come from improperly interpreting the false-positive findings, which could have catastrophic clinical effects on the patient's eyesight. In rare cases, the eye specialist may be unable to assess the performance indicator values utilized by the DL computer program to examine patient information. The technique through which a software program reaches its conclusion and its logic is not always clear. It is probable that a lack of patient trust might be a challenge for remote monitoring (home screening) using DK-powered automated machines. Studies indicate that several patients are more likely to prefer in-person ophthalmologist appointments over computer-aided diagnosis [64,65]. Additionally, there is a chance that physicians would lose their diagnostic skills due to technological addiction. It is vital to create and implement medicolegal and ethical norms in certain specific circumstances, such as when a doctor disagrees with the findings of a DL evaluation or when a patient is unable to receive counseling for the necessary therapy. All of these possible issues demonstrate the need for DL technology to progress over time.

19.7 Future directions

In the future, artificial intelligence will be more prevalent in scientific research, diagnosis, and treatment. When used in ophthalmology, telemedicine allows information to be sent to areas with a dearth of experts but a high need [66]. DR diagnostics already make use of a hybrid approach with high specificity and sensitivity called IDxDR [67–69], which has been classified by the FDA as a moderate-to-low-risk medical device and so helps with the treatment of patients who are in need of an eye specialist. Computers using AI can swiftly sift through mountains of data. On the basis of these studies, AI may investigate possible connections between characteristics of the illness that are not immediately apparent to humans. The ophthalmologist's clinical analysis, bolstered by the findings of the DL analysis, will enhance the personalization of medical therapy [70]. The field of science also benefits greatly from the use of AI. Artificial intelligence may be used to recognize the symptoms of previously unknown eye illnesses [65]. Artificial intelligence algorithms are not restricted to recognizing just clinical aspects, so it is hoped that they will aid in the discovery of novel biomarkers for many diseases. Ongoing studies seek to create self-sufficient software that can detect glaucoma, DR, and AMD, as well as anticipate their development and provide individualized treatment plans [64].

19.8 Conclusion

DL is a new tool that helps patients and physicians alike. The integration of DL technology into ophthalmic care will increase as it develops, relieving the clinician of tedious chores and enabling them to concentrate on enhancing patient care. Ophthalmologists will be able to concentrate their efforts on building patient connections and improving medical and surgical treatment because of DL. Even though medical DL studies have made significant strides and advancements in the realm of ophthalmology, they still confront several obstacles and problems. The advent of big data, the advancement of healthcare electronics, and the public's need for high-quality healthcare are all pushing DL systems to the limit of what they can do in terms of enhancing clinical medical processes, patient care, and prognosis evaluation. In order to employ cutting-edge AI ideas and technology to address ophthalmic clinical issues, ocular medical professionals and AI researchers should collaborate closely with computer scientists. They should also place greater emphasis on the translation of research findings. AI innovations in ophthalmology seem to have a bright future. However, a significant amount of further research and development is required before they may be used routinely in therapeutic settings.

References

[1] D. S. W. Ting, L. R. Pasquale, L. Peng, *et al.*, "Artificial intelligence and deep learning in ophthalmology," *British Journal of Ophthalmology*, vol. 103, pp. 167–175, 2018.

[2] H.-C. Shin, H. R. Roth, M. Gao, *et al.*, "Deep convolutional neural networks for computer-aided detection: CNN architectures, dataset characteristics and transfer learning," *IEEE Transactions on Medical Imaging*, vol. 35, pp. 1285–1298, 2016.

[3] P. S. Grewal, F. Oloumi, U. Rubin, and M. T. S. Tennant, "Deep learning in ophthalmology: a review," *Canadian Journal of Ophthalmology*, vol. 53, pp. 309–313, 2018.

[4] P. Kumar, R. Kumar, and M. Gupta, "Deep learning based analysis of ophthalmology: a systematic review," In *EAI Endorsed Transactions on Pervasive Health and Technology*, p. 170950, 2018.

[5] D. T. Hogarty, D. A. Mackey, and A. W. Hewitt, "Current state and future prospects of artificial intelligence in ophthalmology: a review," *Clinical & Experimental Ophthalmology*, vol. 47, pp. 128–139, 2018.

[6] M. R. K. Mookiah, U. R. Acharya, C. K. Chua, C. M. Lim, E. Y. K. Ng, and A. Laude, "Computer-aided diagnosis of diabetic retinopathy: a review," *Computers in Biology and Medicine*, vol. 43, pp. 2136–2155, 2013.

[7] U. Ishtiaq, S. A. Kareem, E. R. M. F. Abdullah, G. Mujtaba, R. Jahangir, and H. Y. Ghafoor, "Diabetic retinopathy detection through artificial intelligent techniques: a review and open issues," *Multimedia Tools and Applications*, vol. 79, pp. 15209–15252, 2019.

[8] Y. Hagiwara, J. E. W. Koh, J. H. Tan, *et al.*, "Computer-aided diagnosis of glaucoma using fundus images: a review," *Computer Methods and Programs in Biomedicine*, vol. 165, pp. 1–12, 2018.

[9] DGHS, National Programme for Control of Blindness and Visual Impairment (NPCB&VI), Ministry of Health & Family Welfare, Government of India, 2017.

[10] N. Dey, *Classification Techniques for Medical Image Analysis and Computer Aided Diagnosis*, Elsevier Science, 2019.

[11] L. Lu, Y. Zheng, G. Carneiro, and L. Yang, Deep Learning and Convolutional Neural Networks for Medical Image Computing: Precision *Medicine, High Performance and Large-Scale Datasets*, Springer International Publishing, 2017.

[12] Q. Li and R. M. Nishikawa, *Computer-Aided Detection and Diagnosis in Medical Imaging*, CRC Press, 2015.

[13] D. Ghai, S. L. Tripathi, S. Saxena, M. Chanda, and M. Alazab, *Machine Learning Algorithms for Signal and Image Processing*, Wiley, 2022.

[14] M. D. Abramoff, Y. Lou, A. Erginay, *et al.*, "Improved automated detection of diabetic retinopathy on a publicly available dataset through integration of deep learning," *Investigative Opthalmology & Visual Science*, vol. 57, pp. 5200, 2016.

[15] D. S. W. Ting, C. Y.-L. Cheung, G. Lim, *et al.*, "Development and validation of a deep learning system for diabetic retinopathy and related eye diseases using retinal images from multiethnic populations with diabetes," *JAMA*, vol. 318, pp. 2211, 2017.

[16] X. Pan, K. Jin, J. Cao, *et al.*, "Multi-label classification of retinal lesions in diabetic retinopathy for automatic analysis of fundus fluorescein angiography based on deep learning," *Graefe's Archive for Clinical and Experimental Ophthalmology*, vol. 258, pp. 779–785, 2020.
[17] Z. Gao, K. Jin, Y. Yan, *et al.*, "End-to-end diabetic retinopathy grading based on fundus fluorescein angiography images using deep learning," *Graefe's Archive for Clinical and Experimental Ophthalmology*, vol. 260, pp. 1663–1673, 2022.
[18] D. S. W. Ting, C. Y. Cheung, Q. Nguyen, *et al.*, "Deep learning in estimating prevalence and systemic risk factors for diabetic retinopathy: a multi-ethnic study," *npj Digital Medicine*, vol. 2, p. 24, 2019.
[19] A. V. Varadarajan, P. Bavishi, P. Ruamviboonsuk, *et al.*, "Predicting optical coherence tomography-derived diabetic macular edema grades from fundus photographs using deep learning," *Nature Communications*, vol. 11, pp. 130–138, 2020.
[20] M. D. Abràmoff, P. T. Lavin, M. Birch, N. Shah, and J. C. Folk, "Pivotal trial of an autonomous AI-based diagnostic system for detection of diabetic retinopathy in primary care offices," *npj Digital Medicine*, vol. 1, p. 38, 2018.
[21] V. Bellemo, Z. W. Lim, G. Lim, *et al.*, "Artificial intelligence using deep learning to screen for referable and vision-threatening diabetic retinopathy in Africa: a clinical validation study," *The Lancet Digital Health*, vol. 1, pp. e35–e44, 2019.
[22] K. Jin and J. Ye, "Artificial intelligence and deep learning in ophthalmology: current status and future perspectives," *Advances in Ophthalmology Practice and Research*, vol. 2, pp. 100078, 2022.
[23] V. Gulshan, L. Peng, M. Coram, *et al.*, "Development and validation of a deep learning algorithm for detection of diabetic retinopathy in retinal fundus photographs," *JAMA*, vol. 316, pp. 2402, 2016.
[24] H. Liu, L. Li, I. M. Wormstone, *et al.*, "Development and validation of a deep learning system to detect glaucomatous optic neuropathy using fundus photographs," *JAMA Ophthalmology*, vol. 137, pp. 1353, 2019.
[25] J. Chang, J. Lee, A. Ha, *et al.*, "Explaining the rationale of deep learning glaucoma decisions with adversarial examples," *Ophthalmology*, vol. 128, pp. 78–88, 2021.
[26] F. A. Medeiros, A. A. Jammal, and E. B. Mariottoni, "Detection of progressive glaucomatous optic nerve damage on fundus photographs with deep learning," *Ophthalmology*, vol. 128, pp. 383–392, 2021.
[27] F. A. Medeiros, A. A. Jammal, and A. C. Thompson, "From machine to machine," *Ophthalmology*, vol. 126, pp. 513–521, 2019.
[28] Y. Xu, M. Hu, H. Liu, *et al.*, "A hierarchical deep learning approach with transparency and interpretability based on small samples for glaucoma diagnosis," *npj Digital Medicine*, vol. 4, p. 48, 2021.
[29] S. Sun, A. Ha, Y. K. Kim, B. W. Yoo, H. C. Kim, and K. H. Park, "Dual-input convolutional neural network for glaucoma diagnosis using spectral-

domain optical coherence tomography," *British Journal of Ophthalmology*, vol. 105, pp. 1555–1560, 2020.

[30] A. C. Thompson, A. A. Jammal, and F. A. Medeiros, "A review of deep learning for screening, diagnosis, and detection of glaucoma progression," *Translational Vision Science & Technology*, vol. 9, pp. 42, 2020.

[31] H. Fu, M. Baskaran, Y. Xu, *et al.*, "A deep learning system for automated angle-closure detection in anterior segment optical coherence tomography images," *American Journal of Ophthalmology*, vol. 203, pp. 37–45, 2019.

[32] L. M. Zangwill, K. Chan, C. Bowd, *et al.*, "Heidelberg retina tomograph measurements of the optic disc and parapapillary retina for detecting glaucoma analyzed by machine learning classifiers," *Investigative Ophthalmology & Visual Science*, vol. 45, pp. 3144, 2004.

[33] Z. Burgansky-Eliash, G. Wollstein, T. Chu, *et al.*, "Optical coherence tomography machine learning classifiers for glaucoma detection: a preliminary study," *Investigative Ophthalmology & Visual Science*, vol. 46, pp. 4147, 2005.

[34] T. Elze, L. R. Pasquale, L. Q. Shen, T. C. Chen, J. L. Wiggs, and P. J. Bex, "Patterns of functional vision loss in glaucoma determined with archetypal analysis," *Journal of The Royal Society Interface*, vol. 12, pp. 20141118, 2015.

[35] J. C. Wen, C. S. Lee, P. A. Keane, *et al.*, "Forecasting future humphrey visual fields using deep learning,"*PLoS One*, vol. 14, pp. e0214875, 2019.

[36] X. Huang, K. Jin, J. Zhu, *et al.*, "A structure-related fine-grained deep learning system with diversity data for universal glaucoma visual field grading," *Frontiers in Medicine*, vol. 9, pp. 832920–832920, 2022.

[37] Z. Li, Y. He, S. Keel, W. Meng, R. T. Chang, and M. He, "Efficacy of a deep learning system for detecting glaucomatous optic neuropathy based on color fundus photographs," *Ophthalmology*, vol. 125, pp. 1199–1206, 2018.

[38] S. Phene, R. C. Dunn, N. Hammel, *et al.*, "Deep learning and glaucoma specialists," *Ophthalmology*, vol. 126, pp. 1627–1639, 2019.

[39] R. Asaoka, H. Murata, K. Hirasawa, *et al.*, "Using deep learning and transfer learning to accurately diagnose early-onset glaucoma from macular optical coherence tomography images," *American Journal of Ophthalmology*, vol. 198, pp. 136–145, 2019.

[40] B. Y. Xu, M. Chiang, S. Chaudhary, S. Kulkarni, A. A. Pardeshi, and R. Varma, "Deep learning classifiers for automated detection of gonioscopic angle closure based on anterior segment OCT images," *American Journal of Ophthalmology*, vol. 208, pp. 273–280, 2019.

[41] Y.-C. Tham, X. Li, T. Y. Wong, H. A. Quigley, T. Aung, and C.-Y. Cheng, "Global prevalence of glaucoma and projections of glaucoma burden through 2040," *Ophthalmology*, vol. 121, pp. 2081–2090, 2014.

[42] D. S. Kermany, M. Goldbaum, W. Cai, *et al.*, "Identifying medical diagnoses and treatable diseases by image-based deep learning," *Cell*, vol. 172, pp. 1122–1131.e9, 2018.

[43] T. Schlegl, S. M. Waldstein, H. Bogunovic, *et al.*, "Fully automated detection and quantification of macular fluid in OCT using deep learning," *Ophthalmology*, vol. 125, pp. 549–558, 2018.

[44] U. Schmidt-Erfurth, W.-D. Vogl, L. M. Jampol, and H. Bogunović, "Application of automated quantification of fluid volumes to anti–VEGF therapy of neovascular age-related macular degeneration," *Ophthalmology*, vol. 127, pp. 1211–1219, 2020.

[45] G. Moraes, D. J. Fu, M. Wilson, *et al.*, "Quantitative analysis of OCT for neovascular age-related macular degeneration using deep learning," *Ophthalmology*, vol. 128, pp. 693–705, 2021.

[46] Y. Yan, K. Jin, Z. Gao, *et al.*, "Attention-based deep learning system for automated diagnoses of age-related macular degeneration in optical coherence tomography images," *Medical Physics*, vol. 48, pp. 4926–4934, 2021.

[47] G. Zhang, D. J. Fu, B. Liefers, *et al.*, "Clinically relevant deep learning for detection and quantification of geographic atrophy from optical coherence tomography: a model development and external validation study," *The Lancet Digital Health*, vol. 3, pp. e665–e675, 2021.

[48] B. Liefers, P. Taylor, A. Alsaedi, *et al.*, "Quantification of key retinal features in early and late age-related macular degeneration using deep learning," *American Journal of Ophthalmology*, vol. 226, pp. 1–12, 2021.

[49] Z. Xu, W. Wang, J. Yang, *et al.*, "Automated diagnoses of age-related macular degeneration and polypoidal choroidal vasculopathy using bi-modal deep convolutional neural networks," *British Journal of Ophthalmology*, vol. 105, pp. 561–566, 2020.

[50] K. Jin, Y. Yan, M. Chen, *et al.*, "Multimodal deep learning with feature level fusion for identification of choroidal neovascularization activity in age-related macular degeneration," *Acta Ophthalmologica*, vol. 100, 2021.

[51] P. M. Burlina, N. Joshi, M. Pekala, K. D. Pacheco, D. E. Freund, and N. M. Bressler, "Automated grading of age-related macular degeneration from color fundus images using deep convolutional neural networks," *JAMA Ophthalmology*, vol. 135, pp. 1170, 2017.

[52] F. Grassmann, J. Mengelkamp, C. Brandl, *et al.*, "A deep learning algorithm for prediction of age-related eye disease study severity scale for age-related macular degeneration from color fundus photography," *Ophthalmology*, vol. 125, pp. 1410–1420, 2018.

[53] Y. Peng, S. Dharssi, Q. Chen, *et al.*, "DeepSeeNet: a deep learning model for automated classification of patient-based age-related macular degeneration severity from color fundus photographs," *Ophthalmology*, vol. 126, pp. 565–575, 2019.

[54] P. M. Burlina, N. Joshi, K. D. Pacheco, D. E. Freund, J. Kong, and N. M. Bressler, "Use of deep learning for detailed severity characterization and estimation of 5-year risk among patients with age-related macular degeneration," *JAMA Ophthalmology*, vol. 136, pp. 1359, 2018.

[55] Q. Yan, D. E. Weeks, H. Xin, *et al.*, "Deep-learning-based prediction of late age-related macular degeneration progression," *Nature Machine Intelligence*, vol. 2, pp. 141–150, 2020.

[56] C. Y.-l. Cheung, H. Li, E. L. Lamoureux, *et al.*, "Validity of a new computer-aided diagnosis imaging program to quantify nuclear cataract from

slit-lamp photographs," *Investigative Ophthalmology & Visual Science*, vol. 52, pp. 1314, 2011.

[57] E. Long, J. Chen, X. Wu, *et al.*, "Artificial intelligence manages congenital cataract with individualized prediction and telehealth computing," *npj Digital Medicine*, vol. 3, p. 112, 2020.

[58] Z. Li, J. Jiang, K. Chen, *et al.*, "Preventing corneal blindness caused by keratitis using artificial intelligence," *Nature Communications*, vol. 12, p. 3738, 2021.

[59] X. Ye, J. Wang, Y. Chen, *et al.*, "Automatic screening and identifying myopic maculopathy on optical coherence tomography images using deep learning," *Translational Vision Science Technology*, vol. 10, pp. 10, 2021.

[60] T. K. Yoo, I. H. Ryu, G. Lee, *et al.*, "Adopting machine learning to automatically identify candidate patients for corneal refractive surgery," *npj Digital Medicine*, vol. 2, p. 59, 2019.

[61] L. Wang, L. Ding, Z. Liu, *et al.*, "Automated identification of malignancy in whole-slide pathological images: identification of eyelid malignant melanoma in gigapixel pathological slides using deep learning," *British Journal of Ophthalmology*, vol. 104, pp. 318–323, 2019.

[62] R. Gargeya and T. Leng, "Automated identification of diabetic retinopathy using deep learning," *Ophthalmology*, vol. 124, pp. 962–969, 2017.

[63] T. CONSORT-AI and SPIRIT-AS Group, "Reporting guidelines for clinical trials evaluating artificial intelligence interventions are needed," *Nature Medicine*, vol. 25, pp. 1467–1468, 2019.

[64] A. Moraru, D. Costin, R. Moraru, and D. Branisteanu, "Artificial intelligence and deep learning in ophthalmology – present and future (Review)," *Experimental and Therapeutic Medicine*, vol. 12, pp. 3469–3473, 2020.

[65] S. Keel, P. Y. Lee, J. Scheetz, *et al.*, "Feasibility and patient acceptability of a novel artificial intelligence-based screening model for diabetic retinopathy at endocrinology outpatient services: a pilot study," *Scientific Reports*, vol. 8, p. 4330, 2018.

[66] G. W. Armstrong and A. C. Lorch, "A(eye): a review of current applications of artificial intelligence and machine learning in ophthalmology," *International Ophthalmology Clinics*, vol. 60, pp. 57–71, 2019.

[67] N. C. Khan, C. Perera, E. R. Dow, *et al.*, "Predicting systemic health features from retinal fundus images using transfer-learning-based artificial intelligence models," *Diagnostics*, vol. 12, pp. 1714, 2022.

[68] M. Savoy, "IDx-DR for diabetic retinopathy screening," *American Family Physician*, vol. 101, pp. 307–308, 2020.

[69] Commissioner-FDA, *FDA permits marketing of artificial intelligence-based device to detect certain diabetes-related eye problems — fda.gov*, 2018.

[70] A. Consejo, T. Melcer, and J. J. Rozema, "Introduction to machine learning for ophthalmologists," *Seminars in Ophthalmology*, vol. 34, pp. 19–41, 2018.

Chapter 20

Deep learning for biomedical image analysis in place of fundamentals, limitations, and prospects of deep learning for biomedical image analysis

Renjith V. Ravi[1], Pushan Kumar Dutta[2], Pronaya Bhattacharya[3] and S.B. Goyal[4]

Clinical techniques used for the timely identification, observation, diagnostics, and therapy assessment of a wide range of medical problems are just a few examples of how biomedical imaging is crucial in these clinical applications. Grasping medical image analysis in computer vision requires a fundamental understanding of the ideas behind artificial neural networks and deep learning (DL), as well as how they are implemented. Due to its dependability and precision, DL is well-liked among academics and researchers, particularly in the engineering and medical disciplines. Early detection is a benefit of DL approaches in the realm of medical imaging for illness diagnosis. The simplicity and reduced complexity of DL approaches are their key characteristics, which eventually save time and money while tackling several difficult jobs at once. DL and artificial intelligence (AI) technologies have advanced significantly in recent years. In every application area, but particularly in the medical one, these methods are crucial. Examples include image analysis, image processing, image segmentation, image fusion, image registration, image retrieval, image-guided treatment, computer-aided diagnosis (CAD), and many more. This chapter seeks to thoroughly present DL methodologies and the potential for biological imaging utilizing DL, as well as explore problems and difficulties.

20.1 Introduction

Nowadays, medical practice has been utilizing extensive use of biomedical imaging technology. Experts are doing manually analyze biological images and then piece all of the clinical evidence together to get the correct diagnosis, depending on their

[1]Department of Electronics and Communication Engineering, M.E.A. Engineering College, India
[2]Department of Engineering, Amity School of Engineering and Technology, Amity University Kolkata, India
[3]School of Engineering and Technology, Amity University Kolkata, India
[4]City University College of Science and Technology, Malaysia

own expertise. Currently, manual biological image analysis confronts four significant obstacles: (i) Since manual analysis is constrained by human experience, the diagnosis may vary from person to person. (ii) It costs a lot of money and takes years of work to train a skilled expert. (iii) Specialists are under tremendous strain due to the rapid expansion of biological images in terms of both quantity and modality. (iv) Specialists get quickly exhausted by repetitive, tiresome analytical work on unattractive biological images, which might result in a delayed or incorrect diagnosis, putting patients in danger. In some ways, these difficulties make the lack of healthcare resources worse, particularly in developing nations [1]. Medical image analysis with computer assistance is therefore an alternate option.

The use of artificial intelligence (AI) in computer-aided diagnostics (CAD) offers a viable means of increasing the effectiveness and accessibility of the diagnosis process [2]. The most effective AI technique for many tasks, particularly issues with medical imaging, is deep learning (DL) [3]. It is cutting-edge in terms of a variety of computer vision applications. It has been employed in various medical imaging projects, such as the diagnosis of Alzheimer's, the detection of lung cancer, the detection of retinal diseases, etc. Despite obtaining outstanding outcomes in the medical field, a medical diagnostic system must be visible, intelligible, and observable in order to gain the confidence of clinicians, regulators, and patients. It must be able to reveal to everybody why a specific decision was taken in a particular manner in an idealistic situation.

DL tools are among the most often utilized algorithms compared to the machine learning approaches for obtaining better, more adaptable, and more precise results from datasets. DL is also used to identify (diagnose) disorders and provide customized treatment regimens in order to improve the patient's health. The most popular biological imaging techniques for diagnosing patients with the least amount of human involvement include EEG, ECG, MEG, MRI, etc. [4]. The possibility of noise in these medical photographs makes correct analysis of them challenging. DL is able to provide findings that are accurate and precise while also being more trustworthy. Each technology has benefits and drawbacks.

Similar to DL, it has significant limitations, including promising outcomes for huge datasets and the need for a GPU to analyze medical pictures, which calls for more complex system setups [4]. DL is popular now despite these drawbacks because of its capacity to analyze enormous volumes of data. In this chapter, the most recent developments in DL for biological pictures are covered. Additionally, we will talk about the use of DL in segmentation, classification, and registration, as well as its potential applications in medical imaging.

20.2 Biomedical imaging

Several scientists have developed imaging techniques and image analysis approaches to review medical information. However, since medical images are so complex, biomedical image processing needs consistency and continues to be difficult with a wide study scope [5]. There is not a single imaging technique that can meet all radiological requirements and applications. Each kind of medical imaging has limitations caused by the physics of how energy interacts with the physical body, the equipment used, and often physiological limitations. Since Roentgen discovered X-rays in 1895, there has been medical imaging. Later, Hounsfield's realistic computed tomography (CT) machines introduced computer systems into clinical practice and medical imaging. Since then, computer systems have evolved into essential elements of contemporary medical imaging devices and hospitals, carrying out a range of functions from image production and data collection to image processing and presentation [6]. The need for image production, modification, presentation, and analysis increased significantly as new imaging modalities were created [7].

In order to identify disorders, medical researchers are increasingly focusing on DL. As a result of using drugs, drinking alcohol, smoking, and eating poorly, individuals nowadays are predominantly affected by lifestyle illnesses, including type 2 diabetes, obesity, heart disease, and neurodegenerative disorders [4]. DL is essential for predicting these illnesses. For diagnosing and treating any illness using CAD, single photon emission computed tomography (SPECT), positron emission tomography (PET), magnetic resonance imaging (MRI), and other methods, it is preferred in our daily lives. DL may increase the 2D and 3D metrics for more information and speed up the diagnostic processing time. Additionally, it can tackle overfitting and data labeling problems.

20.2.1 Computed tomography

X-ray energy is transmitted to create computed tomography (CT) images, which are then captured from various angles. A thorough cross-sectional view of several bodily images is merged with each image. Invaluable 3D views of certain bodily components, including soft tissues, blood vessels, the pelvis, the lungs, the heart, the brain, the stomach, and bones, are provided by CT images to medical professionals. For the diagnosis of malignant disorders such as lung, liver, and pancreatic cancer, this approach is often used.

20.2.2 Magnetic resonance imaging

Using magnetic resonance imaging (MRI) technology, precise images of the organs and tissues are produced [8] using the dispersion of nuclear spins in a magnetic field after the application of radio-frequency signals [9]. MRI has established itself as a very successful tool of diagnosis by identifying variations in softer types of tissues. The heart, liver, kidneys, and spleen are among the organs in the chest, abdomen, and pelvis that may be assessed using an MRI, as well as blood vessels, breasts, aberrant tissues, bones, joints, spinal injuries, and tendons and ligament problems.

20.2.3 Positron emission tomography

Positron emission tomography (PET) is used in conjunction with a nuclear imaging approach to provide doctors with diagnostic data on the functioning of tissues and organs. This technique may be used to detect lesions that cannot be detected using imaging techniques for the human body, such as CT or MRI [8,9]. PET is used to assess cancer, the efficacy of medicines, cardiac issues, and neurological illnesses, including Alzheimer's and multiple sclerosis.

20.2.4 Ultrasound

The body receives high-frequency sound waves, converts them into images, and then returns them. By mixing sounds and images, medical sonography, ultrasonography, or diagnostic ultrasound, may provide acoustic signals, such as the flow of blood, that let medical experts diagnose the health condition of the patient [8,9]. A pregnancy ultrasound is often used to check for blood vessel and heart abnormalities, pelvic and abdominal organs, as well as signs of discomfort, edema, and infection.

20.2.5 X-ray imaging

Ionizing radiation is used to produce images in the first kind of medical imaging. X-rays operate by directing a beam through the body, which varies in intensity depending on the density of the substance. Additionally, X-ray-type technologies are also covered, along with computed radiography, CT, mammography, interventional radiology, and digital radiography. A device known as radiation treatment utilizes gamma rays, X-rays, electron beams, or protons to destroy cancer cells [8,9]. X-rays are used in diagnostic imaging to assess damaged bones, cavities where items have been ingested, the blood vessels, lungs, and the mammary (mammography).

20.3 Deep learning

DL has emerged as a prominent machine learning (ML) technique because of its better performance, especially in medical image analysis. ML, a branch of AI, includes DL. It relates to algorithms that are inspired by the design and operation of the brain. With the help of several hidden processing layers, it enables computational methods to learn from the representations of the dataset [4]. The concept of feature transformation and extraction is addressed at these levels. The results of the preceding layer are loaded into the one after that. In this method, the predictive analysis may be automated. Additionally, it may function in both unsupervised and supervised strategies.

Figure 20.1 illustrates the operation of the DL method. This method selects the dataset and specific DL algorithm for which the model will be created first. The findings of extensive experiments are created and examined in subsequent phases.

20.3.1 Artificial neural network

The nervous system of humans serves as the basis for the organization and operation of an artificial neural network. The perceptron, the first artificial neural network, is based on a biological neuron. Specifically, it consists of an input layer and an output layer, with a direct connection between the two. Simple tasks like

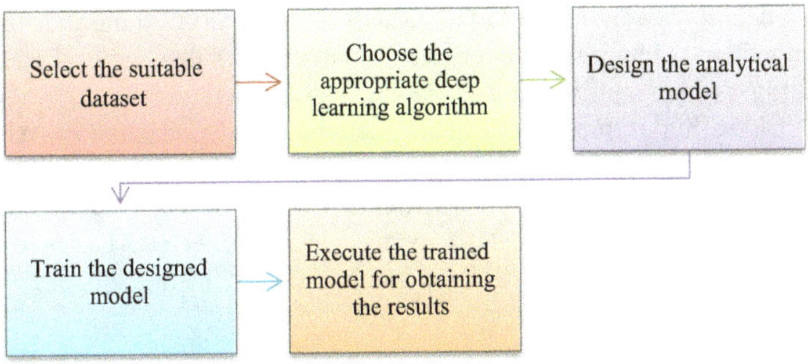

Figure 20.1 The different steps in a deep learning

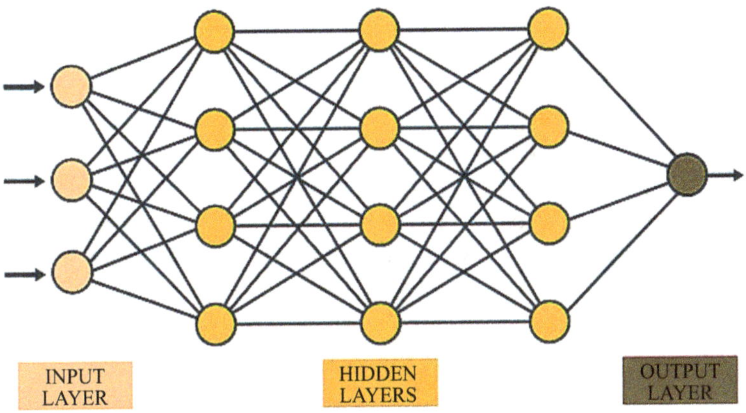

Figure 20.2 The construction of an artificial neural network

categorizing linearly separable patterns may be accomplished with it [10]. In order to solve problems, a "hidden layer" of neurons with more connections was added between the input and output layers. A neural network's fundamental task is to accept input, analyze it, and then send the output to the layer below it. Figure 20.2 depicts the construction of a neural network. The input is initially received by each neuron in the network, which then sums the weighted input, adds the activation function, computes the output, and sends it on to the next layer.

The next subsections address a variety of DL models that have been created, including convolutional neural networks (CNN), deep belief networks (DBN), recurrent neural networks (RNN), etc.

20.4 DL models with various architectures

The complexity of the issue increases with the number of layers in the neural network. There are many hidden layers in a deep neural network (DNN). At the moment, a neural network may have thousands or even tens of thousands of layers. When trained on a huge amount of data, a network of this scale is capable of remembering every mapping and producing insightful predictions [10]. DL has therefore had a big influence on fields including voice recognition, computer vision, medical imaging, and more. Among the DL techniques used in research are DNN, CNN, RNN, deep conventional extreme learning machine (DC-ELM), deep Boltzmann machine (DBM), DBN, and deep autoencoder (DAE). In computer vision and medical imaging, the CNN is gaining greater traction.

20.4.1 Deep neural network

A DNN has more than two layers that allow for complicated non-linear interactions. Regression and classification are two uses for it. Despite its high accuracy, the training procedure is difficult due to the error's propagation backward through the layers, which makes it smaller [10]. The model's learning rate is likewise relatively poor. Figure 20.3 displays a DNN with an input layer, an output layer, and hidden layers.

20.4.2 Convolutional neural network

For two-dimensional data, CNN are effective. It consists of convolution layers that have the ability to transform two-dimensional content into three-dimensional content [10]. This model performs well, and it also has a quick

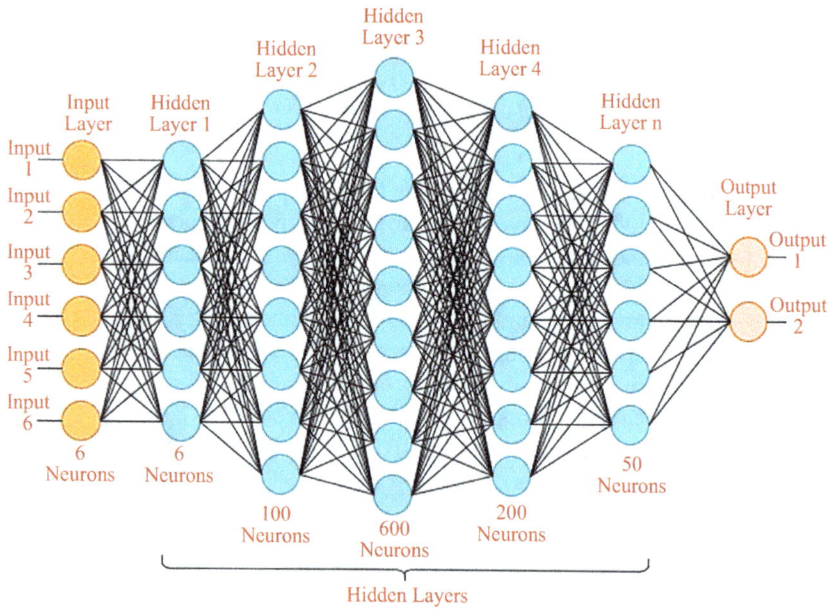

Figure 20.3 Deep neural networks

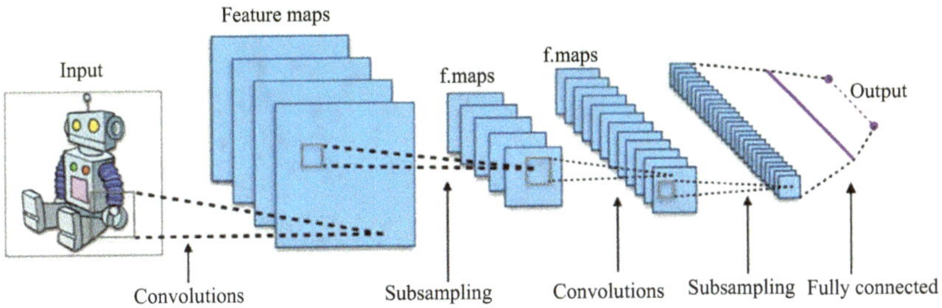

Figure 20.4 Architecture of CNN

learning mode. However, this approach has a disadvantage in that it necessitates a large number of datasets for classification. An example of a CNN is shown in Figure 20.4.

As part of a DL approach, CNN may accept an image as input, assign weights to individual image objects, and impart biases to groups of related items. Compared to other classification methods, CNN requires less pre-processing. CNN's structure is encouraged by the visual cortex's design [4]. In this, a single neuron in the constrained area of the receptive field is responsible for reacting to inputs (visual field). These fields are grouped together to fill the whole visual field. There are several convolutional layers. While further layers adjust to produce a general comprehension of the images, which would be incorporated into the dataset, the initial layer collects low-level characteristics (Figure 20.4).

20.4.3 Recurrent neural network

RNNs are capable of learning sequences, and all phases and neurons share the same weights. It can simulate temporal dependencies as well. Long- and short-term memory (LSTM), BLSTM, HLSTM, and MDLSTM are only a few of its numerous varieties. In many other NLP tasks, including voice recognition, character identification, and others, RNN offers great accuracy [10]. However, due to gradient vanishing, this model has several problems and also requires huge datasets. A representation of a RNN is shown in Figure 20.5.

20.4.4 Deep convolutional extreme learning machine

A deep convolutional extreme learning machine (DC-ELM) is created by combining extreme learning machines and CNNs [10]. Fast training features are derived from the extreme learning system, and feature abstraction performance is derived from the convolution network. For sampling local connections in this network, a Gaussian probability distribution is used. An example of a deep convolutional extreme learning machine (DC-ELM) is shown in Figure 20.6.

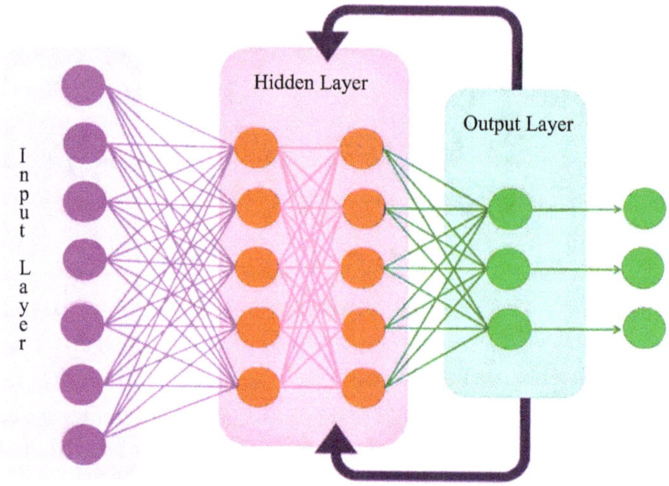

Figure 20.5 Recurrent neural network architecture

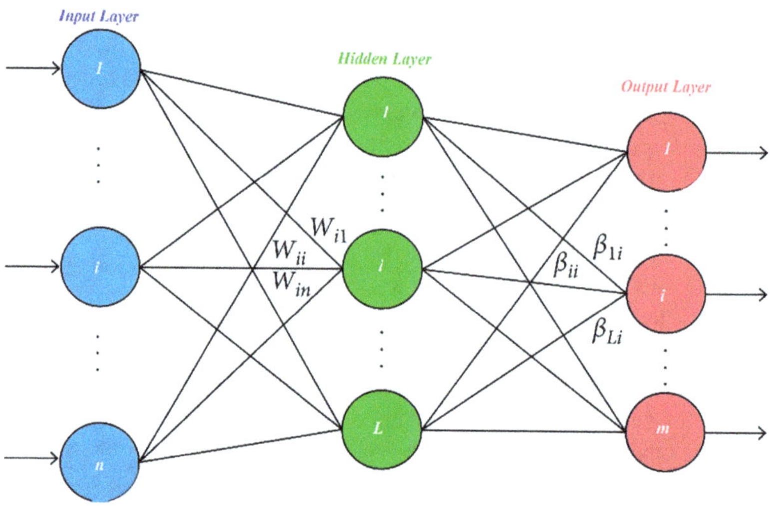

Figure 20.6 Structure of an extreme learning machine [11]

20.4.5 Deep Boltzmann machine

All hidden layers in this model are connected in a single direction and are founded on the Boltzmann formula group. Because it incorporates top-down feedback and ambiguous content, this model produces robust inference [10]. However, this methodology does not allow for parameter optimization for huge datasets. A deep Boltzmann machine (DBM) with two hidden layers is shown in Figure 20.7.

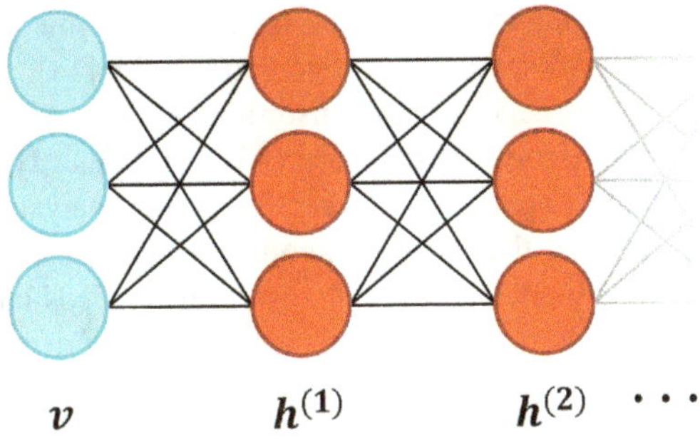

Figure 20.7 Deep Boltzmann machine with two hidden layers [12]

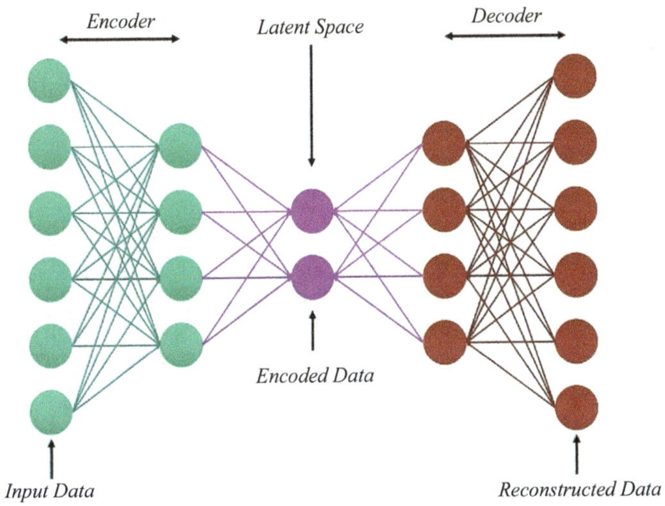

Figure 20.8 Structure of a deep autoencoder [13]

20.4.6 Deep autoencoder

When utilized in supervised learning, DAE is primarily intended for feature extraction and dimensionality reduction. The range of outputs and the quantity of inputs are equal. There is no requirement for labeled data [10]. DAEs come in a variety of forms, including conventional autoencoders for increased sturdiness, de-noising autoencoders, and sparse autoencoders. It requires a step before training. Its training might be harmed by disappearing. The structure of a DAE is shown in Figure 20.8.

20.5 DL in medical imaging

Humans often choose professions involving machines or computers because they are quicker and more precise than people. CAD and automated medical image processing are desirable, though not essential, choices in the medical sciences [4]. CAD is also important in disease progression modeling [14]. A brain scan is essential for several neurodegenerative disorders (NDD), including strokes, Parkinson's disease (PD), Alzheimer's disease (AD), and other types of dementia. Detailed maps of the brain's areas are now accessible for analysis and illness prediction. We may also include the most common CAD applications in biomedical imaging, cancer detection, and lesion intensity assessment. CNN has gained greater popularity in recent years as a result of its incredible performance and dependability. The effectiveness and efficacy of CNNs are shown in an overview of CNN techniques and algorithms where DL strategies are employed in CAD, shape predictions, and segmentations as well as brain disease segmentation.

It may be especially difficult to differentiate between various tumor kinds, sizes, shapes, and intensities in CAD while still employing the same neuroimaging method. There have been several cases when a concentration of infected tissues has gathered alongside normal tissues. It is difficult to manage various forms of noise, such as intensity-based noise, Ricardian noise effects, and non-isotropic frequencies in MRI, using simple machine learning (ML) techniques. These data issues are characterized using a unique method that blends hand-defined traits with tried-and-true ML techniques.

Automating and integrating characteristics with classification algorithms is possible with DL techniques [15,16]. Because CNN has the ability to learn more complicated characteristics, it can handle patches of images that focus on diseased tissues. In the field of medical imaging, CNN can classify TB manifestations using X-ray images [17] and respiratory diseases using CT images [18]. With hemorrhage identification in color fundus images, CNN can identify the smallest and most discriminatory regions in the pre-training phase [19]. The segmentation of isointense stage brain cells [10] and the separation of several brain areas from multi-modality magnetic resonance images (MRI) [20] have both been suggested by CNN. Several hybrid strategies that combine CNN with other approaches have been presented. For instance, a DL technique is suggested in [21] to encode the characteristics of distorted models and the procedure for segmenting the left ventricle of the heart from short-axis MRI. The left ventricle is identified using CNN, and its morphology is inferred using DAE.

This method seeks to assist computers in identifying and characterizing data that may be relevant to a particular problem. Many machine-learning algorithms are based on this idea. Increasingly complex models that are built on top of one another transform input images into results. For image analysis, CNNs are preferable. They classify and categorize data that could be relevant to a certain problem. Many ML algorithms are based on this idea. Increasingly complex models that are built on top of each other transform input images into responses. For image analysis, CNNs are a superior model [22]. The CNNs analyze the input using many filter layers. By subjecting them to a variety of input representations, such as three-dimensional data, DL techniques are usually used in the medical profession to familiarize themselves with contemporary architecture. Due to the size of 3D convolutions and the extra restrictions they imposed, CNNs previously ignored dealing with the high amount of interest.

20.5.1 Image categorization

The categorization of medical images, one of the key jobs in DL, is focused on examining clinical-related concerns, including earlier patient treatment. Multiple image input; single diagnosis (yes or no to diseases). When compared to the number of test models and sample points used in computer vision, a medical imaging technique or software program typically employs fewer diagnostic tests overall. According to [23], feature extraction, or a little adjustment, seems to have performed better, attaining 57.6% accuracy in determining the presence of knee osteoarthritis as opposed to 53.4%. The CNN extraction of features for cytology categorization seems to produce accuracy values ranging from 69.1% to 70.5% [24].

20.5.2 Image classification

The categorization of medical images into different categories in order to diagnose diseases or aid researchers in their ongoing studies is a critical aspect of image recognition. Medical images may be classified by extracting key images from them, which can then be used to create classifiers that categorize the images from the datasets.

When computer-aided design (CAD) was not as prevalent as it is now, physicians often used their expertise to extract and identify medical image elements. Normally, this is a difficult, tiresome, and time-consuming task [4]. DL addresses the issue of accurate prediction, allowing it to forecast more quickly and accurately than humans. Additionally, it can handle a variety of patient-specific datasets. Medical imaging technologies have greatly benefited studies in recent years, not only in terms of helping physicians with their problems. The objective is still beyond our ability as researchers to effectively fulfill. If researchers could categorize illnesses effectively and quickly, it would be a huge help to doctors in making diagnoses.

20.5.2.1 Object classification

For patients with a higher degree of interest, the object categorization is put on tiny, targeted portions of the medical image. These pieces allow for projections into two different classes. Local information as well as multilateral data are essential for getting more precise findings. According to the findings in [22], an image with objects of various sizes was corrected using three CNN DL techniques. Following the use of these three techniques, the ultimate image feature matrices were computed.

CAD serves as an aid in biomedical image diagnosis and interpretation by offering a second objective or supplementary perspective. Numerous studies and kinds of studies have shown recently that using CAD systems accelerates and improves image diagnosis while lowering inter-observer variability [25,26]. For clinical advice like a biopsy, CAD improves the quantitative backing [18]. Selection of features, extraction, and classification are crucial processes that are often used while building CAD for tumor identification [14,26]. The categorization of malignant and healthy cells has been suggested using a variety of ML and DL algorithms [23]. The primary difficulty is shrinking the features without losing crucial information. A significant problem in DL is the size of the dataset; a relatively small dataset reduces the ability to forecast particular cases with the lowest possible risk of over-fitting [27]. The researchers have offered a wide variety of lesion categorization options. However, the majority of them achieve feature space minimization by creating new features under supervision, picking existing features, or deriving small feature sets.

20.5.3 Detection

In the last several decades, academics have paid a lot of attention to object detection tasks. Researchers have started to think about applying object detection processes to healthcare to increase the effectiveness of doctors by using computers to aid them in the detection and diagnosis of images. DL methods are still making significant advancements, and object detection processes in healthcare are popularly used in clinical research as part of the AI medical field [28]. From the perspective of regression and classification, the problem of object identification in the medical profession is difficult. Because of their importance in CAD and detection procedures, several researchers are adapting object detection methods to the medical profession.

A typical objective is finding and recognizing minor illnesses within the entire image domain. Many researchers have conducted a thorough investigation in this regard. By recognizing various medical images, computer-aided detection methods have a long and sordid history and are intended to increase detection performance or shorten the reading time for individual professionals. For pixel (or particle) classifications, CNNs continue to be used in most reported deep-learning object identification approaches [29]. This is proceeded by some image capture to provide image recommendations.

20.5.3.1 Organ or region detection

Organ and region detection is a crucial task in medical imaging, particularly for cancer and neurological illnesses. It is feasible to identify the kind of illness and its phases when the organ deformation activities are captured by MRI or other modalities [30]. The diagnosis of malignancy in a tumor or malignant tumor is crucial for clinical assessment. The analysis of all open and transparent cells for accurate detection is a major difficulty in microscopic image assessment. Cell-level data, however, allows for the distinction of the majority of illness grades [31]. Researchers and academicians employed CNN to successfully identify and segment cells from histo-pathological imaging [32], which is widely used for the diagnosis of cancer.

20.5.3.2 Object and lesion detection

An important stage in the diagnostic procedure that takes a lot of time for physicians to complete is the identification of the required items or lesions in the medical image. Finding the little lesion in the large medical images is part of this endeavor. Computer-aided automatic lesion identification systems are being studied in this field; they improve detection accuracy while speeding up physicians' ability to evaluate medical images. In 1995, a proposal for the first automatic object-detecting system was made. To find nodules in X-ray images, it employed a CNN with four layers [33]. The majority of research on DL-based object identification systems conducts first-pixel classification with CNN before obtaining object candidates via post-processing. Multi-stream CNNs may also incorporate 3D or context information into medical images [34].

Detecting and categorizing objects and lesions are comparable with classification. The main distinction is that in order to identify lesions, we must first conduct a segmentation task, and only then can we classify or forecast a disease [35]. Currently, DL offers encouraging outcomes that allow for the correct timing of early diagnosis and therapy for the patient [36,37].

20.5.4 Segmentation

Segmentation is essential for disease/disorder prediction by dividing an image into several parts and associating them with testing outcomes [38]. The most extensively used framework for 3D image segmentation lately has been the CNN framework, which is broadly utilized in segmentation. Segmentation is an essential part of medical image analysis, in which the image is broken up into smaller parts based on shared characteristics such as color, contrast, grey level, and brightness.

20.5.4.1 Organ and substructure segmentation

Organ substructures must be divided up by volume in order to do a quantitative examination of clinical factors like shape [39]. In the automatic detection procedure, it is the first stage. Finding the group of voxels that make up an object's interior is known as "segmentation." The segmentation method is the most often used subject in DL in medical imaging.

Segmentation activities are carried out by researchers and medical professionals to determine the stage and severity of the illness [40]. Given the prevalence of cancer nowadays, it is often employed in cancer diagnosis. However, brain surgery is the most common form of therapy for brain tumors. The pace of tumor development is slowed by additional therapies like radiation and chemotherapy. The brain's structural and functional makeup is revealed by MRI. The enhanced diagnosis, tumor growth rate, tumor size, and treatment may all be aided by tumor segmentation using MR images, CT images, or other diagnostic imaging modalities [41]. Meningiomas, for example, may be segmented with ease. Due to their low contrast and long tentacle-like features, gliomas are difficult to segment [42]. Tumor segmentation's main goals are to identify the tumor's location, find the tumor's expanded territory (when cancer cells are present), and make a diagnosis by contrasting the damaged tissues with healthy tissues [43].

20.5.4.2 Lesion segmentation

Lesion segmentation combines the difficulties associated with object recognition with organ and substructure segmentation. The global and local environments are necessary for precise image segmentation. There are several cutting-edge methods for segmenting lesions. In contrast, CNN yields the most encouraging outcomes in

2D and 3D biological data [32]. Applying convolution and deconvolution techniques, Yuan suggested a lesion segmentation approach [33] for the automated identification of melanoma from nearby skin cells. CNN and other DL approaches are employed to diagnose different malignant cells because they provide more accurate findings more quickly.

20.5.5 Data mining

During the segmentation procedure, parts of the body, such as organs and structures, are removed from medical imaging [22]. It is used to assess the patient's clinical features, a heart or brain examination, for instance [44]. Additionally, it serves a variety of purposes in CAD. By identifying the specifics that comprise the subject of interest, the digital image may be defined. Selecting a particular dataset from each layer and moving it below, mixes upsamples with downsamples. By combining layer de-convolution samples and convolution points, it linked these two processes [45].

20.5.6 Registration

A frequent activity in the image analysis process is registration, also known as spatial alignment, which involves calculating a common coordinate to align a certain object in the images [46]. A unique kind of parameterized transform is assumed, iterative registration is employed, and then a set of matrices is optimized. Lesion recognition and segmentation are two of the more well-known applications of DL, but researchers have shown that DL also produces the greatest results in registration [47]. Two deep-learning registration algorithms are now being extensively used in research [48]. The first one involves evaluating the similarities between two pictures in order to derive an iterative optimization approach, while the second one involves using a DNN to forecast transformation characteristics.

20.5.7 Other aspects of DL in medical imaging

In medical imaging, there are a large number of additional activities that, when completed, improve the overall image quality and provide more accurate disease identification. The following subsections will provide basic descriptions.

20.5.7.1 Content-based image retrieval

Another emerging technique that might aid radiologists in improving image interpretation is content-based image retrieval (CBIR). The capacity of CBIR to seek and identify similar pictures may be useful for a variety of image and multimedia applications [49]. Using CBIR applications in multimedia instead of laborious, unstructured searching might save user's time. While CBIR can be used for similarity indexing, it may also provide support for CAD based on details of the image and other data related to medical pictures, which might be highly beneficial in medicine [50]. Since it has been so successful in other medical fields, CBIR seems to have had a minimal impact on radiology [51]. Current research in computer vision, biomedical engineering, and information extraction may considerably increase CBIR's application to radiology practice.

20.5.8 Image enhancement

In medical imaging, DL has traditionally concentrated on segmenting, forecasting, and readjusting reconstructed images. DL has recently made progress in MR image capture, noise removal [52,53], and super-resolution [54,55], among other lower-level MR measuring methods or procedures [4].

20.5.9 Integration of image data into reports

DL's pre-processing of a large amount of data produces superior findings, assisting the radiologist in the diagnosis of illness and future study [56,57]. In order to make good decisions, the subjects' reports include information about what happened around them and how likely it is that the disease's symptoms will show up [4].

20.6 Summary of review

Over the course of the last several years, capabilities for DL have been established. DL techniques are now reliable for use in practical cases, and the structures build on this benefit. This is used to apply to the setup of medical imaging. It will make significant progress in the future. DL-enabled machines are capable of drawing the conclusions required for medicine delivery. Patients will benefit from it since it is a crucial component of our study. A significant component of the DL solution is confirming that the machines are being used to their full capability. The classification, categorization, and enumeration of patterns of illness are made possible by the DL algorithms employed in medical image analysis. Additionally, it makes it possible to investigate the limits of analytical objectives, which aids in the creation of treatment prediction models. These issues, including the use of DL in healthcare services, are being considered by researchers in the imaging profession. As DL becomes more prevalent in various other industries, including healthcare, it is advancing quickly.

20.7 Challenges of DL in medical imaging

The application of DL to diagnostic instruments has been the most innovative technical advance since the emergence of digital imaging. The fundamental advantage of DL in medical imaging is the discovery of hierarchical correlations in image pixel data. This information can be discovered theoretically and algorithmically, eliminating the need for time-consuming manual feature creation. DL is being used to advance many important academic fields, including classification, segmentation, localization, and object recognition. The medical business has dramatically expanded the usage of electronic records, which has contributed to the vast volumes of data needed for precise deep-learning algorithms. Recognizing the severity of symptoms from psychiatric assessment, brain tumor identification and segmentation, biomedical image analysis, digital pathology, and diabetic self-management are major uses of DL in medical imagery. A fascinating and expanding area of study, the use of DL-based techniques in the medical industry is currently being slowed down by a number of obstacles [4,58,59]. The next subsection goes through each of them.

20.7.1 Large amount of training dataset

DL techniques need a significant quantity of training data in order to obtain the desired level of precision. The quantity and quality of the dataset heavily influence the performance of the DL model in every application, including regression, classification, segmentation, and prediction. One of the biggest difficulties in using DL clinical imaging is the dearth of training datasets [60,61]. Medical professionals must put in a great deal of effort and struggle with the development of such massive volumes of medical image data. Furthermore, a lack of trained specialists or enough examples of uncommon illnesses may make it hard to annotate each projected disease.

20.7.2 Legal and data privacy issues

When real images are used for DL in biomedical imaging, the privacy issue becomes far more complex and difficult to solve [62,63]. Data privacy is a social and technological problem that has to be addressed. The governments had already established guidelines for healthcare providers to follow in order to protect, limit the sharing, and use of patient confidential information. It also grants people legal rights over their personal details and medical records. When personal information is lost, it is more challenging to connect the data to a specific individual. However, by utilizing association algorithms, privacy violators may quickly locate sensitive data. Because they could have a detrimental effect from an ethical and legal perspective, privacy concerns must be resolved as soon as possible. Reduced information content caused by limited and constricted data availability may affect the accuracy of DL.

20.7.3 Standards for datasets and interoperability

One of the main obstacles is the lack of dataset standards and interoperability due to the varied nature of training data [64,65]. Because different hardware settings provide different types or natures of data, there is significant variance in medical imaging owing to things like national standards and sensor types. Since DL in the field of medical imaging demands a lot of training data, merging numerous diverse datasets to improve accuracy is basically necessary [66]. In the health industry, interoperability is a crucial quality, but implementing it is still difficult. To improve the accuracy of DL, the data from the health industry must be standardized [67]. Many standardization organizations, including HL7 and HIPAA, are working on this problem to specify certain rules and protocols that will lead to increased interoperability.

20.7.4 Black box problem

Numerous medical imaging applications were launched by DL in this field, opening up new opportunities. Despite having excellent performance across a wide range of applications, including segmentation and classification, it may sometimes be challenging to describe the judgments it takes in a manner that the typical person can comprehend. This is referred to as a "black box issue" [68–70]. The DL approaches take in a lot of data, find features, and create prediction models, but since their underlying workings are not well understood, they are sometimes difficult to comprehend.

20.7.5 Noise labeling

Accurate algorithm design is complicated by noise, even when the data being analyzed are labeled by medical professionals, as in the identification of nodules in lung CT using the LIDC-IDRI dataset. In this dataset, respiratory nodules are annotated separately by several radiologists [53]. This assignment did not need universal agreement, and it turned out that there were three times as many nodules that they did not all agree were nodules as there were nodules [61]. Extra caution should be used while training a DL algorithm on such data in order to account for the presence of noise and uncertainty in the reference standard.

20.7.6 Images of abnormal classes

Finding images of aberrant classes in the realm of medical imaging could be difficult [71]. For instance, a tremendous quantity of mammography data has been collected globally as a consequence of the screening program for breast cancer has been collected globally as a consequence of the screening program for breast cancer [72]. However, most of these mammograms are unremarkable. In building these DL systems, accuracy and efficiency might therefore be key study fields. Using clinical data effectively is another difficulty for DL. In addition to using medical images, doctors may draw better conclusions by using a variety of information on patient data, age groups, and demographics. DL networks incorporating medical images often use this information to enhance their performance; however, the results were not as encouraging as anticipated [73]. In order to keep the imaging characteristics, separate from the different clinical characteristics and avoid the clinical manifestations from being drowned out, this is one of the main challenges in DL.

20.8 The future of DL in biomedical image processing

The health industry will soon enter a new age when biomedical imaging and data will be crucial. As DL is applied to large datasets, the number of scenarios will grow in lockstep with the human population. The issue of the big dataset will automatically be solved when the number of instances being recorded rises. The essential necessity for every topic is that appropriate care be given as soon as possible. We might infer from this that the availability of enormous datasets presents both enormous potential and problems.

According to several studies, CAD can handle multiple instances at once and is more precise than humans at diagnosing diseases. Therefore, in today's technological age, CAD accessibility and dependability are no longer problems. Due to the availability of several data-driven medical imaging technologies that allow autonomous feature construction and minimize human interaction during the operation, DL has supplanted pattern recognition and traditional machine learning in recent years. It is advantageous for a variety of health informatics issues. In the end, DL reinforces unstructured data from diagnostic imaging, bioinformatics, and health informatics quickly and in a progressive direction. The majority of the developed DL applications for medical imaging analyze unprocessed health data. However, structured data contains a wealth of information. This provides comprehensive details on the history, care, pathology, and diagnosis of the topic. The cytological comments in situations of cancer identification in medical imaging provide details regarding the stage and distribution of the tumor. Such details are essential since they are needed to assess the patient's illness or condition. With AI, DL improves the dependability of medical decision-support systems.

20.9 Conclusion

Our lives have been significantly impacted by recent developments in DL algorithms that have automated most procedures. Compared to conventional machine learning methods, they have shown improvements. Researchers predict that DL will replace the majority of human labor in the next 15 years, and autonomous robots will handle most of our daily duties. This prediction is based on the pace at which performance is improving. In contrast to other real-world issues, the adoption of DL in the healthcare sector is rather gradual because of the responsiveness of the domain. In this chapter, we have highlighted the issues that are impeding the development of DL in the healthcare sector. The use of DL in the analysis of medical images has also been covered. The report gives an idea of the broad range of applications for DL in medical imaging, even if the list of possible applications is never exhaustive. DL has received excellent reviews for all of its applications to date in all other fields, but because of the sensitive nature of the medical imaging field, DL has only had a limited impact there. As a result, it can be said that DL's usage in this field is restricted.

References

[1] P. Zhang, Y. Zhong, Y. Deng, X. Tang, and X. Li, "A Survey on Deep Learning of Small Sample in Biomedical Image Analysis," 2019, *arXiv preprint arXiv:1908.00473*.

[2] A. Singh, S. Sengupta, and V. Lakshminarayanan, "Explainable deep learning models in medical image analysis," *Journal of Imaging*, vol. 6, p. 52, 2020.

[3] F. Altaf, S. M. S. Islam, N. Akhtar, and N. K. Janjua, "Going deep in medical image analysis: concepts, methods, challenges, and future directions," *IEEE Access*, vol. 7, p. 99540–99572, 2019.

[4] M. Jyotiyana and N. Kesswani, "Deep learning and the future of biomedical image analysis," in *Studies in Big Data*, Springer International Publishing, 2019, p. 329–345.

[5] M. A. Haidekker, *Advanced Biomedical Image Analysis*, John Wiley & Sons, 2010.

[6] P. M. de Azevedo-Marques, A. Mencattini, M. Salmeri, and R. M. Rangayyan, *Medical Image Analysis and Informatics: Computer-Aided Diagnosis and Therapy*, CRC Press, 2017.

[7] S. Renukalatha and K. V. Suresh, "A review on biomedical image analysis," *Biomedical Engineering: Applications, Basis and Communications*, vol. 30, p. 1830001, 2018.

[8] A. Maier, *Medical Imaging Systems An Introductory Guide*, Springer Nature, 2018, p. 259.

[9] V. I. Mikla and V. V. Mikla, *Medical Imaging Technology*, Elsevier, 2013.

[10] S. Tanwar and J. Jotheeswaran, "Survey on deep learning for medical imaging," *Journal of Applied Science and Computations*, vol. 5, p. 1608–1620, 2018.

[11] T. L. Qinwei Fan, "Smoothing L0 regularization for extreme learning machine," *Mathematical Problems in Engineering*, vol. 2020, Article ID 9175106, 2020.

[12] H. Manukian and M. D. Ventra, "Mode-assisted joint training of deep Boltzmann machines," *Scientific Reports*, vol. 11, p. 19000, 2021.

[13] S. Latif, M. Driss, W. Boulila, *et al.*, "Deep learning for the industrial Internet of Things (IIoT): a comprehensive survey of techniques, implementation frameworks, potential applications, and future directions," *Sensors*, vol. 21, Article 7518, 2021.

[14] H.-P. Chan, L. M. Hadjiiski, and R. K. Samala, "Computer-aided diagnosis in the era of deep learning," *Medical Physics*, vol. 47, no. 5, p. e218–e227, 2020.

[15] D. Nie, H. Zhang, E. Adeli, L. Liu, and D. Shen, "3D deep learning for multi-modal imaging-guided survival time prediction of brain tumor patients," in *Medical Image Computing and Computer-Assisted Intervention – MICCAI 2016*, Springer International Publishing, 2016, p. 212–220.

[16] T. Xu, H. Zhang, X. Huang, S. Zhang, and D. N. Metaxas, "Multimodal deep learning for cervical dysplasia diagnosis," in *Medical Image Computing and Computer-Assisted Intervention – MICCAI 2016*, Springer International Publishing, 2016, p. 115–123.

[17] Y. Cao, C. Liu, B. Liu, *et al.*, "Improving tuberculosis diagnostics using deep learning and mobile health technologies among resource-poor and marginalized communities," in *2016 IEEE First International Conference on Connected Health: Applications, Systems and Engineering Technologies (CHASE)*, 2016.

[18] M. Anthimopoulos, S. Christodoulidis, L. Ebner, A. Christe, and S. Mougiakakou, "Lung pattern classification for interstitial lung diseases using a deep convolutional neural network," *IEEE Transactions on Medical Imaging*, vol. 35, p. 1207–1216, 2016.

[19] M. J. J. P. van Grinsven, B. van Ginneken, C. B. Hoyng, T. Theelen, and C. I. Sanchez, "Fast convolutional neural network training using selective data sampling: application to hemorrhage detection in color fundus images," *IEEE Transactions on Medical Imaging*, vol. 35, p. 1273–1284, 2016.

[20] W. Zhang, R. Li, H. Deng, *et al.*, "Deep convolutional neural networks for multi-modality isointense infant brain image segmentation," *NeuroImage*, vol. 108, p. 214–224, 2015.

[21] M. R. Avendi, A. Kheradvar, and H. Jafarkhani, "A combined deep-learning and deformable-model approach to fully automatic segmentation of the left ventricle in cardiac MRI," *Medical Image Analysis*, vol. 30, p. 108–119, 2016.

[22] S. Panda and R. Kumar *Dhaka,* "Application of artificial intelligence in medical imaging," in *Machine Learning and Deep Learning Techniques for Medical Science*, 2022, p. 195–202.

[23] J. Antony, K. McGuinness, N. E. O'Connor, and K. Moran, "Quantifying radiographic knee osteoarthritis severity using deep convolutional neural networks," in *2016 23rd International Conference on Pattern Recognition (ICPR)*, 2016.

[24] E. Kim, M. Corte-Real, and Z. Baloch, "A deep semantic mobile application for thyroid cytopathology," in *SPIE Proceedings*, 2016.

[25] S. Singh, J. Maxwell, J. A. Baker, J. L. Nicholas, and J. Y. Lo, "Computer-aided classification of breast masses: performance and interobserver variability of expert radiologists versus residents," *Radiology*, vol. 258, p. 73–80, 2011.

[26] R. Liu, H. Li, F. Liang, *et al.*, "Diagnostic accuracy of different computer-aided diagnostic systems for malignant and benign thyroid nodules classification in ultrasound images," *Medicine*, vol. 98, p. e16227, 2019.

[27] A. Anaya-Isaza, L. Mera-Jiménez, and M. Zequera-Diaz, "An overview of deep learning in medical imaging," *Informatics in Medicine Unlocked*, vol. 26, p. 100723, 2021.

[28] Y. Shou, T. Meng, W. Ai, C. Xie, H. Liu, and Y. Wang, "Object detection in medical images based on hierarchical transformer and mask mechanism," *Computational Intelligence and Neuroscience*, vol. 2022, p. 1–12, 2022.

[29] J. Moorthy and U. D. Gandhi, "A survey on medical image segmentation based on deep learning techniques," *Big Data and Cognitive Computing*, vol. 6, p. 117, 2022.

[30] A. S. Lundervold and A. Lundervold, "An overview of deep learning in medical imaging focusing on MRI," *Zeitschrift für Medizinische Physik*, vol. 29, p. 102–127, 2019.

[31] K. A. Tran, O. Kondrashova, A. Bradley, E. D. Williams, J. V. Pearson, and N. Waddell, "Deep learning in cancer diagnosis, prognosis and treatment selection," *Genome Medicine*, vol. 13, Article no. 152, 2021.

[32] K. Lee, J. H. Lockhart, M. Xie, *et al.*, "Deep learning of histopathology images at the single cell level," *Frontiers in Artificial Intelligence*, vol. 4, p. 754641–754641, 2021.

[33] S.-C. B. Lo, S.-L. A. Lou, J.-S. Lin, M. T. Freedman, M. V. Chien, and S. K. Mun, "Artificial convolution neural network techniques and applications for lung nodule detection," *IEEE Transactions on Medical Imaging*, vol. 14, p. 711–718, 1995.

[34] S. P. Singh, L. Wang, S. Gupta, H. Goli, P. Padmanabhan, and B. Gulyás, "3D deep learning on medical images: a review," *Sensors*, vol. 20, p. 5097, 2020.

[35] M. A. Abdou, "Literature review: efficient deep neural networks techniques for medical image analysis," *Neural Computing and Applications*, vol. 34, p. 5791–5812, 2022.

[36] C. Ieracitano, N. Mammone, M. Versaci, *et al.*, "A fuzzy-enhanced deep learning approach for early detection of Covid-19 pneumonia from portable chest X-ray images," *Neurocomputing*, vol. 481, p. 202–215, 2022.

[37] N. Mahendran and M. Durai Raj Vincent P, "A deep learning framework with an embedded-based feature selection approach for the early detection of the Alzheimer's disease," *Computers in Biology and Medicine*, vol. 141, p. 105056, 2022.

[38] R. Wang, T. Lei, R. Cui, B. Zhang, H. Meng, and A. K. Nandi, "Medical image segmentation using deep learning: a survey," *IET Image Processing*, vol. 16, p. 1243–1267, 2022.

[39] J. Harms, Y. Lei, S. Tian, *et al.*, "Automatic delineation of cardiac substructures using a region-based fully convolutional network," *Medical Physics*, vol. 48, p. 2867–2876, 2021.

[40] M. Aljabri and M. AlGhamdi, "A review on the use of deep learning for medical images segmentation," *Neurocomputing*, vol. 506, p. 311–335, 2022.

[41] M. Bardis, R. Houshyar, C. Chantaduly, *et al.*, "Deep learning with limited data: organ segmentation performance by U-Net," *Electronics*, vol. 9, p. 1199, 2020.

[42] Q. Liu, K. Liu, A. Bolufé-Röhler, J. Cai, and L. He, "Glioma segmentation of optimized 3D U-net and prediction of multi-modal survival time," *Neural Computing and Applications*, vol. 34, p. 211–225, 2021.

[43] T. Magadza and S. Viriri, "Deep learning for brain tumor segmentation: a survey of state-of-the-art," *Journal of Imaging*, vol. 7, p. 19, 2021.

[44] F. Behrad and M. S. Abadeh, "An overview of deep learning methods for multimodal medical data mining," *Expert Systems with Applications*, vol. 200, p. 117006, 2022.

[45] T. H. Jaware, K. S. Kumar, R. D. Badgujar, and S. Antonov, *Medical Imaging and Health Informatics*, Wiley, 2022.

[46] S. Abbasi, M. Tavakoli, H. R. Boveiri, M. A. M. *et al.*, "Medical image registration using unsupervised deep neural network: a scoping literature review," *Biomedical Signal Processing and Control*, vol. 73, p. 103444, 2022.

[47] X. Chen, X. Wang, K. Zhang, *et al.*, "Recent advances and clinical applications of deep learning in medical image analysis," *Medical Image Analysis*, vol. 79, p. 102444, 2022.

[48] D. Sengupta, P. Gupta, and A. Biswas, "A survey on mutual information based medical image registration algorithms," *Neurocomputing*, vol. 486, p. 174–188, 2022.

[49] M. A. Dhaygude and S. Kinariwala, "A literature survey on content-based information retrieval," *Journal of Computing Technologies*, vol. 11, pp. 1–6, 2022.

[50] R. Vishraj, S. Gupta, and S. Singh, "A comprehensive review of content-based image retrieval systems using deep learning and hand-crafted features in medical imaging: research challenges and future directions," *Computers and Electrical Engineering*, vol. 104, p. 108450, 2022.
[51] J. Janjua and A. Patankar, "Comparative review of content based image retrieval using deep learning," in *Intelligent Computing and Networking*, Springer Nature Singapore, 2022, p. 63–74.
[52] S. Kaji and S. Kida, "Overview of image-to-image translation by use of deep neural networks: denoising, super-resolution, modality conversion, and reconstruction in medical imaging," *Radiological Physics and Technology*, vol. 12, p. 235–248, 2019.
[53] D. Karimi, H. Dou, S. K. Warfield, and A. Gholipour, "Deep learning with noisy labels: exploring techniques and remedies in medical image analysis," *Medical Image Analysis*, vol. 65, p. 101759, 2020.
[54] W. Ahmad, H. Ali, Z. Shah, and S. Azmat, "A new generative adversarial network for medical images super resolution," *Scientific Reports*, vol. 12, p. 9533, 2022.
[55] W. Muhammad, M. Gupta, and Z. Bhutto, "Role of deep learning in medical image super-resolution," in *Advances in Medical Technologies and Clinical Practice*, IGI Global, 2022, p. 55–93.
[56] A. L. Appelt, B. Elhaminia, A. Gooya, A. Gilbert, and M. Nix, "Deep learning for radiotherapy outcome prediction using dose data – a review," *Clinical Oncology*, vol. 34, p. e87–e96, 2022.
[57] N. Subramanian, O. Elharrouss, S. Al-Maadeed, and M. Chowdhury, "A review of deep learning-based detection methods for COVID-19," *Computers in Biology and Medicine*, vol. 143, p. 105233, 2022.
[58] V. Saraf, P. Chavan, and A. Jadhav, "Deep learning challenges in medical imaging," in *Algorithms for Intelligent Systems*, Springer Singapore, 2020, p. 293–301.
[59] S. N. Saw and K. H. Ng, "Current challenges of implementing artificial intelligence in medical imaging," *Physica Medica*, vol. 100, p. 12–17, 2022.
[60] N. Tajbakhsh, L. Jeyaseelan, Q. Li, J. N. Chiang, Z. Wu, and X. Ding, "Embracing imperfect datasets: a review of deep learning solutions for medical image segmentation," *Medical Image Analysis*, vol. 63, p. 101693, 2020.
[61] M. J. Willemink, W. A. Koszek, C. Hardell, *et al.*, "Preparing medical imaging data for machine learning," *Radiology*, vol. 295, p. 4–15, 2020.
[62] Y. Y. M. Aung, D. C. S. Wong, and D. S. W. Ting, "The promise of artificial intelligence: a review of the opportunities and challenges of artificial intelligence in healthcare," *British Medical Bulletin*, vol. 139, p. 4–15, 2021.
[63] G. A. Kaissis, M. R. Makowski, D. Rückert, and R. F. Braren, "Secure, privacy-preserving and federated machine learning in medical imaging," *Nature Machine Intelligence*, vol. 2, p. 305–311, 2020.
[64] P. M. A. van Ooijen, "Quality and curation of medical images and data," in *Artificial Intelligence in Medical Imaging*, Springer International Publishing, 2019, p. 247–255.

[65] H. Harvey and B. Glocker, "A standardised approach for preparing imaging data for machine learning tasks in radiology," in *Artificial Intelligence in Medical Imaging*, Springer International Publishing, 2019, p. 61–72.

[66] C. Park, S. C. You, H. Jeon, C. W. Jeong, J. W. Choi, and R. W. Park, "Development and validation of the radiology common data model (R-CDM) for the international standardization of medical imaging data," *Yonsei Medical Journal*, vol. 63, p. S74, 2022.

[67] F. Prior, J. Almeida, P. Kathiravelu, *et al.*, "Open access image repositories: high-quality data to enable machine learning research," *Clinical Radiology*, vol. 75, p. 7–12, 2020.

[68] I. Castiglioni, L. Rundo, M. Codari, *et al.*, "AI applications to medical images: from machine learning to deep learning," *Physica Medica*, vol. 83, p. 9–24, 2021.

[69] J. Petch, S. Di, and W. Nelson, "Opening the black box: the promise and limitations of explainable machine learning in cardiology," *Canadian Journal of Cardiology*, vol. 38, p. 204–213, 2022.

[70] G. S. Handelman, H. K. Kok, R. V. Chandra, *et al.*, "Peering into the black box of artificial intelligence: evaluation metrics of machine learning methods," *American Journal of Roentgenology*, vol. 212, p. 38–43, 2019.

[71] J. M. Johnson and T. M. Khoshgoftaar, "Survey on deep learning with class imbalance," *Journal of Big Data*, vol. 6, 2019.

[72] G. Murtaza, L. Shuib, A. W. A. Wahab, *et al.*, "Deep learning-based breast cancer classification through medical imaging modalities: state of the art and research challenges," *Artificial Intelligence Review*, vol. 53, p. 1655–1720, 2019.

[73] E. Tasci, Y. Zhuge, K. Camphausen, and A. V. Krauze, "Bias and class imbalance in oncologic data—towards inclusive and transferrable AI in large scale oncology data sets," *Cancers*, vol. 14, p. 2897, 2022.

Index

www.ingramcontent.com/pod-product-compliance
Lightning Source LLC
LaVergne TN
LVHW011838200926
848484LV00002B/371